Insulating Modernism

Kiel Moe
Insulating Modernism
Isolated and Non-isolated Thermodynamics in Architecture

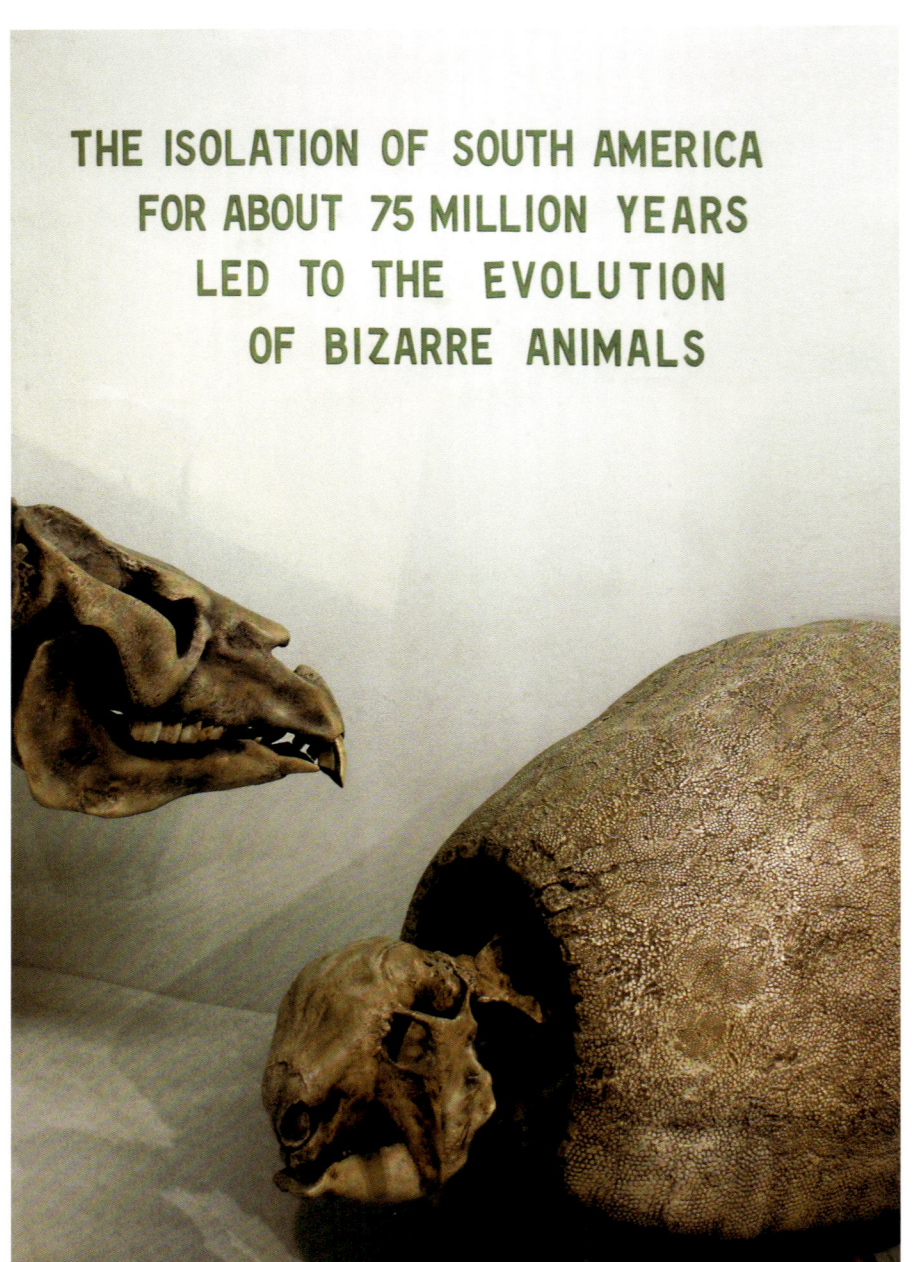

Glyptodont Exhibit, Harvard Museum of Natural History

"It seems that architects build in an isolated, self-contained, ahistorical way. They never seem to allow for any kind of relationships outside of their grand plan."[1]
Robert Smithson, 1973 "Entropy Made Visible"

"… the only rooms that can still be locked from the inside are reserved for isolates, fetishists, lost stumblers-in out of the occupation who need loneliness like the dopefiend needs his dope …"[2]
Thomas Pynchon, 1973 Gravity's Rainbow

"… the ideology of consumption, far from constituting an isolated or successive moment of the organization of production, must be offered to the public as the ideology of the correct use of the city."[3]
Manfredo Tafuri, 1973 Progetto e Utopia

"If we now consider instead of an isolated system, a system in contact with an energy reservoir … we necessarily are confronted with open systems in which the exchanges with the external world play a capital role…
In all these phenomena, an ordering mechanism not reducible to the equilibrium principle appears. For reasons to be explained later, we shall refer to this principle as order through fluctuations. One has structures which are created by the continuous flow of energy and matter from the outside world. Their maintenance requires a critical distance from equilibrium, i.e. a minimum level of dissipation. For all these reasons we have called them 'dissipative structures.'"[4]
Ilya Prigogine & René Lefever, 1973 "Theory of Dissipative Structures"

"Instead of the confusion that comes from the western civilization's characteristic educational approach of isolating variables in tunnel-vision thinking, let us here seek common sense overview which comes from overall energetics."[5]
Howard T. Odum, 1973 "Energy, Ecology, and Economics"

Notes

1 Robert Smithson, "Entropy Made Visible," (1973) in Jack Flam, ed., *Robert Smithson: The Collected Writings*, Berkeley: University of California Press, 1996. p. **309**. Originally published in Alison Sky, ed., *On Site #4*, **1973**.
2 Thomas Pynchon, *Gravity's Rainbow*, New York: Penguin Classics, **1973**. p. **686**.
3 From the English translation of Manfredo Tafuri, *Progetto e Utopia: Architettura e Sviluppo Capitalistico*, Bari: Laterza & Figli, **1973**. p. **78**.
4 Ilya Prigogine and René Lefever, "Theory of Dissipative Structures," *Synergetics*, **1973**. pp. **124–125**.
5 Howard T. Odum, "Energy, Ecology, and Economics," *Ambio*, vol. **2(6)**, **1973**. p. **220**.

Layout, cover design, and typography
Miriam Bussmann, Berlin

Library of Congress Cataloging-in-Publications Date
A CIP catalogue record for this book has been applied for at the Library of Congress, Washington D.C., USA.

Bibliographic information published by the Deutsche Nationalbibliothek
The Deutsche Nationalbibliothek lists this publication in the Deutsche Nationalbibliografie; detailed bibliographic data are available on the Internet at http://dnb.dnb.de.

This work is subject to copyright. All rights are reserved, whether the whole or part of the material is concerned, specifically the rights of translation, reprinting, re-use of illustrations, recitation, broadcasting, reproduction on microfilms or in other ways, and storage in databases. For any kind of use, permission of the copyright owner must be obtained.

© 2014 Birkhäuser Verlag GmbH, Basel
P.O. Box 44, 4009 Basel, Switzerland
Part of Walter de Gruyter GmbH, Berlin/Boston

Printed on acid-free paper produced from chlorine-free pulp. TCF ∞
Printed in Germany

ISBN 978-3-03821-539-4

9 8 7 6 5 4 3 2 1

www.birkhauser.com

Contents

Foreword
Iñaki Abalos 8

Buildings are Non-isolated, Transient Structures of Dissipation:
A Reckoning in the Form of an Introduction 10

A History of Heat Transfer in Buildings 54

A Material History of Insulation in Modernity 126

Physiology, Insulation, Climate, and Pedagogy 188

The Architecture of Dissipation 228

A Dissipative Epilogue: Breathing Walls, by *Salmaan Craig* ... 272

Conclusion:
The Metabolic Rift, Gift & Shift of
Architecture's Necessary Excess 288

Acknowledgements 310
Illustration Credits 312
About the Author 314
Index of Persons, Firms, and Institutions 315
Index of Buildings and Objects 319

Foreword

Iñaki Abalos

There is, apparently, nothing more irrelevant in a building than the layers of insulating material that, only over the past few decades, has been filling chambers and sheathing façades all over the world with supposedly beneficial universal effects. This book presents a surprisingly ambitious proof that the assumption that these insulating hidden materials are innocent and irrelevant is one of the biggest mistakes that have been made by architects, educators, and historians over the past fifty years. Not only does it debunk the innocence of these few interior centimeters with irrefutable technical reasoning — in a way that shows the pedagogic and communicational skills of the author — but this technical discussion is just the starting point in the construction of an architectural agenda for the coming decades whose theoretical and critical precision, historic timeliness, and futuristic vision — aspects that are progressively unveiled as one reads the book — are simply astonishing. With his new book, Kiel Moe manages to make a qualitative leap forward in the ambitions already expressed in previous books such as *Convergence* or *Thermally Active Surfaces in Architecture* by proposing a politically critical vision of the material culture of our time, highlighting the impossibility for architects to construct a consistent discourse because of their inability to understand the material, tectonic, and thermodynamic facets of which buildings are made. He thus proposes a new design culture that reclaims an integrated — "convergent," in his own terms — knowledge which will allow the architect to not only address form, matter, and energy with new tools but also to address the increasingly dramatic requirements of society through ascertaining precisely the appropriateness and creativity of these instruments. The different thermodynamic scales of the building, from millimeters to decameters, give way to comprehensive views of the city and the territory, delimiting the systemic character of a thermodynamic conception of architecture that does not dwell upon the technical deception of minimizing consumption through maximizing the building's insulation; an activity which conceals both a dark business and a voluntary short-sightedness in trying to address the energy problem in our buildings and cities with physiocratic attitudes. By doing this, Moe shows us the importance of welcoming, in a holistic way, a new form of understanding between architecture and thermodynamics. *Insulating Modernism* is a decisive step in the theoretical and critical establishment of what is known as thermodynamic materialism, a step that aims at recovering the knowledge — and therefore the authority — from the more general thermodynamic principles, avoiding the dominant technocratic fascination and bringing back the historic dimension of the discipline, which holds the promise of a radical revision of modernism. In this trend of thought one can identify the presence of a bloodline of authors amongst which Reyner Banham and Howard T. Odum become indispensable references. *Insulating Modernism* unifies different cultural, historical, and technological approaches to architecture from a deep understanding of the scales at which the laws of thermodynamics operate, exposing them in a concise and pedagogical way that will certainly have an important effect on the way in which the future generations will design. This book definitively insulates modernism and its long and underground influence in order to celebrate the coming of a new architecture that will be anchored both to historical time and to the entropic time of thermodynamic processes.

1973 Petro-Pentecostalism: Calvinist Thermodynamics, Born Again
Amid the oil embargo fervor, this vacant Potlatch, Washington gas station was readily converted to a religious meeting hall.

Saving energy, saving money, saving the planet, and saving souls share a deeply rooted form of rhetoric. "The Protestant Ethic and Spirit of Capitalism" of the crude energy efficiency/conservation discourse overlays linear cultural and economic concepts on the non-linear behavior of thermodynamics. Calling, frugality, and atonement are routinely the impetus of these practices. However, you cannot make energy more or less efficient as all energy is always and only conserved; as unambiguously stated in the first law of thermodynamics. The energy efficiency discourse of consuming less and minimizing dissipation ultimately discloses little about the role of people, buildings, and design in the thermodynamic evolution of urbanization but does finally amplify many neoliberal dynamics. To address the non-isolated, non-equilibrium, and non-linear thermodynamics that float the operations of buildings and of life itself, architects by now need a radically different epistemology — a different ethic of work in these systems — for energy and the exergy designs that will engender maximal entropy production futures for civilization.

Buildings are Non-isolated, Transient Structures of Dissipation: A Reckoning in the Form of an Introduction

1973

To continue to pretend that the 1973 OPEC oil embargo did anything other than perpetuate and intensify modern, problematic intellectual habits and path dependencies about energy in architecture would continue to overlook the power of far more cogent contemporaneous observations on energy systems. So, instead of yet another recitation of the 1973 OPEC oil embargo as a pivot in the energy praxis of architecture, this book instead begins with a quartet of 1973 observations about the concept of isolation from Robert Smithson, Thomas Pynchon, Manfredo Tafuri, Howard T. Odum, and Ilya Prigogine. Though divergent from one another, all of these quotations depart sharply from a persistent modern preoccupation with isolation. Together, these contemporaneous scientific, urban, and artistic observations reflect minor, but poignant, non-isolating habits of mind that strongly contrast with isolating intellectual habits that, even now, continue to coddle the thermodynamic platitudes of energy efficiency and energy conservation. These platitudes persist in architecture despite profound transformations in thermodynamics that, more than a century ago, imparted a far more accurate and powerful characterization of energy system dynamics. In regards to these transformations, the persistent focus on efficiency and conservation is but reverb in the oil embargo echo chamber of architecture's modernist discourse on energy.

Worse still, the oil embargo narrative of energy in architecture is at best a dog's dinner: a sloppy serving of Calvinist economic notions about frugality and calling applied to the decidedly non-economic behavior of energy systems; nascent neoliberal economic dynamics unwittingly guiding the hands and voices of architects; a broadened professionalization of a managerially-driven agenda for efficiency and expertise; and the deadening conflation of energy and fuel — rather than work — that together routinely misconstrues the thermodynamics of architecture. The narrative of the oil embargo only intensified this muddled praxis, a predicament that today continues to occlude a more complete and compelling account of thermodynamics in architecture. So as to avoid the confusions and conflations inherent in this misconstrued embargo litany, let us instead, to quote Odum, "seek common sense overview which comes from overall energetics."[1]

This book is about the history of insulation. But far more importantly, this history of insulation is the perfect alibi to help explicate the non-isolated thermodynamics of architecture: its overall energetics of buildings and its implications for design. Insulation, it turns out, is the consummate counter-concept to the overall energetics and thermodynamics of buildings. This explication of insulation provides the means to move beyond the crippling praxis of energy efficiency and energy conservation that determines and delimits much of the contemporary discourse on energy.

As a start, the selected quartet of quotations offers one way to see around architecture's own corner when it comes to energy and its thermodynamics. These quotations help architects to begin to imagine the otherwise absent implications of thermodynamics in architecture. They thus help architects grasp century-old thermodynamic concepts that position designers to finally fulfill their architectural, urbanistic, and ecological potential.

A history of insulation is now necessary, largely because the energy discourse in architecture generally lacks any sense of irony about the role of isolation in its mangle of energy practices. Without some sense of irony regarding its own contingencies and entanglements, little can be known about a practice as pervasive as insulation. Without its history, one will not know much about how or why current insulation practices and pedagogies are structured the way they are, much less the isolating concepts that systemically over-determine many ideas and practices about energy in general. Throughout modernity, insulation has been a central, emblematic protagonist in the praxis of energy in architecture. How did insulation come to shape so much of what architects think energy is, how it behaves, and what they think they can do with it?

This is a history of a most common material/energy practice in architecture. But this history aims for decidedly uncommon outcomes for architecture: fulfilling the non-isolated thermodynamics of architecture. To reach this aim, in the course of this history, we find just how "isolated, self-contained, ahistorical" architects have built, especially with respect to energy. We will find that architects "never seem to allow for any kind of relationship outside of their grand plan," especially when it comes to energy and everything that modernist architects externalized in their construal of energy systems in architecture. Likewise, in the course of this book, we will find plenty of "isolates, fetishists, lost stumblers-in out of the occupation" who locked out the thermodynamic potential of architecture, often from inside the research laboratories and the corollary lecture halls in architecture schools. Likewise, in ideological *and* thermodynamic terms, the role of consumption and dissipation is not an "isolated or successive moment of the organization of production" but rather constitutive of the open thermodynamic systems of architecture. Thus, from a thermodynamic perspective, consumption is "the ideology of the correct use of the city." In short, a range of isolated energy concepts in the past century led to the evolution of modern architects as somewhat bizarre thermodynamic and ideological animals.

Alternately, we will also consider what architecture stands to gain by designing "dissipative structures" and "order through fluctuations" that characterize the non-isolated, non-equilibrium thermodynamics of buildings, cities, and the universe. This book explains the isolating habits of insulation, but only so as to reveal the potential of a non-isolated praxis in architecture that confronts the vitality of its "overall energetics," leading to an evolved, non-modern thermodynamic and ideological species of architect.

No matter how tacit, out of sight, and benign it might seem, insulation was one of architecture's master modernization agents. For this reason alone, it needs reconsideration. What might initially seem to be a simple matter of heat flux management in turn reveals how inadequate, as a point of departure, the concepts of both management and insulation are when it comes to the actual thermodynamic vitality of buildings. This book is indeed about insulation, but, more importantly, it is also a necessary *contournement* around the misguiding and obdurate concept of isolation that continues to impede the thermodynamic evolution of architecture.

Definitive Isolation

Thermal insulation in the context of buildings commonly refers to the capacity of a material or material assembly to purposively resist the flow of heat in a building's exterior envelope. The concept of insulation, however, has played a much more beguiling epistemological role in the total assemblage of matter and energy in modern buildings, one that reaches far beyond its purported function as a neutral and neutralizing thermal boundary between warm and cold. One of its most paradoxical roles has been the resistance to and occlusion of more cogent — non-equilibrium, non-isolated — thermodynamic concepts and practices. Instead, insulation, it turns out, was focused on quite a different set of isolated, linear thermodynamic principles applicable only to isolated thermodynamic systems.

The concept of insulation — as well as the theories and practices based on this concept — is rooted in the Latin word *insula*, meaning *to isolate or set in a detached condition*. Ancient Romans, for example, isolated themselves from the tumult and filth of early Roman street life by occupying the interior cells of large apartment blocks called *insulae*. Likewise, the Romans isolated and confined their sick to an island, the Insula Tiberina, replete with its own Temple of Aesculapius. The same clear emphasis on isolation in the etymology of the Latin word for island — *insula* — appears in the contemporary word insulation:

—
to insulate is to isolate.

Isolation unquestionably constitutes a clear and recurrent ambition when it comes to energy in architecture. The ostensibly clear — but ultimately obfuscating — role of isolation as a concept has a short but pervasive history in modern architecture. For instance, the Latin etymology of isolation appears in the promotional material of a 1909 cork insulation board manufacturer: "The word insulation is derived from a Latin word meaning island…In insulating a cold storage room, what the engineer tries to do is to make it an island in the ocean of heat."[2] Similarly, consider a *Popular Mechanics* headline from 1924 that also unambiguously expresses this recurrent concept: "Homes Are Turned into Thermos Bottles."[3] While the concept of an isolated interior in an ocean of heat was superb fodder for marketing insulation, the stubborn laws of thermodynamics grant no sustained islands in an ocean of heat and literal isolation is thermodynamically impossible to achieve in the context of buildings. Buildings, it turns out, are not thermos bottles nor would any well considered thermodynamic design aim for that kind of isolation.

Yet this clear and recurrent ambition to isolate has proved to constrain and befuddle how architects understand what actually constitutes energy and its actual thermodynamic behavior in architecture, not to mention its connection to life itself. The impossibility of isolation is even more apparent when both the short and long-term energy cycles of a building are fully considered *together*. Throughout modern architecture, the thermal/exergy flux of a wall has been isolated from the emergy flux embedded in that wall *and especially* from the flux of energy not present in the wall but inextricable from it. The resultant externalizations have engendered a crippling epistemology about energy and a corresponding lack of thermodynamic depth in architecture. For architects claiming an interest in energy, these manifold forms and scales of energy matter and their relative power hierarchy must always be situationally established. But in contemporary practice, a gaping rift persists between the consideration of short-term thermal energy flux through a wall and the longer-term energy flux of matter that presupposes that wall: the total thermodynamic exchanges inherent in architecture. The fetters of this enabling rift have been a central characteristic of insulation practices in modernity and persist today. Energy, when considered at all in architecture, is typically reduced to short cycles of thermal and electrical flux. This is *at best* an incomplete account of energy in architecture. The thermodynamic reality is that energy readily and spontaneously cycles from the molecular to the territorial and from short to long-term energy cycles. So should architects and so must a book on the subject.

How did architects acquire such a peculiarly narrow understanding and practice of energy? Isolation, and the insulating apparatus built on it, has been a false project with false origins, aims, and means. The insulation apparatus in architecture hunted the shadow of isolation rather than the substance of dissipation. Insulation taught modern architects to resist the flow of energy with astonishing and over-reaching effectiveness. The problem, however, is that this is the exact opposite of how architects should engage energy. Instead, a non-modern architect in this century will come to know how to capture, channel, intercept, store, accelerate, and modulate the total energetic dissipation of building *through design*. In doing so, they will finally begin to maximize the entropy production of a building over multiple simultaneous scales, formations, and dissipations of energy. The evolution and engenderment of this kind of non-modern architect is a primary goal of this book.

Temps

As Michel Serres has noted, "by chance or wisdom, the French language uses a single word, *temps*, for the time that passes and for the weather outside."[4] Time, in short, is inextricable from the vortex of swirling energy and matter set in motion by the superabundant influx of solar radiation that drives nearly all earthly processes, not the least of which is architecture, urbanization, and weather. The *temps* and tempos of a building, a city, or climate far exceeds the beat of diurnal heat transfer, and instead pushes towards the rhythmic cycling of the mega-pulsing of energy over very short *and* very long periods and paradigms. Lewis Mumford, for one, chronicled whole phases of civilization by these large spatial and temporal scales of energy.[5] With regard to the thermodynamics of a building and its manifold systems, the shrewd brevity of *temps* as a linguistic convergence of time, space, and energy — when even faintly grasped — is wildly productive, to say nothing of the results when its near boundlessness is deeply plumbed. Buildings are but integrating centers, the hardened core of multiple spatial and temporal scales of energy flux, without which there will be no architectural agenda for energy. *Temps* provides a better and more totalizing starting point for the temporality of thermodynamics than *insula* ever could.

So while isolation, and by extension insulation, is perhaps a clear and compelling colloquialism, its very origin introduced a fundamental irreconcilable thermodynamic conundrum into the core of the modernist agenda for energy: it aimed to physically isolate what it could never isolate. In doing so, insulation *did* isolate architects from more relevant, if not mixed, thermodynamic concepts and practices. In doing so, insulation *did* deeply isolate and externalize long-term energy flux from architects, even as it never fully dealt with small-scale heat flux. This epistemological isolation of long flux from short energy flux has become so deeply structural that architects cannot even envision or readily accept these larger fluctuations of energy that are so essential to the thermodynamics and ecologies of a building, a city, and life itself: the now "imperceptibly large" flux of energy that will become increasingly familiar to non-modern designers.[6]

The persistent focus on isolation in modernity has proved problematic from a thermodynamic point of view, yet it still dominates contemporary notions of energy in architecture. This book documents how this misleading concept of isolation thoroughly conditioned modern architects' understanding not only of insulation, but of energy more generally. Most importantly, once the problems of isolation are firmly established, this book aims to show what architects stand to gain by evolving a non-isolated, non-modern praxis of energy.

Energetic Distortion

My primary claim in this book is that no other concept has disturbed and disfigured our understanding of energy more than the seemingly innocent idea of isolation. Moreover, no material and energy practice in architecture has more completely instilled and reinforced notions of isolation than insulation. So, whence these distorting theories and practices and how to move beyond them?

The core of the problem is this: precisely when architects began to systemically engage heat flux in early modernity, the pedagogies and practices of modern architecture, paradoxically, became epistemically insulated from the architectural and urban vitality of thermodynamics through the prevalence of isolation as a concept. The early preoccupations of insulation fostered a systemic culture of isolation that left few practices of energy untouched in the twentieth century. The concept of isolation is as constrained, from a thermodynamic point of view, as the pedagogies and practices it engendered. If insulation initially aimed to resist heat transfer, historically its habits of mind have also resisted basic thermodynamic realities and possibilities.

Given the persistence *and* impossibility of this perennial isolating ambition, the developments of insulation in modern architecture did much more than attempt to isolate the warm from the cool. In unwarranted and inexplicable ways, isolation privileged certain aspects of heat flux in buildings over others. Ultimately, it also isolated architects from adjacent thermodynamic and architectural possibilities that now deserve consideration in contemporary architecture. This systemic and epistemic oversight is no longer valid for any architect interested in the role of energy in architecture.

Thermodynamic Depth

In insulation practices, isolation conspicuously obfuscates a necessary totalizing perspective on energy over small and long-term temporal cycles as well as over small and large spatial scales. Throughout this book I will characterize insulation theory and practices, like many energy practices in architectures, as lacking thermodynamic depth. Thermodynamic depth is a measure of the complexity, the capacity for complexity, in a system.[7] Complexity in this context refers to the non-linear adaptive feedback that is a central characteristic of life itself. Complexity in this context *does not* refer the trivial, additive, linear, technical complicatedness that has characterized much of contemporary buildings and energy practices. Life strains against mindless complication and thrives on the vital complexities and feedbacks of thermodynamic depth. This thermodynamic depth requires knowledge, methods, and outcomes based on the energy that fluxes in long and short time spans in the smallest and largest of spatial scales. But methodological connections in both practice and the academy between short/long, small/large energy cycles are today as rare as they are necessary. Isolation, in my view, is the central epistemological antagonist between current praxis and a far more powerful, non-modern energy praxis in architecture.

In this century, there is no ecological or thermodynamic justification or purpose for considering short-term energy cycles as isolated from long-term energy cycles. There is no reason that architecture should resist the sublimity of this thermodynamic depth. This thermodynamic depth is what fundamentally drives the energetics of buildings, ecologies, and our planet, yet it has been systemically occluded from modern architecture. Architects thus need to consider this thermodynamic depth in order to make valid observations and claims about energy in buildings. Isolated perspectives on energy are ecologically and thermodynamically troubled and will ultimately be haunted by their energetic isolations. Short and long-term energy cycles are far more connected in reality than most contemporary thermal simulations, codes, or pedagogies will suggest. Why? Because these cycles are also far more connected than what the development of heat transfer and insulation in the context of architecture has suggested for architects.

In terms of contemporary energy dynamics, the lack of thermodynamic depth in architecture remains beholden to the limitations of originating concepts such as *isolation*. As Gregory Bateson observed, "There is an ecology of bad ideas, just as there is an ecology of weeds and it is characteristic of the system that basic error propagates itself."[8] Isolation's weedy ecology of bad ideas and basic errors persists and propagates in contemporary praxis. It persists not only in terms of heat transfer in walls but moreover in many backward, reductive ideas about energy systems and the thermodynamics of architecture.

The Future can no Longer be a Colony of Today

Misguiding propagations and permutations of isolation are evident in standing heat transfer practices for building envelopes, as well as in contemporary agendas for "sustainable," "self-sufficient," "energy-independent," "autonomous," and, most impossibly, "net-zero energy" buildings. In each of these perhaps well-intentioned autarkical claims, isolation serves as the motivation, the means, and the endgame of a series of architectural decisions that rarely have actual ecological correspondence in reality. These problematic concepts are only possible if large portions of a building's energy hierarchy are isolated and externalized from its actual, complete energy systems. If you consider the total energy hierarchy of a building, how could any building, anything consist of net-zero energy? Such a claim requires a most distorted and opportunistic system boundary, one that is difficult to reconcile with that which it externalizes. The future can no longer be a colony of that which is isolated today.

In each of the above agendas and cases, the motivating isolation not only is impossible to achieve but, if achieved, would strain against the large-scale, non-linear, far-from-equilibrium thermodynamic constitution of buildings and cities. Further, and most importantly, *if the full energy system is considered, one would never want to achieve these zero-sum, equilibrium ends because such ends direly minimize ecological engagement and potential complexity through isolating ambitions*. Such claims fundamentally and problematically isolate buildings from their contingent energy hierarchies and engenderment; this is a profound ecological liability, not an asset. A zero-energy building is not only constitutively impossible without opportunistic system boundary selections but would forfeit the great ecological potential inherent in architecture's necessary excess. Buildings are vast vortices of matter and energy. The only relevant question is how that matter and energy can yield the greatest power and vitality, through design.

Isolation will never support a powerful thermodynamic and ecological perspective on energy in architecture. So a primary aim for this book is to help question this characteristic managerial focus on isolation in the energy discourse of architecture. To base so much of energy praxis in architecture on the concept of isolation is most certainly a basic error that cannot be propagated one generation — or even one minute — longer.

Instead of isolating concepts of classical thermodynamics, this book is about the non-equilibrium thermodynamics that drive every building, city, and form of life. In this context, the primary aim of any thermodynamic system is to yield the maximum entropy possible. One strong thermodynamic indicator of maximal entropy is that the intake, transformation, and feedback of a thermodynamic system consumes as much exergy as possible and thus re-radiates energy at its lowest possible level.[9]

Isolated, Non-isolated

The subject of this book is the history of heat transfer and insulation as a practice. However, the salient lesson it seeks to impart is the respective implications of isolated and non-isolated perspectives on energy in architecture: design modalities that alternately emerge from classical thermodynamics in the former and the non-equilibrium thermodynamics that float the operations of buildings, cities, and life itself in the latter.

Today the concept of isolation persistently characterizes architecture's perspectives on energy: from the vain aspiration to thermally isolate interior from exterior, to homeostatic views of sustainability, to net-zero energy claims. Yet autarkical premises are again impossible thermodynamic propositions. So, although this book documents the outwardly innocuous technical, material, and historical development of insulation in modernity, it, again, more importantly documents the habitual and problematic impulse *to isolate* when it comes to energy and the thermodynamics of architecture.

The centrality of an isolating habit of mind is fundamentally flawed, because buildings are anything but isolated in either short or long-term energy cycles. To be emphatically clear and to speak in strict thermodynamic terms — despite pervasive and persistent assumptions to the contrary in modernity —

buildings ARE NOT, EVER:

– isolated
– self-sufficient
– autarkical
– self-sustaining
– homeostatic
– energy-independent
– autonomous
– closed systems
– net-zero energy

Rather, buildings ARE:
– non-isolated, open systems
– transient
– structures of massive energy dissipation

When I say that "buildings are non-isolated, transient structures of massive energy dissipation," I state this in a strictly thermodynamic sense. *Buildings are fundamentally non-isolated thermodynamic systems.* In technical terms, they are always and only open to the exchange of energy and material. In other words, any energy system an architect will engage in practice will be a non-isolated thermodynamic system, for which isolating concepts and practices are rarely valid. Buildings are but the temporary, hardened edge of great energy and material flux that can only be characterized as non-isolated, non-equilibrium yet coherent structures. Buildings are an expression of non-isolated formations of energy captured in matter, around space, for finite durations. Any temporary, instrumental abstraction of an isolated phenomenon in a laboratory or simulation cannot be confused for that which it abstracts and should dissimulate. The conception of a building as an isolated, potentially autarkical system is errant. It is the product of an excessively parochial, isolated view of the thermodynamic systems of buildings.

Open

By open — a more technical way to identify a system as non-isolated — I mean that buildings and cities are open to energy and material exchange between their system and their surroundings. This is an absolute, irreducible, and uncontestable reality and must be the basis of any non-modern agenda for energy in architecture in the future.

"Not only are these systems open," Ilya Prigogine and Isabella Stengers note, "but also they exist only because they are open. They feed on the flux of matter and energy coming to them from the outside world…They form an integral part of the world from which they draw sustenance, and they cannot be separated from the fluxes that they incessantly transform."[10] In other words, buildings are not isolated, autonomous, or self-sustaining systems. They are radically contingent, gradient-dependent, dissipative systems. They only exist — like any form — to dissipate and degrade massive and manifold gradients of available energy, ideally in the most powerful ways possible. If design intervenes on these thermodynamics, it should be in the direction of the maximum power of these dissipative systems.

Buildings intake and dissipate great amounts of material and energy as a constitutive function of their construction, operation, use, and maintenance. They only persist, and can only thrive, because of this intake/dissipate constitution. Again, the isolated, counterproductive notion of a zero-energy building in a constitutively open system, even if possible, would severely constrain a building's potential ecological contribution back into a much more complex and powerful flux of energy.

Occasionally, one might provisionally simplify a system spatially and temporally, as well as reduce system and state variables, sufficiently to treat the system as an isolated and/or steady-state system. In the right hands, this may yield elegant insight into the behavior of a system for design. In the wrong hands, this unwitting abstraction will yield a distorted view of reality and misconstrue the aim of design. It will lead to a practice of isolated objects rather than contingent systems. In all cases, the abstraction of a more thermodynamically finite system should always and only be placed back into its non-isolated system context and energy hierarchy for the best understanding of system behavior and design. The nested energy hierarchies of a building might be provisionally construed as nested steady-state components, but their non-isolated inputs and outputs ultimately have to be re-situated and evaluated in its energy hierarchy.

In too many cases, designers and engineers treat buildings as increasingly "efficient" isolated systems without any regard for the larger energy hierarchies of a building. In doing so, the effect and purpose of the exergy spent on the "energy efficiencies" remain unknown and abstract. In my view, such problematic thermodynamic characterizations are constitutive of an epistemology entangled in the concept of isolation and are thus one of the primary motivations for this book.

These epistemic problems are prevalent in the contemporary architectural discourse on energy. They are problematic because, in their isolation, they treat large and largely indeterminate phenomena as determinant phenomena. In the terms of Karl Popper, they assume "clouds" to be "clocks" wherein open systems are construed as closed, a common enabling fiction for false certitudes.[11] More than problematic, Popper characterized such methodological, epistemic mistakes as a "deterministic nightmare:"

> By a physically closed system I mean a set or system of physical entities, such as atoms or elementary particles or physical forces or fields of forces, which interact with each other — and only with each other — in accordance with definite laws of interaction that do not leave any room for interaction with, or interference by, anything outside that closed set or system of physical entities. It is this 'closure' of the system that creates the deterministic nightmare.[12]

Contemporary buildings, both in their conceptualization and in reality, are too often characterized as closed and deterministic objects from a thermodynamic point of view. The temptation to reduce thermodynamic phenomena to a few factors or forces, while at times necessary, inadequately limits what a building or city is and what it can do. Without the less determinant, more non-linear aspects of thermodynamics that the insulation apparatus routinely isolates, little will be known of thermodynamics. This is a wickedly persistent conundrum of an architectural agenda for energy: how to continuously toggle between multiple spatial and temporal gradient dissipations of energy to ensure that relevant and situated — that is to say open, non-isolated — dissipations of energy are the focus of design.

Contrary to much energy discourse in architecture, I claim that — because buildings are open, non-equilibrium systems — the necessary excess of buildings in open systems is the source of their great ecological and thermodynamic potential, not the source of ecological liability. In this way, buildings should be designed to be far from equilibrium, not at zero or equilibrium. Architects today need a sufficient sense of thermodynamic irony to recognize that buildings are not only massive accumulations of matter and energy, but that this is their greatest ecological and thermodynamic asset in the non-linear, open systems of life.

Transient-oriented Development

In strict terms, buildings must generally be understood as transient, non-steady-state systems. A system is a steady-state system if its state variables do not change over time. In both short and long-term energy cycles, buildings will occasionally exhibit quasi-stable steady-state properties, but only under certain circumstances and system boundaries, and for certain periods of time. In some cases, for the time required for a heat transfer process to occur, for instance, the system might be sufficiently close to steady-state to characterize the limits of the system as such. But even this local, temporary steady-state condition is enabled by many non-steady-state processes. So this local, temporary consideration should, then, recursively include the more general non-steady-state condition of a building for both short and long periods. Buildings exhibit both transient and steady-state behaviors, often simultaneously, at different scales.

The state variables and boundary conditions of a building change with time. Energy fluxes through buildings as boundary conditions vary from seconds to centuries. In shorter-term energy cycles, the sun shines, people occupy spaces, computers and stoves are activated and produce heat, and heating systems are activated, each aperiodically affecting the state variables and boundary conditions of a building.

In longer-term cycles, generally excluded from energetic considerations, buildings are anything but static or steady-state. Buildings are extracted, built, used, maintained, and discarded in a non-stable flow. Each of these states involves vast quantities and qualities of energy generally externalized in modern architecture. Buildings are highly contingent on non-steady-state flows of energy and this flux must be part of a non-anthropocentric perspective on energy. To paraphrase John Oschendorf, contemporary buildings are designed as but waste in transit. This contingency is essential to any productive, homeorhetic understanding of sustainability. Do not let your anthropocentrism fool you: buildings are but one state in a real flow with varying velocities. *All buildings have a velocity as dissipative structures.*

As another instantiation of the non-steady states of building, one can also consider cities such as Rome, wherein generations of people occupy the same buildings in differing ways and with differing uses over the long durée. This also constitutes several different states in another kind of flow with its own shifting boundary conditions. This transience, so important to an ecological program for architecture, is severely lacking in architecture today.

In short, designers must contend, in multiple and synchronous ways, with these simultaneous transient and steady-state dynamics. With respect to these fundamentally transient energy systems, scientist James J. Kay observes the following:

There are three ways to cope with a changing environment.
1. Take control of the environment.
2. Isolate the system from the environment.
3. Adapt the system to the changed environment by:

a. changing the behavior and role of elements of the system;
b. changing the elements of the system;
c. changing the interconnections between elements.[13]

By insulating modernism, architects generally adopted the second option in a vain attempt to achieve the first option. In contrast, this book presents some of the architectural possibilities inherent in the third option when buildings and cities are properly construed as agents in open, non-equilibrium thermodynamics that, while approaching steady-state behaviors at certain scales, are nonetheless constitutively transient, exchange-oriented systems and structures.

Isolation in the context of buildings and buildings envelopes — while conceptually clear and conducive to steady-states characterizations — is itself often but an adiabatic dream, especially when multiple spatial and temporal scales are considered together. Steady-state scenarios are no doubt a valid and necessary concept for isolated study, but that isolated study must always be resituated back in the open realities of its larger energy hierarchy. Given the number of variables in its thermodynamic system, it is better to begin with the understanding that a building is fundamentally a transient system in short and long cycles with fluctuating multi-scalar variables. Working the system the other way around — conceptualizing buildings as steady-state objects seeking an isolated construal of equilibrium — is clearly not working.

In short, buildings might occasionally reflect near-steady-state conditions, but their transient reality in short and long-term energy cycles must be considered as part of an architectural agenda for energy. The pulsing of material, energy, construction, use, operation, and maintenance cycles of a building are all ultimately non-steady-state phenomena, even if they exhibit steady states in the course of a building.

Structures of Dissipation

By structures of dissipation, again in strict technical terms, I mean that buildings are fundamentally sources and sinks of dissipating energy. Since anything, any form exists to dissipate available energy gradients, buildings are well construed as thermodynamic pumps for the dissipation of energy. In short-term energy cycles, buildings are sources and sinks of thermal energy, dissipating exergy, information, material, and dissipated solar energy. Further, the design, construction, and operation of a building are sources of enormous dissipation and act as important means of maximizing power and entropy production through design. Finally, architecture — generally some more highly ordered embellishment of building — requires additional dissipations to maintain its order, appearance, and functions. To cripple these massive dissipations and their latent power with Calvinist guilt — by aiming to minimize energy consumption as the task of energy design — *is to fail to recognize the thermodynamic and ecological purpose and potential of architecture.*

The design implications of the non-equilibrium thermodynamics of a building's ecology are vast and not always immediately apparent, but the fundamental principles and indicators are absolutely clear. As Eric Schneider and James J. Kay note,

> Ecosystems develop in a way which increases the amount of exergy that they capture and utilize. As a consequence, as ecosystems develop, the exergy of the outgoing energy decreases as ecosystems develop. It is in this sense that ecosystems develop the most power, that is, they make the most effective use of the exergy in the incoming energy while at the same time increasing the amount of energy they capture.[14]

Designs for the most powerful dissipative structures — buildings and cities — must thus mind these multiple, non-isolated flows of energy, tracking various exergy and emergy relationships in short and long cycles.

In long-term energy cycles, buildings are storage sinks of massive emergy flows and conductors of other massive emergy flows such as human occupation, use, and stored information. In all of these short and long-term cycles, dissipation plays a vital role in the energy dynamics of a building. In the Calvinist ethos of frugality that characterizes the modernist discourse on energy in architecture, dissipation is something to be minimized through efficiency measures. But this problematic application of a fundamentally cultural/economic habit of frugality on energy and dissipation systems occludes the thermodynamic potential of buildings as structures of dissipation as a productive process. The massive sinks, storages, and flows of energy inherent to a building as a dissipative structure need not necessarily be understood as bad or problematic, for when designed correctly, this abundance is an ideal starting point for designing great ecological exuberance and maximal entropy production.

As such, in this book, I maintain the technical meaning of "dissipative structures" that Ilya Prigogine introduced in 1973 and, more famously, later to a broader audience with Isabelle Stengers in the 1979 book *La Nouvelle Alliance*.[15] For this scientist, the dissipative structure of reality means that inevitable entropic production in open systems at times nonetheless produces new levels of order and organization. In short and long periods, novel orders appear through dissipative fluctuations and mutualities in the system. As such, Prigogine and Stengers were preoccupied with what they described as "the constructive role of dissipative processes" that produces "transformation from disorder, from thermal chaos, into order."[16] As in good exergy design where one function's "waste" energy is another function's exergy, productive order does in fact arise from dissipation so as to maximize power and entropy production. Precisely because of architecture's necessary excess, the productive, projective design of dissipation is a key task for architects in this century. An account of the constructive role of dissipation is absolutely essential to an accurate and powerful architectural agenda for energy.

Far from the notions of equilibrium that characterize isolated perspectives on energy, Prigogine and Stengers remark that "both the biosphere as a whole as well

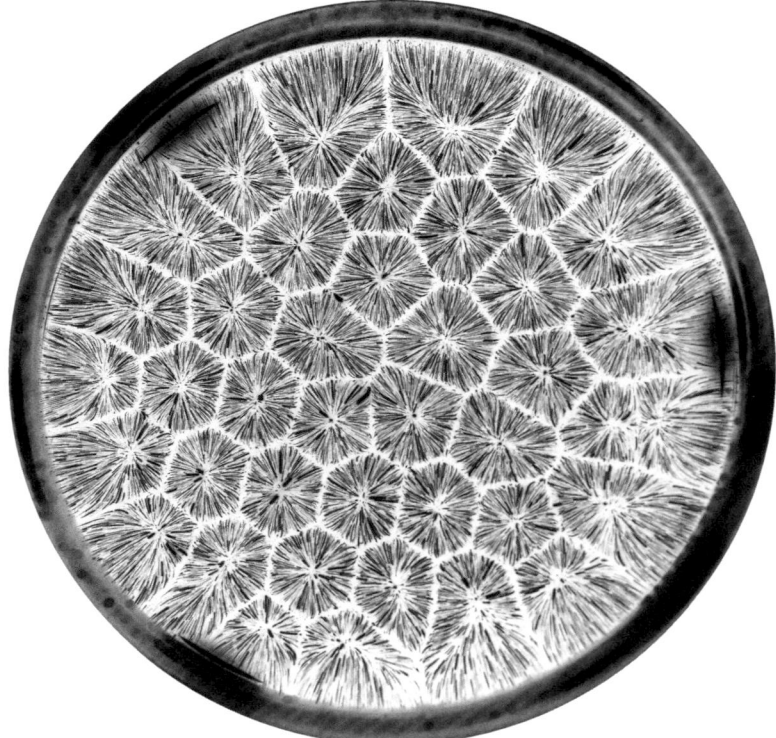

Plate Tectonics, Clouds, Cities, and Buildings are all Material and Energy-convective Flow Structures
Above: Rayleigh-Bénard convection cells / Right: Beni Isguen, Algeria
Grasping the multi-scalar temporal/spatial flow structures that organize and configure all the matter of life is essential to a thermodynamically valid engagement with energy in architecture.

as its components, living or dead, exist in far-from-equilibrium conditions."[17] As a concise example of these dissipative systems, Prigogine and Stengers state that the metabolism of a city "can only survive as long as it is a center of inflow of food, fuel and other commodities and sends out products and wastes."[18] What Prigogine calls "non-linear interactions" in this type of system are what an ecosystem scientist like Howard T. Odum would describe as "feedback reinforcements."[19] Both are key concepts for the latent thermodynamic depth of architecture and its constitutive dissipative structure.

In longer-term energy cycles, buildings and cities share some of the self-organizing properties of dissipative structures. The appearance of pattern in Rayleigh-Bénard convection cells, and the convection pattern inherent in the material, use, structure, and typology of buildings in a city/region are fundamentally and literally

related as dissipative structures. As with Rayleigh-Bénard convection cells, the formation of cellular solids (buildings) in a city from largely cellular solid matter (such as wood and ceramics) fundamentally follows the principles of energy laws and non-linear thermodynamics. The turning over and structured shaping of matter in both examples by the flux of energy follows similar dynamic principles and in both examples exists to dissipate energy in the most powerful way possible. While convection cells and buildings in cities no doubt operate on vastly different temporal and spatial scales, current thermodynamic thinking — such as Adrian Bejan and Sylvie Lorente's Constructal Law — "places the occurrence of design and pattern in nature on the basis of a law of physics."[20] Whether considering the dendrites of neurons, a lung, a tree, or a river system, Bejan and Lorente see no difference in the operative dynamics in the designs and patterns of the dissipative structures that emerge over time. This perspective on the dissipative structures of buildings and cities is, in my mind, absolutely essential to an architectural agenda for both thermal flux and energy in general. Today architects must make readily apparent connections between long and short-term energy systems to make any coherent and valid claim about energy in architecture. In other words, universal energy laws suggest that architects can no longer have isolated views of buildings as energy systems. Dissipation should more fundamentally and optimistically structure architectural agendas for energy.

In the course of the book, I will articulate the implications of a non-isolated agenda on dissipations in architecture through a more totalizing perspective than isolation, by definition, could ever afford. The circumstances of architecture in this century suggest that architects need a more totalizing, thermodynamically accurate, ecologically powerful, and architecturally ambitious agenda for energy than insulation offers through its definitive isolation. Such an agenda would not aim to isolate or minimize energy flux, but rather engage it more directly and fully so as to maximize its relative dissipations, velocities, and ultimately its power.

As a dissipative structure, the ultimate ecological/thermodynamic ambition for building design, thus, is maximum entropy production. As Schneider and Kay observe,

> we expect a more mature ecosystem to degrade the exergy content of the energy it captures more completely than a less developed ecosystem... if a group of ecosystems are bathed in the same amount of incoming energy, we would expect that the most mature ecosystem would reradiate its energy at the lowest exergy level, that is, the ecosystem would have the coldest black body temperature.[21]

How architects might design accordingly must include not only aspects of the building but its very becoming, its functions in a city and civilization, and its eventual fate. Imagining relationships that would push architecture towards maximum entropy production for any given quantity of energy is daunting, to say nothing of designing them, but it is nonetheless an absolute indicator of ecological and thermodynamic capacity. What is most compelling, however, is the way in which these maximal power designs reinforce architectural ambitions, as they always have in the non-modern canon of architecture.

Entanglement

In less strict scientific terms but in more strict historical terms, I also assert that buildings are non-isolated, transient, dissipative structures of social, technical, cultural, economic, and ecological relations. While technical topics in architecture are routinely framed in tidy and supposedly rational ways, history shows that any technical endeavor is fraught with endless entanglements and contradictory demands that make them anything but tidy, rational subjects. Technology is not an ahistorical, autonomous agent but rather a highly contingent cultural artifact, subject to enmeshed demands and persuasions.

Insulation, for example, is not as neat a subject as a textbook on the topic or manufacturer marketing material might suggest. Any technology or technical assemblage of parts and systems is in fact a mangle of practices, procedures, and histories.[22] As anthropological narratives of technology observe, even the most idiosyncratic and novel buildings emerge from practices that are not unique to any single building but instead are highly contingent on a vast history of practices, procedures, and products.[23]

The mixed cultural *and* scientific factors inherent to the entangled development of insulation cannot be overestimated. There is considerable merit in understanding the particular, and at times, peculiar, culture of science that produced heat transfer equations and insulation products. Likewise, there is considerable merit in understanding the thermodynamic cycles that drive cultural and information systems. Both the cultural and scientific are necessary, mutually reinforcing ways to understand the historical phenomena of insulation. One without the other would be incomplete and preclude the compelling mixtures they produce. One without the other would isolate what in reality is non-isolated.

Architects today would immensely benefit from a much greater sense of irony about technological systems in general and energy specifically. Technological practices, procedures, and products reflect a very conditional agglomeration of telluric, technical, economic, social, and cultural systems. These mixtures routinely cause as many problems as they are intended to solve and they rarely determine history, despite much hand-waving to the contrary.[24] From a historical point of view, buildings and building-related practices are decidedly non-isolated and open in this sense as well. Buildings and building practices are radically contingent mixtures of the larger cultural, political, economic, and intellectual conditions that presuppose them.

The technical and historical basis of a non-isolated agenda on energy systems demands that the conceits of otherwise separate technology and history studies in architecture must be merged. In reality, these aspects of architecture are merged, despite what the administrative conveniences and contrivances of academic faculties might otherwise represent. The neat categorization of architecture into isolated topics serves the administration of instruction and tenure dossiers well but it obscures the messy mixing of their respective interests in practice. To this end, this book is at once technical and historical. A non-isolated argument for an inherently non-isolated topic demands non-isolated treatment.

Assemblage, Dispositif, Apparatus, Instrument

The development of insulation in modernity did not occur in isolation, but as the product of many interests, motivations, and agendas. Insulation as a concept and as a practice will not be well understood — or known at all — without some larger grasp of its developmental context. My focus is on the larger assemblage of insulation as a technical implement and my means build on the framework of other related observations about technological systems. It is worth disclosing a few of the primary frameworks that are at the core of this book. For instance, the necessity of apprehending the larger assemblage of any technical implement or practice is emphatically stated by Gilles Deleuze and Félix Guattari:

> the principle behind all technology is to demonstrate that a technical element remains abstract, entirely undetermined, as long as one does not relate it to an assemblage it presupposes.[25]

For the discipline of architecture, insulation, while seemingly familiar, has long remained abstract and undetermined. In other words, insulation has (unsurprisingly, given its contradictory etymological origins and entombment in not only walls) remained isolated from full view and a full account of its role in modern architecture. A major aim of this book is to explicate insulation; to relate insulation to the assemblage it presupposes. Insulation — as one emblematic indicator of how an architect understands technique and energy — should today be far less abstract and undetermined. This is my aim, because insulation and isolation continue to constrain architects today: not only about energy as a grossly underestimated and unconsidered topic, but also about broader possibilities for architecture inaccessible to an isolating habit of mind.

Thinking likewise about the implications of assemblages, Michel Foucault described any assemblage, such as insulation, as a *dispositif*:

> a thoroughly heterogeneous ensemble consisting of discourses, intuitions, architectural forms, regulatory decisions, laws, administrative measures, scientific statements, philosophical, moral and philanthropic propositions — in short, the said as much as the unsaid.[26]

In this book I consider insulation, from its earliest scientific observations to contemporary habits of mind, as a response not only to heat transfer in buildings, but to a larger set of pressures that Foucault indicates in his explanation of the *dispositif*. Thus, I am interested in the "discourses, intuitions, architectural forms, regulatory decisions, laws, administrative measures, scientific statements, philosophical, moral and philanthropic propositions" inherent to insulation. But even more so, I am *keenly* interested in not only the "said" in the discourse on insulation, but also the "unsaid." I am inordinately interested in the various missions *and* omissions of the heterogeneous protagonists and agents involved with the insulation apparatus.

For example, what is included *and* what is excluded in an outwardly innocuous heat transfer equation — and how its results are projected as a simulation of reality — are as important as the solution itself. The certainty with which contemporary numerical models of the thermal reality of buildings are abstracted, codified, and received is a central organizing mechanism of the insulation apparatus and must be routinely questioned if it is to be of any use at all. In this assemblage, certain ideas frequently prevail over others, but rarely for scientific or ecological reasons. Other concerns intervene and, as such, our understanding of insulation is incomplete and inadequate without knowledge of the assemblages that presuppose it.

Regarding the selection of one idea or practice over another in any assemblage, the philosopher Giorgio Agamben builds directly on Foucault's *dispositif* and his reflections are constructive here. In his essay "What is an Apparatus?," Agamben further describes an apparatus as

> literally anything that has in some way the capacity to capture, orient, determine, intercept, model, control, or secure the gestures, behaviors, opinions, or discourses of living beings.[27]

I will show that insulation as a set of concepts and practices is most certainly an apparatus of this type, and has had the capacity to determine, direct, and control the behaviors of architects and the building industry in both productive and unproductive ways. In this way, the missions and omissions of this apparatus are central to the narrative of insulation and demand articulation today. When one considers the range of environmental and epistemological problems that face this century, the specter of the unsaid, the omitted, and the externalized looms large in the history of insulation and its isolating premise. Through these missions and omissions, insulation engendered new subjects and objects, and thereby conditioned the relationship between the apparatus of insulation and that of living beings. It produced a new system of objects and subjects in the modern architectural apparatus. It produced new, often capitulating, roles for architects: new experts, new demands, and new, more "conditioned" occupants in buildings. It was, in short, a master modernization process in architecture. What might otherwise seem to be a simple obligation of energy codes actually coded multiple aspects of contemporary architectural practice and life itself.

Peter Sloterdijk has articulated the social implications of an insulated apparatus in his discussion of "'*Insulierungen*' (insulations)."[28] It is not a mere coincidence that Sloterdijk employs insulation as a metaphor for any isolated social apparatus that grants itself both autonomy and authority. The apparatus of insulation, as discussed in this book, aptly fits the contours of Sloterdijk's observations:

> Just as every group, whether it means to or not, produces its own seclusion in an auditory world proper to itself, as if it were hidden behind a fence of incomprehensibility, so every cultural unit insulates itself primarily and spontaneously through its modus vivendi, or its normative constitution. I am talking about a state of affairs for which there exists no simple and convincing general term, but on which differently tinged customs, cultures, rights, laws, rules, relations of production, language games, forms of life, institutions, and habituses rely.
>
> All insulated human groups, asserting themselves through processes of generation and thereby existing each in their proper time (Eigenzeit), partake in a mystery of stability that has been little investigated but

> without which their continuation could hardly be understood. They generate within themselves a normative architecture that exhibits a sufficiently supra-personal, imposing, and torsion-resistant character to be regarded by its users as valid law, as an apparatus of obligatory principles, and as a coercive normative reality.
> This ethical ether has, to borrow from Hegel, the characteristics of objective spirit: it precedes the individuals as something that stands against — and is unaffected by — what they hold to be right. It is handed down from one generation to the next in a stable (or only imperceptibly modified) form, in the same manner as the names of gods, the myths, and the rituals of a tribe.[29]

The insulation apparatus — the insulated human group par excellence — most certainly fosters "a mystery of stability that has been little investigated but without which [its] continuation could hardly be understood." The foundational equilibrium of this stability is fundamentally questioned in this book. Within this apparatus, a "normative architecture" of insulation standards — reductive rules of thumb, standards, and best practices — cultivates a "coercive normative reality" and a "torsion-resistant character" through its "apparatus of obligatory principles." Worse, the "objective spirit" of the R-value, for instance, projects the moralistic certitude of such an apparatus: any "value" always implies more than mere quantification or arithmetic. One can only hope that when the inadequacies of the R-value concept and metric are articulated and called into question, such values — "handed down from one generation to the next in a stable (or only imperceptibly modified) form, in the same manner as the names of gods, the myths, and the rituals of a tribe" — would also have to be called into question. A less reductive, equilibrium-prone — and more secular — epistemology and practice offers an alternative non-modern model for far less normative architectures: a radically different tribe of architects.

Focusing more directly on the apparatus of science, Peter Galison has observed that the very instruments of any apparatus are never neutral and the instruments of the laboratory merit as much attention as more abstract theory and experimentation.[30] The instruments in this book — from Jean Baptiste Joseph Fourier's early conductive metal rod to the hot-plate apparatus to full-scale, *in-situ* building experiments — are included in this historical narrative because there emerges in the insulation apparatus a sharp distinction between how heat transfer is construed in theory through abstraction in closed systems, and how it is conducted and evaluated *in situ* in open systems. As Galison notes,

> The daily activities of instrument builders and experimentalists and their perspective on their discipline often differ strikingly from each other, and from the view of their theoretical colleagues in the upstairs office. Material culture and the accompanying practices of the laboratory are not vulgarized versions of high theory, nor its primitive building blocks.[31]

Inherent to each of these observers of cultural/technological systems is a lack of faith in the isolated, bounded domains of knowledge that characterize orthodox modernism. This book certainly shares that agnosticism. Indeed, the notion of *boundary* is absolutely essential to the apparatus of insulation and yet it is so poorly construed from thermodynamic and intellectual points of view. The nonsensical vanity that an exterior might be isolated from interior, as in early insulation advertisements, is but a physical manifestation of a larger modernist faith in bounded separations: the separate forms of knowledge, expertise, and practice that characterize the pedagogies and practices of modern architecture. Likewise, that insulation might be reduced to a neat, simple, rational parameter (such as the R-value) reflects a belief in other equally pervasive and partitioned forms of knowledge and practice.

In an effort to move beyond these parochial modern beliefs, Bruno Latour has long cultivated an alternative epistemology, aptly summarized in the following observation from his recent *An Inquiry into Modes of Existence: An Anthropology of the Moderns*. Latour suggests that we:

> need a very different investigative tool, one that takes account of the fact that a border indicates less a dividing line between two homogenous sets than an intensification of crossborder traffic between foreign elements.[32]

Whether in strict thermodynamic terms of more exacting system boundaries, or in the cultural constitution of insulation and its apparatus, the relationship between thermodynamics and architecture indeed too needs a very different investigative tool. Such a tool diverges from what the modern domains of design, building science, and — least and most backwardly — sustainability currently afford as praxis and proxies of energy in architecture. To grasp anything of merit about the energetics of buildings is to grasp the hybrid flux of multiple forms of energy that freely dissipate through multiple spatial and temporal scales: a paradigm of cellular solidarity of co-isolated but communicating forms of energy. More than most domains of architecture, thermodynamics cannot sustain the isolated impulses and false purities of "modern constitutions".[33] There is no room in this century for committed isolates — *committed Latourian Moderns* — when it comes to energy.

> *In short, architects today need a non-modern, non-equilibrium praxis of energy.*

A Partial Science

To understand the history of insulation, one must first grasp that the historical narrative of insulation is not always a narrative of rational scientific development. Thus, its present state cannot simply be assumed complete or to be an optimal response to the thermodynamic reality of buildings, to the physiology of human bodies, or much less to the ecology of urbanization. The constitutive practices of insulation, in fact, address only a few of these necessary scales of energy systems: an inadequate praxis in as much as it is incomplete and partitioned. The praxis of insulation has been, and remains, based on a partial science.

By partial, I mean that the modernist apparatus of insulation concerned itself primarily with a partial understanding of heat flux *and*, on account of that partial view, the modernist apparatus was also partial to certain forms of energy, materials, equations, citations, practices, explanations, pedagogies, industries, and regulations. What is left unsaid in insulation — from the physiological to the ecological — makes it a partial science. What is precluded from prevalent views of insulation — transient behaviors or large scale energy dynamics, for instance — also makes it a partial science.

While partial, it must be noted that insulation is, of course, very well-established and broadly implemented on its own terms. These terms are sound and in many cases benefited and advanced built environments, yet they remain incomplete and inadequate. As noted, in routine pedagogies and practices, the terms of insulation theory and practice isolate certain important thermal parameters (such as thermal conductivity) for attention regarding heat transfer in the context of buildings while also isolating equally powerful, if not complicated, parameters (such as diffusivity or maximum entropy production) from attention. Likewise, insulation theory and practice isolated certain formations of energy (typically short-term forms of energy) for consideration while externalizing others (typically long-term forms of energy). In each case, this partiality has constrained habits of mind about buildings and energy that persist even now. A partial view of the thermodynamics of buildings fails built and un-built environments.

The isolation and abstraction of heat flux in equations is by definition partial: the selection and abstraction of certain parameters and properties over others — a seemingly simple act of rationality — is not neutral, but is in fact a decidedly ideological choice. Given the non-linear thermodynamics of reality, any reductive numerical model can serve to answer only isolated inquiries and, as such, the derivation of any numerical model must raise large questions. The opposite is generally true today: architects routinely grant great authority to numerical models that are assumed to address large questions and thus elicit only isolated inquiry. Numerical models are thus overburdened with expectations the tool is not designed to deliver and cannot ultimately bear. In this admittedly necessary process of abstraction and isolation, what is left out of contemporary numerical accounts of heat flux is as important as what is quantified. A non-modern agenda for numerical models and simulations is needed to advance praxis.

Thusly, this book will cover what has been said and quantified about heat transfer in buildings as it emerged through this apparatus in the twentieth century. This book also attempts to uncover what has not been said, not quantified, about energy exchange in buildings. Any apparatus conceals as much as it reveals and this is certainly true in the case of insulation.

Early modern architectural attempts to modulate heat flow in walls are my starting point, not only because this apparatus reveals one way in which architects are structurally blind to large-scale, non-isolated energy flows but because, paradoxically enough, *study of this apparatus reveals how it occluded full understanding of the heat flux in walls itself!*

To present this at times incongruous historical narrative of insulation, this book is organized into a series of chapters that consider the apparatus of insulation from different, but not isolated points of view. Throughout, my focus is largely on the early, formative years of insulation in modernity and how that period determined a series of theories and practices. In certain instances, I will follow typically more minor theories and practices of insulation through to contemporary states. In successive chapters, I consider when and how certain concepts took hold, why certain material practices prevailed, how a pedagogy of heat transfer and energy systems developed in architecture, and ultimately what would constitute a non-isolated agenda of energy flux in architecture.

Early History of Heat Transfer

The first chapter considers the development of the science of heat transfer in buildings. It tracks the development of key scientific observations, thermodynamic concepts, equations, and standards, a narrative which identifies primary insights and oversights as well as the quantified and unquantified aspects of heat flux in buildings. Thus, I am equally interested in both rigorous scientific description and the looser accounts of heat transfer afforded by material manufacturers that were the primary source of heat transfer education for practicing architects at the time.

Central to this history is the undue historical preoccupation with certain modes of energy at the expense of others: namely, a preoccupation with thermal conduction. On account of this preoccupation, North American insulation theories and practices themselves became isolated from a range of adjacent possibilities and even from the full specificity of heat transfer in buildings. The assumptions embedded in the technological momentum of conductivity continue — largely unquestioned — into the twenty-first century context as evident in bureaucratic standards, energy codes, and absurdly incomplete checklists that frequently characterize the topic of energy, environment, and ecology in the practice of architecture. In this way, insulation is positioned in this book as anything but a neutralizing and isolating component in a building envelope. Rather, in modernity, insulation became a highly active physical, conceptual, epistemological, and historical co-determinant in the habits of twentieth century architectural design and its associated construction practices.

A central aspect of the first chapter is a focus on how the emergence of insulation theory in the context of North American buildings was largely based on prior developments in the realm of refrigeration. While refrigeration was perhaps an obvious and necessary point of departure for early building scientists, in retrospect it is difficult to argue that buildings are refrigerators. Some vessels — refrigerators, thermos bottles — are better suited to theories of heat transfer based on ideas of isolation than buildings. Buildings have radically different thermodynamic contexts and radically different opportunities and capacities for heat modulation and the velocity of energy in a system.

The research and industrial culture of insulation emerged directly from the simpler thermodynamic domain of refrigeration, and the insulation apparatus continues to mime the techniques and technologies of this chilly antecedent. For instance, if you consider a contemporary refrigerator and a double-skin high-rise building, whether in Berlin or Dubai, the operative principles of refrigeration and building science remain all too resonant: a co-determinant approach of envelope and heat pump as the basis of their thermal design and construction. This is undoubtedly a logical outcome of the insulation and refrigeration apparatus. The designs of the refrigerator and the contemporary high-rise are both resultants from strategies that isolate interior from exterior through envelope design, coupled with a heat pump to manage internal convection. But the isomorphism of the refrigerator and the high-rise, or even a Passivhaus, is difficult to resolve given the multifarious obligations and opportunities of architecture when it comes to energy. Buildings are not refrigerators and should consider far more than isolated short-term energy cycles alone. Today architects need to move towards a more ecologically and architecturally powerful manifestation of convergent evolution for buildings than that which the co-determinants of early refrigeration and insulation research engendered for modernity.

Refrigeration and insulation are bonded in a history of technological momentum: a co-determinism of technological momentum that seated modern architecture into particular trajectories. One does not persist well without the other; they are mutually amplifying systems. As refrigerators were scaled up to the size of buildings — first as cold storage facilities — the stage was set for refrigeration-based transformation of buildings: how they were built, who determined their new science, how they were designed, and how they were regulated (both thermally and politically).

In the North American context in particular, the large debt to early refrigeration research largely accounts for its preoccupation with conduction and convection. It also reveals why the R-value concept, for example, so forcefully emerged in tandem with air-conditioning in the mid-twentieth century building market. More directly than the reciprocal of conductivity (a persistent heat transfer platitude which in fact is only valid under certain, narrow circumstances), the R-value concept is the reciprocal of mid-century air-conditioning practices when one looks at the apparatus of insulation and its role in the larger building industry in the twentieth century. To reconsider insulation practices presumes a reconsideration of air-conditioning practices as well.

The R-value concept is perhaps the primary product of the North American insulation apparatus in modernity. Therefore an explication of the R-value concept offers much about the said and the unsaid in insulation. The reductions of this all too neat and curiously incomplete "value" are deeply problematic today — primarily on account of its great commercial success. Its cultural and commercial success is its ultimate technical and ecological failure for architects and contemporary buildings. The R-value concept was derived to sell insulation, not to advance the thermodynamic or ecological efficacy of buildings. The saturation of the R-value concept in the pedagogies and practices of architecture reduces and trivializes an architect's capacity to engage the topic of energy. The more an architect knows about the R-value concept, the more the "value" of its moniker is known to be not a definitive quantity, but an entrepreneurial variable of primarily economic value and motivation rather than ecological value.

A central liability of the R-value concept is that it assumes steady-state conditions. Buildings may operate in steady-state conditions but only in certain circumstances. As one begins to think about the flux of energy over longer periods (minutes, hours, days, weeks, months, years, centuries), many other factors can and should temper an architect's engagement with energy and dissipation. In both short and long-term energy cycles, transient boundary conditions are certainly more complex, but that is precisely their primary attribute: their complex capacity for exchange and feedback in the system. The oversimplified, steady-state narrative of heat transfer through R-values conceals too much from architects. Throughout the development of insulation, a persistent will to simplify the complex dynamics of heat transfer routinely undermined a broader, more convergent approach not only to heat, but to energy at large.

As great as the temptation for abstraction and purification was in modernity — as evidenced in the construal of thermal resistance (R-value) — the messy realities and contingencies of heat transfer resist this simplification and reduction. Further, attempts to isolate particular heat transfer phenomena and resistances are just as futile as the attempt to isolate heat transfer from other cultural constructions. Indeed, heat transfer became a cultural construction in modern architecture. To understand heat transfer without its cultural attachments would be to further pursue in an isolated way that which is fundamentally non-isolated. Thus, understanding heat transfer demands that we not further isolate its aspects but rather the necessity of understanding how and when isolation was attempted, how the isolates succeeded only by knowing what they ignored in their mad pursuit, and that, rare as they might be, we identify the idiosyncratic instances that sought non-isolated mixtures of heat, matter, climate, physiology, and design.

Despite the authority granted to reductionist, numerical accounts of reality in modernity — such as the R-value — any quantitative account of heat flow is incomplete. While these abstractions often serve as necessary shorthand for certain physical processes, they do not correspond to reality in all its compelling complexities, vitality, and varying boundary conditions. This thermodynamic shorthand cannot be mistaken for anything but an abbreviated account of energy, which so often appears in the building industry. The complex dynamics of energy strain against the simple habits of measure, quantification, and simulation. Yet it is precisely the vital complexity of energy exchange — as opposed to heat transfer — that must be seen as an opportunity for architects, not an imposed liability or something to be reduced. After all, heat exchange itself is not all that complex but what it can achieve in its most totalizing manifestation is truly complex in the most vital and productive sense possible. If thermodynamics teaches architects anything, it is a perspective on the non-linear, non-isolated energy systems of buildings and cities that are not at equilibrium.

To summarize, the first chapter ultimately considers how the dutiful apparatus of insulation isolated itself, paradoxically, from the reality of heat flux in buildings — precisely in order to study the reality of heat flux in buildings!

Material History

The second chapter, building on the partial sciences of the first chapter, articulates the material history of insulation: the materials, products, and practices that contribute to the application of insulation theory in architecture. Hardly a product of sound science alone, the material history of insulation has been continuously subject to a range of pressures whose presence neither guarantees sound heat flux in walls nor the energy cycling in the biosphere at large.

The material history of insulation — from archaic pelts to modern iatrogenic products and contemporary foams — indexes more than just material facts about insulation. Notably, insulation materials are but the hardened expression of a range of pressures and forces that determined their history. The history of insulating materials reveals the otherwise isolated motivations that shifted insulating materials to insulating products.

In the short history of insulation materials — especially as they became insulation *products* — it is immediately apparent that insulation was introduced to modulate not only heat flows but, of course, fiscal flows as well. Insulating materials were introduced to architects in the twentieth century through a mixture of scientific, economic, and corporate rhetoric. Insulation corporations or associations, in fact, typically funded early educational materials about insulation. They also described the topic of heat flux in almost exclusively economic terms — rhetoric about fuel and cost savings. In this fiscally-driven milieu of early insulation, architects quickly became *prosumers* of insulation materials. Again, insulation is anything but a neutral and neutralizing material layer in a building. It cannot be considered as isolated from the formative socio-economic circumstances of its material culture that shaped so much of our contemporary practices.

More generally, the long history of insulation matter is a history of human beings discovering and exploiting cellular solid materials. Entrained air cavities in fibers, foams, and fabrications at multiple scales are the fundamental variable in this material history. The materials and motivations vary, but the fundamental parameter remains the same: relative cavity-to-matter ratios in the consistently astonishing class of materials known as cellular solids.

The use of asbestos, as one example in the material history of insulation, reflects how the isolating habit of mind that is so present in insulation theories and practices proves iatrogenically counterproductive when other factors enter the otherwise isolated consideration of this particular material. Why isolate them from long-term well-being as accounted for in emergy analysis? Thus from this material perspective, this chapter on the material culture of insulation articulates how, in the heroic rifts and hubristic separations of modernity, insulation was isolated from great liabilities and potentials. Asbestos best exposes the strange separation of insulation practice from physiological well-being, a topic expanded in the third chapter of the book.

Physiology and Pedagogy

The third chapter articulates how insulation theory was partial in another regard: that of human physiology and architectural pedagogies. Insulation theory rarely leaps beyond the domain of material science and conductivity to consider the role of physiology, even if human comfort is purportedly a primary purpose for considering the heat flux of walls. How a body perceives heat in short-term energy cycles is a non-trivial matter but is rarely present in the development of insulation. From human comfort to thermal delight to new theories of formation based on the thermodynamic figuration of architecture, physiology has been an under-represented agent in the history of modern architecture. Thus, the aim of the third chapter is to expand upon aspects of heat flux from the point of view of human physiology, as one way not only to ameliorate energy flux in architecture, but to enrich architecture and life itself in buildings.

With the human body and human comfort in mind, the third chapter acknowledges that insulation theory and practice does not only address heat flux but the flux of moisture and vapor as well. The mixture of heat, moisture, condensation, and vapor in walls became necessary, and thus better understood, only as construction materials became lighter and thinner. It must be noted, however, that insulation materials and their related construction assemblies only became increasingly sensitive to moisture and vapor as interior air conditions became more controlled and as material became thinner and lighter. This co-determinant tautology is a primary reason why older building materials and insulation methods were systematically dismissed in the course of the twentieth century. In related ways, however, the more massive and diffusivity-based approaches to the thermal modulation of buildings, and the massive, diffusion-based modulation of moisture and vapor have much in common. They both point to an alternate paradigm in which one, or a few, low-transformity layers of construction do the work of many contemporary, high-transformity layers. When this paradigm is coupled with a long-term energy cycle agenda, an altogether different perspective on energy in architecture emerges.

Perhaps the most revealing way to track the co-development of these techniques regarding human physiology is in the introduction and evolution of the "environmental control" courses that entered architectural pedagogies after World War II. In this period, new systems and requirements were introduced to architecture, occasionally through architects, but most commonly through engineers teaching courses on "environmental control" systems. This is a key pedagogical turn in the architectural habitus regarding energy. Architects in this period were not only taught about certain principles and systems, but they were simultaneously taught how to think about those principles and systems. With a few notable exceptions the topic of energy largely entered the pedagogies and practices of architecture from outside of the discipline during this consequential period of development. It is not surprising, then, that architectural agendas for energy and heat flux have been few and far between. Again, what is left out of architectural pedagogies on energy is as important as what has been included.

Dissipation & Design: The Dionysian's Guide to De-isolating Architecture

In response to the first three chapters, the fourth chapter considers the role of insulation — and non-isolation — in larger energy flows, a topic rarely part of any historical or contemporary discussion on insulation. Understanding the relative position of insulation practices of heat flux in larger energy hierarchies and dissipative structures brings the concept and science of insulation closer to the non-isolated, open, dissipative structure of energy in reality. The long-term energy memory — emergy — inherent in insulation cannot be isolated from the short-term energy consideration of a building envelope. The current inflated preoccupations with very discrete modes of heat flux in walls appear inexplicable, considering the magnitude of emergy flux in architecture. Heat flux in walls should remain a central concern for architects, but that concern must both be made more specific and put in a larger context of energy, exergy, and emergy exchange. A less isolated and modern perspective — a non-isolated, non-modern agenda for energy — is necessary for architecture in this century.

In both scientifically and culturally open systems, perhaps the most salient task for building design today is to account — in the fullest degree possible — for the implications of the fact that buildings are non-isolated, transient structures of dissipation. In a non-isolated agenda for energy in architecture, both short and long-term energy cycles provide specific opportunities and obligations for design. A modernist epistemology alone will thus be insufficient. What is designed (the object, the system, or both), how it is designed, and why it is designed as such, all point to distinct opportunities, obligations, and outcomes when a non-isolated and dissipative understanding of buildings orients design. The "order through fluctuations" in non-isolated buildings needs to be captured and channeled in novel and productive ways today, by design.

Thus throughout this book, insulation is the subject of study. But insulation is merely a gauzy fiberglass veil that, under sufficiently close scrutiny, finally permits us to peer into far more consequential and contingent systems: the non-isolated energetic systems and dynamics of buildings. To be emphatically clear, this non-isolated view of architecture in energy systems should extend well beyond our modernist knowledge and practices regarding heat exchange in building envelopes, for walls and roofs are but one type and scale of heat transfer that is innate to the energy exchanges that presuppose architecture. Again and again, architecture's history with regard to insulation reveals just how isolated our understanding of energy became through modernism: how the mineral wool was systematically pulled over the eyes of modern architects. As such, this history serves as an astute and emblematic way to understand why architects and engineers think about energy the way they do today, why they practice energy the way they do today, and how else we could think about it and practice it.

From Heat Transfer to Energy Transformities

Rather than accepting an isolated paradigm of resistance to *heat transfer* — too often reduced to a uni-dimensional, uni-directional, uni-temporal flow — this book suggests that architects instead adopt a more expansive paradigm of *energy exchange* modulation in an open system: multi-dimensional, multi-directional, multi-temporal fluxes of matter and energy. As strange as it may initially seem, the task of energy design in this more thermodynamically cogent, if also complex, milieu is ultimately the design of relative velocities of energy and mass in built and un-built environments. To achieve much more consequential and powerful energy velocities requires a decidedly non-isolated view of the world.

These observations about isolation and non-isolated systems set the terms of this book. In due course, the development and limitations of the isolated perspective on energy in architecture will be tracked through the scientific, material, and pedagogical practices of insulation that emerged in the very late nineteenth and early twentieth century. Why are buildings insulated the way that they are? Why, in certain cases, are they insulated at all? Most importantly, why do we have such isolated concepts and practices when it comes to energy? The aim of this book is to address these basic but unfamiliar questions that have few, if any, studied answers.

The insights, and oversights, evident in the development of insulation in North America have had a profound, and not always empowering, impact on the practices, pedagogies, and performance of modern buildings and modern architects. From the earliest scientific observations to contemporary energy codes, this book discusses the roles of the agents inherent in the technological momentum of modern isolation concepts and insulation practices: architects, engineers, scientists, materials, equations, thermodynamics, codes, test standards, entrepreneurs, academics, supply and value chains, marketing, and consumers. Throughout this narrative of insulation in modernity, a dominant preoccupation with isolation supersedes the non-isolated, open, dissipative reality of buildings.

Insulation theories and practices have no doubt advanced aspects of architecture. Yet, as with any technology, the path determinacy of these theories and practices often dispossesses architecture and society of adjacent possibilities. In this case, a prevalent preoccupation with conductivity — best applied to limited, steady-state, lightweight conditions in buildings — focused on the resistance to dissipation. While a valid concern, when considered in isolation it obfuscated and delimited adjacent practices that would have otherwise proven efficacious in long and short-term energy cycles. Operating as a wild expansion of conductivity considerations, what would an alternate, more totalizing agenda based on the diathermic, dissipative reality of short and long-term energy cycles look like?

Rather than isolation, a more totalizing approach to the energy exchange — rather than the heat flux — of buildings can begin, conceptually, with the absolute inevitability of dissipation as a starting point. Rather than futilely resisting heat flux and dissipation as a primary objective, an alternate approach puts dissipation and diffusion to work in thermodynamically *and* architecturally productive ways. The most compelling work on energy in buildings today begins with this perspective.

Although not named, this approach was also in fact the primary self-organized paradigm of thermal modulation and other energy dynamics for thousands of years of construction in most parts of the world. It was the industrializing amnesia of the nineteenth and twentieth centuries that diminished thermal modulation based on dissipation and diffusion systematically into conduction concerns, by way of modernist entrepreneurship and design. One aim of this book is to revisit certain self-organized approaches — and advance novel non-modern approaches — to the topic of energy exchange and dissipation in architecture.

To neglect the history of insulation, and by extension isolation, would be to let insulation persist as pure technique, an unchecked and crippling epistemology of energy. While it is common in architectural pedagogies and practices to separate technical practices from historical and formal practices as seemingly neutral reserves of technique, I see this as a grave, chronic error embedded in the institutions and practices of architecture.

The foremost of these errors is the assumption that modernist technique and habits of mind will somehow ameliorate the ills of modernist technique and habits of mind. Especially when it comes to our culture's greatly inflated expectations and exaggerated hopes for technology and simulation, technique is never a simple matter and routinely conceals vital possibilities. As the philosopher George Grant observed, "We can hold in our minds the enormous benefits of a technological society, but we cannot so easily hold the ways it may have deprived us, because technique is ourselves."[34] In the case of insulation, there have clearly been enormous benefits, but it is now important to discern how it has also deprived buildings, and us, of (i) other capacities for energy in short and long-term cycles, (ii) other forms of comfort and delight, (iii) other epistemologies that can engender other worlds, and, not least, (iv) other architectural/ecological potentials and capacities.

Given its reluctance to address the epistemologies of its own practices and techniques, the autonomy of modernist building science is as problematic as, if not more than, the knowing conceits of modernist formalist counterpart. With its reduction of knowledge to isolated systems and techniques, building science was condemned to the same fate of autonomous formal agendas. Rather than an incursion into the "autonomy of architecture," a non-modern thermodynamic formation of a building finally fulfills the terms of architecture's most persistant definitions of architecture, from Vitruvius and Alberti onwards. A rich, ambitious non-modern formalism that is indissociable from thermodynamic science would otherwise position technique, design methodology, and simulation/analysis towards radically different and more powerful ends.

What is non-modern? In simple terms it is that which is not modern in its posture, habits, and practices.[35] The non-modern generally resists the linear, equilibrium-seeking, positivist, mechanistic, techno-bureaucratic managerial posture associated with the modern constitution. This would include pre-modern modalities, but non-modern practices could never be construed as Luddite recidivism because such practices are much more projective, keenly pursuing more powerful futures for architecture and life itself.

System Boundary

Do not let unquestioned system boundaries fool you. It is an enormous thermodynamic error to associate the thermal exchanges of a building primarily with its building envelope. Vast forms of dissipated heat riddle the energy hierarchies of a building. Without any scale analysis, it is impossible to know which form of thermal dissipation matters most in a building or, more importantly, in its larger energy hierarchy and ecology.

In this non-isolated perspective on energy systems, perhaps the most critical and persistent question for a designer today must be the question of system boundary. What system boundary matters for a given design? How is that system boundary situated back into a nested hierarchy of total energy? These questions must become a central habit of mind for the thermodynamically-minded designer. This book finds the neat boundary between the cultural and scientific accounts as more of a mixed-up collusion in reality. Likewise, the system boundary of buildings must be radically questioned today, for buildings are but the manifestation of much larger systems.

Whether in terms of their energetic, urban, ecological, or formal ambition; architects frequently select delusional system boundaries. To paraphrase Thomas Pynchon, architects currently lock themselves inside peculiar rooms, inside peculiar system boundaries. The near-sightedness of this false autonomy is as architecturally unambitious as it is thermodynamically impaired. More ambitious, thermodynamically accurate system boundaries and, thus, forms of architecture are necessary today not only for the vitality of life on this planet, but for the vitality of the discipline of architecture itself. A non-isolated, radically contingent perspective is absolutely critical equipment to avoid the constraints of these rooms and boundaries locked from the inside.

This book aims to unlock these unreflective, systematically taught system boundaries. No system boundary deserves to be blasted wide open more than the assumption that the building envelope is the system boundary of a building. Once these parochial boundaries finally dissipate, the only thing that penultimately matters is the relative velocity and temperature of the system in short and long-term cycles. The only thing that *then* matters is how architects temper the relative rate of available and dissipated energy mixing in the system through design.

Passiv-aggressive

If the recent Passivhaus paradigm is perhaps the ultimate logical extension of the received insulation apparatus in architecture, the fourth chapter thus concludes with a consideration of a *Massivhaus* paradigm. Massivhaus constitutes a broader non-isolated, non-steady-state, dissipative way to think about and practice energy in architecture. How architects might otherwise design heat flux, shift boundary conditions, and evaluate emergy accumulation elicits great architectural potential in this paradigm. By the end of this chapter, I ask the reader to consider a mode of architecture that, rather than operating with an illusory notion of isolation through insulation, fundamentally accepts and makes productive the inevitabilities of dissipation in the building envelope, building structure, and whole civilizations, so as to maximize the use of available energy. This approach is thermodynamically more cogent as a starting point for practicing both short and long cycles of energy that are highly relevant to the work architects need to do in this century. It also offers unique formal, structural, spatial, and material opportunities that can serve to amplify the ambitions of architects and ecologies today.

Ultimately, in these chapters I point to an alternate apparatus of energy exchange in buildings, one that reaches far beyond the heat transfer in a wall. As such, this book oscillates between cultural accounts and technical topics to pose broader questions of knowledge and the design of buildings. Far from effacing technical accounts of building science, the aim of this book is to extend the science of buildings into much larger domains, for buildings are, again, ultimately non-isolated, non-steady-state, dissipative objects and assemblages. Diffusion and dissipative structures are concepts of utmost importance to energy systems at multiple spatial and temporal scales and boundary definitions. I offer diffusion and dissipative structures in the context of architecture as the most operative units of understanding about energy in this expanded view. The story of insulation in modernity in this book merely serves to reveal at once much larger and much more nuanced lessons about energy in the dissipative domain of buildings.

Excerpts from an Atlas of Non-isolated, Non-modern Architecture

Once our attention about the manifold energy dissipations inherent to buildings becomes less isolated and singular in focus, it is productive to consider how agendas for the thermal milieu of buildings converge with other obligations and opportunities in architecture through design. So, to punctuate each of these chapters, I offer a few non-modern building case studies to further illustrate the specific architectural implications of each chapter.

The buildings included in these chapters as examples are excerpts from a larger atlas of non-isolated, non-modern architecture. They illustrate important alternative concepts and departures from the insulation apparatus chronicled in the historical chapters. Some of the examples delve into non-isolated thermal exchanges more directly; others consider the otherwise externalized thermodynamics of modern architecture. The best of the examples do both, through design.

Among the examples is an early in-situ thermal test on building materials in Norway. Another example considers the repeatedly misunderstood thermal architecture of Frank Lloyd Wright, as evident in the Jacobs House. Also frequently inadequately characterized, the architecture of ancient Roman baths offers one way to see the questions of matter, space, energy, body, and form converged through design in non-isolated ways. The final example, the Rauch House, considers architecture as an act of "controlled erosion," pointing towards much larger and often much more powerful forms of dissipation inherent in architecture. This building is coupled with Sal Craig's own contribution on the role of contra-flow design as an envelope-scaled consideration of non-isolated design.

Collectively, the intricacies and specificity of each of these non-modern examples far exceed the crippling simplifications of R-value characterizations of dissipation and its role in architecture. These cases help articulate a most important thermodynamic and thus ecological principle: that through design, buildings that maximize energy intake, deploy exergy matching and use, and use feedback reinforcement as part of a maximum power, maximum energy velocity approach to design, should prevail.

In each example used in this book, the thermal strategy engages not only the short energy cycle of the building but simultaneously aims to maximize the power of its larger energy hierarchy and dissipations. More than specific techniques, what is fundamentally at stake in the following projects are non-modern architectural and ecological habits of mind that reflect a very different agenda for energy than the one that shaped insulation in the twentieth century. The modern insulation apparatus never fully disclosed the energetics of architecture to architects. These examples help articulate some, but only some, of the implications of the dissipative structure of reality in built forms.

What is at stake in the case studies, and in the book as a whole, is not merely a more subtle, yet more expansive understanding of energy exchange in buildings. *What is ultimately at stake is a radical shift to a non-isolated habit of mind about energy in architecture.* Insulation is not an isolating boundary between interior and exterior. It never performs that thermodynamically impossible ambition. Instead, the insulation apparatus has been more consistently successful in creating boundaries between simultaneously more nuanced and broader concepts of energy exchange in architecture. Much more than walls were insulated in modernity: in the course of modernity, insulation became a highly active physical, conceptual, and historical agent in the determinant habits of twentieth century architectural design and its associated construction practices.

To understand the apparatus of insulation is to come to better understand the larger flux of energy in architecture. To rethink the premises and developments of insulation also aids in rethinking a range of systemic energy concepts and practices in architecture. The flux of energy is not an isolated technical matter, as it is so often treated in the current pedagogies and practices of architecture. Buildings, as non-isolated, non-steady-state, dissipative structures of matter and energy, demand much greater facility with the flux of energy than insulation affords architects.

This book ultimately projects a non-isolated agenda for energy systems and dissipation. As such, aspects of this book may appear foreign to an orthodox, more isolated, building science, architecture, or engineering perspective. This might be the case because such orthodoxies have long only partially considered the role of energy dissipation in short-term energy cycles. I invite the reader to see around the corner of insulation's deep-seated orthodoxies and platitudes in order to look instead towards the "overall energetics" of architecture. More than the intricacies of any particular scale of energy behavior, what architects today need to see are the full implications of the energy system dynamics of buildings: the total thermodynamics of ecosystems. Many of the theories, techniques, and technologies of building science remain important, of course, but these concerns become so much more powerful and ecological when the thermodynamics of large scale energy systems and ecosystems are coupled with the insights of short-term energy cycle concerns: when the dissipative structure of buildings is finally considered in full. When short and long-term energy cycles recursively shape architectural agendas, the basis for more sane and powerful buildings/urbanization is in place.

Today architects need an expanded view of thermodynamics that exceeds what the managerial tautologies of R-values, air-conditioning calculations, energy simulations, embodied energy values, life-cycle analyses, and energy codes can disclose about thermodynamics alone. Ecologically powerful, if not fecund, modes of building design and construction are possible, but extant concepts and tools must be augmented with other, admittedly less familiar, concepts, methods, models, and materials. This non-modern approach to energy flux — a mongrel of the greater archaic intuitions about the energetics of life and architecture coupled with certain techniques, analytics, and methods available in contemporary modalities — is one way around the limits of either pre-modern recidivism or the many fetters of a modernist, autonomous apparatus.

As pre-modern "designers" intuited and contemporary system ecologists know, *architects must not aim to minimize the impact of buildings on the environment.* Rather they must become much more powerful and productive protagonists of abundant and vital environments that result in maximum productive impact. Their impact must be substantive, and amplify the abundance of non-isolated world dynamics. This book aims to help enhance and extend this effort.

Temperare

To conclude this introduction, this book in sum presents a greater mixture of forces and fluxes that have characterized insulation in buildings in the last century: a mixture that the isolated characterizations of heat transfer in a textbook will never offer. This cultural/technological mixture is decidedly different today than in the thousands of self-organized years of construction that yielded such powerful systems before the industrialization of building and the professionalization of architects. But whether archaic, modern, or ideally, non-modern, the apparatus of energy exchange in buildings is always a multifarious mixture that cannot be ignored or externalized. So, at the conclusion of this introduction, this inevitable mixture of forces and fluxes recalls another etymological root that is compelling to consider in the thermodynamic milieu of this book. As historian of thermodynamics Ingo Müller notes:

> Temperature — also *temperament* in the early days — measures hot and cold and the word is, of course, Latin in origin: *temperare* — to mix… The passive voice is employed — the "*-tur*" of the present tense, third person singular — which indicates that some liquid *is being mixed* with another one.[36]

Much more than just hot and cold is mixed in the temperature modulation of modern buildings. Thus, not only does the nuanced and complex mixture of energy exchange deserve our attention, but the equally idiosyncratic and complex mixture of *how* heat flux in buildings has come to be characterized and modulated by now also deserves our attention. Further, architects would certainly benefit from a greater epistemological mixture of pre-modern and non-modern approaches to energy in architecture. The heat flux in walls is inordinately nuanced and compelling, but so is the larger apparatus of insulation and even more so the delirious dissipative thermodynamics that float the operations of life. As manifestations of both thermodynamics and social dynamics, temperatures in buildings are always and only complex mixtures of far more than hot and cold.

This book — at once a technical, material, and cultural history — aims to look more directly at aspects of this thermal mixture that has always been less clear, less isolated than its technocratic treatments suggest. To fulfill architecture's own terms in this century, the divergent implications of isolated and non-isolated, of modern and non-modern perspectives on the dissipative structure of buildings are of profound consequence for architects. To help articulate the significance and substance of the latter for architects is a primary ambition in this book.

The Latin root of temperature — *being mixed* — is another astute starting point for understanding energy systems and their ceaseless technical and cultural exchanges. Any energy system that designers will encounter is an open system: constantly mixing, exchanging vortices, velocities, and discourses over short and long temporal cycles. Any particular isolated spatial or temporal scale — perhaps provisionally necessary — must always be synthesized back into its larger, non-isolated energy hierarchy, back into the mix through a recursive methodology. Contemporary building scientists and architects do not yet do this. Any powerful understanding of energy systems in architecture must begin with a grasp of the endless mixing of open systems and how buildings fit with the resulting cascade of energy in an energy hierarchy and how to maximize exergy use, thereby maximizing entropy production. Without that, designers hazard many enabling fictions — false system boundaries and curiously a-contextual quantifications — that ultimately restrict design to a number of thermodynamic platitudes.

This book is intended to be a handbook for explicating and expanding the boundaries of architecture locked from within the discipline's routine working habits and assumptions: its most parochial system boundaries, whether physical, professional, or epistemological. A shift from heat transfer in closed system thermodynamics to energy exchange in open systems — the shift from isolated perspectives on energy to non-isolated perspectives on energy, and from a modern to a non-modern intellectual posture — is the purpose of this book. The thermodynamic, ecological, and intellectual vitality of this shift can no longer be isolated from contemporary architecture. An architectural agenda for energy is long overdue and no other topic reveals more about this agenda than an explication of the insulation apparatus of modern buildings.

In closing this introduction, it should be noted that insulation is but one aspect of architecture's many routine procedures and practices that are taken for granted. Thus, the present study is emblematic of many other unexamined techniques in architecture. In this sense, this book is only one contribution to a larger, and largely unwritten, history of practices that questions architecture's most habitual working procedures as the site of evolutionary, liberating novelty.

Notes

1. Howard T. Odum, "Energy, Ecology, and Economics," *Ambio*, vol. 2 (6), 1973. p. 220.
2. Armstrong Cork Company Insulation Department, *Nonpareil Corkboard Insulation: for cold storage warehouses, abattoirs, breweries, ice plants, fur storage vaults, dairies, creameries, candy factories, bakeries, fish freezers, canneries, refrigerators, freezing tanks, and generally wherever refrigeration is employed.* Pittsburgh, Pennsylvania. 1909. p. 9.
3. *Popular Mechanics Magazine,* June 1924. p. 868.
4. Michel Serres, *The Natural Contract,* Ann Arbor, MI: The University of Michigan Press, 1995. p. 27.
5. Lewis Mumford, *Technics and Civilization*, New York: Harcourt, Brace & World, 1963.
6. John May, lecture, Harvard Graduate School of Design, April 20, 2012.
7. Seth Lloyd and Heinz Pagels, "Complexity as thermodynamic depth," *Annals of Physics,* vol. 188, issue 1, 15 Nov. 1988. pp. 186–213.
8. Gregory Bateson, *Steps Towards and Ecology of Mind,* New York: Ballantine, 1972. p. 484.
9. Eric D. Schneider and James J. Kay, "Complexity and thermodynamics: Towards a New Ecology," *Futures*, vol. 26(6) 1994, p. 626–647.
10. Ilya Prigogine and Isabelle Stengers, *Order Out of Chaos: Man's New Dialogue with Nature*, New York: Bantam Books, 1984. p. 127.
11. Karl Popper, *Objective Knowledge: An Evolutionary Approach*, Oxford: Oxford University Press, 1972. pp. 206–255.
12. Ibid. p. 219.
13. James J. Kay, "Complexity theory, exergy, and industrial ecology," in Charles J. Kilbert, Ja Sendzimir, and G. Bradley Guy, eds., *Construction Ecology: Nature as the basis of green buildings*. London: Spon Press, 2002. pp. 94–5.
14. Schneider and Kay. pp. 637–39.
15. Ilya Prigogine and Isabelle Stengers, *La Nouvelle alliance: Métamorphose de la Science*, Paris: Gallimard, 1979. (English translation: Ilya Prigogine and Isabelle Stengers, *Order Out of Chaos: Man's New Dialogue with Nature*, see footnote 10)
16. Prigogine and Stengers, *Order Out of Chaos: Man's New Dialogue with Nature*. p. 12
17. Ibid. p. 175
18. Grégoire Nicolis and Ilya Prigogine, *Self-Organization in Nonequilibrium*

Systems: From Dissipative Structure to Order through Fluctuations, New York: John Wiley & Sons Inc., **1977**. p. **4**.

19   For an elaborated discussion on the role of feedback reinforcement in energy systems, see multiple references in: Howard T. Odum, *Environment, Power, and Society for the Twentieth century: The Hierarchy of Energy,* New York: Columbia University Press, **2007**.

20   Adrian Bejan and Sylvie Lorente, "The constructal law of design and evolution in nature," *Philosophical Transactions of the Royal Society, Biological Sciences*, **2010**, **365**. p. **1335**.

21   Eric D. Schneider and James J. Kay, "Order from Disorder: The Thermodynamics of Complexity in Biology," in Michael P. Murphy, Luke A. J. O'Neill, eds., *What is Life: The Next Fifty Years. Reflections on the Future of Biology*, Cambridge, UK: Cambridge University Press, **1995**. pp. **169–70**.

22   Andrew Pickering, *The Mangle of Practice: Time, Agency, and Science*, Chicago: University of Chicago Press, **1995**.

23   Merrit Roe Smith, ed., *Does Technology Drive History?*, Cambridge, MA: The MIT Press, **1994**. For an example, consider Jim Johnson (Bruno Latour), "Mixing Humans and Nonhumans Together: The Sociology of a Door-Closer," *Social Problems: Special Issue: The Sociology of Science and Technology*, vol. **35**, no. **3**, Jun., **1988**. pp. **298–310**.

24   Ulrich Beck, *World at Risk*, London: Polity Press, **2009**.

25   Gilles Deleuze and Félix Guattari, "A Treatise on Nomadology," in *A Thousand Plateaus: Capitalism and Schizophrenia*, Minneapolis: University of Minnesota Press, **1987**. pp. **397–398**.

26   Michel Foucault, "The Confession of the Flesh" (**1977**) interview, in Colin Gordon, ed., *Power/Knowledge: Selected Interviews and Other Writings 1972–1977*, New York: Pantheon Books, **1980**. p. **194**.

27   Giorgio Agamben, *"What Is an Apparatus?" and Other Essays*, Stanford, CA: Stanford University Press, **2009**. p. **14**

28   Peter Sloterdijk, *Sphären III: Schäume, Plurale Sphärologie,* Berlin: Suhrkamp Verlag, **2004**.

29   Peter Sloterdijk, "The Nomotop: On the Emergence of Law in the Island of Humanity," *Law and Literature*, vol. **18**, no. **1** (Spring **2006**). pp. **1–2**.

30   Peter Galison, "History, Philosophy, and the Central Metaphor," *Science in Context*, vol. **2**(1), **1988**. pp. **197–212**.

31   Ibid. p. **209**.

32   Bruno Latour, *An Inquiry into Modes of Existence: An Anthropology of the Moderns*, Cambridge, MA: Harvard University Press, **2013**. p. **30**.

33   For more on the "modern constitution," see Bruno Latour, *We Have Never Been Modern*, Cambridge, MA: Harvard University Press, **1993**.

34   George Grant, "A Platitude," in *Technology and Empire*, Toronto: Anansi, **1969**. pp. **137–43**.

35   For one expanded consideration of the non-modern see, Latour, *We Have Never Been Modern*.

36   Ingo Müller, *A History of Thermodynamics: The Doctrine of Energy and Entropy*, Berlin: Springer-Verlag, **2007**. p. **1**.

"... by confidently positing techno-scientific management as the solution to life's uncertainties, modernization has exposed flaws in its own conception of 'the natural.'"[1]

John May

A History of Heat Transfer in Buildings

In this chapter, I track the historical development of heat transfer science in relation to buildings: from its origins in early scientific observations about heat to contemporary assumptions and codes regarding insulation in buildings. This account articulates the roles of the various agents inherent in the technological momentum of insulation and its associated practices: architects, engineers, scientists, materials, equations, thermodynamics, codes, test standards, entrepreneurs, academics, marketing, and consumers.

The complexity of heat transfer in buildings and building assemblies is itself an antagonistic, somewhat elusive actant in its own history. In the development of insulation theory and practice, the intricacies of heat transfer frequently resisted thorough study and, in turn, routinely favored reductive characterizations for practice. While physicists and engineers have elegantly accounted for the conjugate behaviors of thermodynamics and/or fluid dynamics at multiple scales in this history, neither the elegance of the analytics nor the breadth of behaviors are commonly manifest in the practices and pedagogies of architects. The occlusion of important heat transfer mechanisms and scales thereby privileged a particular set of reductive insulation theories, practices, and products in the insulation apparatus. This tendency towards reductive, violent abstractions directly contributed to the technological momentum of certain theories and practices over other, more complete and thus much more powerful approaches.

Further, as more reductive characterizations came to determine how architects thought about energy, even the most elegant and rigorous characterizations of heat transfer in a building still externalized other forms of energy and its inevitable dissipation: the larger scales of dissipation that presuppose and enable the building. An architectural grasp of the compound diffusions and dissipations of heat at multiple scales — the thermodynamics of architecture in aggregate — is absolutely essential to a relevant design agenda for energy today, if not just to know their relative importance in any given circumstance and the scale of analysis that buildings in fact represent in the energy flux of their energy hierarchies.

But even when considered as small-scale thermodynamic objects, overly reductive characterizations of heat transfer nonetheless emerged and constrained the thermodynamic imagination of architects. Buildings — as non-isolated energy systems — are in reality almost always characterized by a nuanced set of transient behaviors with shifting boundary conditions. While they can be considered steady-state for finite periods and treated as such in various heat transfer equations, the larger transience of a building cannot be discounted or ignored. Buildings are lodged in simultaneous, seemingly imperceptible multi-scalar fluxes and hierarchies of energy. Both are absolutely essential to an architectural agenda for energy. One without the other is at best an incomplete, non-recursive consideration of the energetics of architecture; at best a modernist conceit rather than a design for thermodynamic depth.

A Bifurcation from Reality

In the course of the twentieth century, the recurrent tendency was to simplify and purify concepts of heat transfer, typically reducing it to less complex matters of heat conduction in walls while entirely ignoring much larger dissipations of heat. This desire for abridged cognition is somewhat understandable, for as building assemblies and systems became more industrialized and complicated in the course of the twentieth century, so did the actual complexities of heat transfer at the wall or building scale. These complicated thermal behaviors only amplified a will for their abstraction and reduction into rudimentary analysis and quantification. Mounting market pressure favored, and ultimately implemented, an abridged approach to and characterization of heat transfer. This served market interest foremost and heat transfer knowledge — and the potential of architecture — less. The result was a bifurcation from reality: heat transfer became more abstract and isolated so that a partial science could address a broad market. This is a central problem of the insulation apparatus: the only way to get broad saturation for heat transfer in practice was to isolate more nuanced and complicated behaviors. The abstractions and externalizations of heat transfer, not to mention the larger, enabling externalizations of the modern insulation apparatus, thus shaped not only reductive assumptions about insulation but made its externalizations systemic.

Today the looming externalities of wall or building-scale energy flux haunt this reductive modernist agenda for energy. As such, a few questions were the impetus for the research of this historical account of heat transfer in buildings. These questions are applicable equally to agents (people, equations, etc.) in this history and to contemporary architects, academics, and techniques. They include: What is your system boundary? What do you externalize? What do your quantifications elegantly simplify and what do they oversimplify?

Hardly a *Pièce de Résistance*

With such questions in mind, this narrative of the technological momentum of certain heat transfer practices, especially in the context of North America, is particularly revealing, because it helps explain how and why a steady-state understanding of heat transfer — as exemplified in the now ubiquitous R-value concept — could come to serve as such a crude yet prevalent proxy for the actual heat transfer behavior of buildings. As oversimplified as it is pervasive, the concept of the R-value of insulation dominates the building industry in North America, as evident in building codes, building science pedagogies, and green building certification programs. R-values have come to represent all of what many architects and engineers know about heat transfer in practice. Given its current technological momentum, the assemblage of scientific, technical, cultural, and entrepreneurial factors that structured the development of insulation in North America now warrants explication.

Thermal resistance — the R-value of a material — ostensibly indexes the resistive capacity of a material. But more than relative thermal flux in a material, what the R-value concept resists most is a complete and accurate description of heat transfer in the minds of architects and other agents in the building industry. The R-value concept introduced heat-resistance concepts and materials into the market, and in doing so, it also introduced resistance to more thorough understandings of heat transfer in that same market. One aim of this chapter is to establish a pre-history of the R-value concept to help articulate its original purposes and clear limitations.

The historical narrative of insulation and heat transfer in the building industry thus presents a core conundrum: the actants involved in the development of building insulation occasionally had motivations to isolate themselves from the realities of heat transfer. They did so in order to simplify the study of heat transfer and so construed a particular, limited understanding of heat transfer. The motivations were multiple, often simultaneous, and rational on their own terms: to resolve prior scientific problems, set national material and performance standards, and increase financial gain and/or market share. A self-reinforcing dynamic of scientific rigor, a recurrent will to abstraction, market pressure, corporate cant, and consumer choice all characterize the history of insulation in the twentieth century building industry. This chapter charts the development of insulation's unremittingly complicated technological momentum and heat transfer logics through the terms of this core conundrum.

Insulation & Heat Transfer

Again, the word *insulation* is derived from the Latin *insula,* which means to isolate or set in a detached condition. In this very name, then, from the beginning a false and thermodynamically impossible ambition is embedded in the root concepts and vocabulary of insulation: to isolate interior from exterior. While perhaps conceptually clear, this nomenclature triggers a cascading set of implications and associations that constitute the first of many problems associated with the development of insulation in modernity: insulation theories and practices do not always align in name or in reality with the physics of heat transfer in buildings. Not only is the idea of insulation problematic, it has problematically conditioned how many architects, engineers, and building scientists think about energy in general: as a sport of quantities played within narrow system boundaries.

Unfortunately, decades of architects, engineers, and professors now think of the energy systems of buildings as a sport of energetic quantities, minimized ostensibly to be less bad or to avert the specter of thermodynamic death. But buildings are open thermodynamic systems, so the question of energy in architecture cannot be posed in linear interpretations of the first law alone and the peculiar concepts of efficiency and conservation. The profound implications of the second law must drive architecture not away from thermodynamic death but towards thermodynamic depth. From the beginning, however, thermodynamic death and pessimistic views of dissipation motivated the discourse on insulation, in which it became a phenomenon to be minimized.

Other problematic concepts emerged early in this history. For instance, one early, emblematic textbook on insulation presents insulation as a "non-conductor" of heat.[2] While this, too, might seem clear and straightforward, it is also false from a physical point of view: every material conducts heat (except at zero Kelvin). This common view of insulation as a non-conductor and isolator of heat belies many of the conceptual and physical problems that have beset the history of insulation practices and architecture's thermodynamics. Why not characterize materials more directly as conductors with different properties, capacities, and velocities? A more astute view would focus on the relative efficacy of the inherent dissipative mechanisms, not just for heat transfer, but also for the total exergy exchanges in buildings. In reductive accounts, though, the inevitable flow of heat in matter is seen as problematic rather opportunistic. The "isolator" and "non-conductor" lexicon for heat transfer fundamentally mischaracterized the topic from the outset.

The "non-conductor" perception of insulation nonetheless became prevalent throughout the twentieth century. The means and ends of insulation were thereby quickly reduced to a focus on lowest conductivity in a steady-state condition. Yet the transfer of heat, of course, occurs in three modes — conduction, convection, and radiation — and most often in simultaneous modes with shifting boundary conditions in either transient or periodic modalities. Conduction is certainly a primary concern in heat transfer in the context of buildings, but all three of these modes of heat transfer are often present, mixed in simultaneous and transient conditions. While a primary concern, conduction should rarely be the only concern. So whence the preoccupation with conductivity?

The twentieth century preoccupation with lower thermal conductivity in architectural insulation practices was based on a then extant body of scientific knowledge of thermal conductivity, mostly as manifest in contemporaneous refrigeration research. This refrigeration research was based on now famous mid-nineteenth century scientific developments regarding heat and thermodynamics. These initial thermodynamic observations in turn emerged from earlier characterizations of heat and temperature. Early scientific observation and derivation of thermal conductivity formed the basis of this chain of observations leading to twentieth century insulation practices in buildings and is thus the starting point of this account.

Thermal Conductivity

The mathematical characterization of thermal conduction — the molecule-to-molecule flow of heat through a material or a material assembly — was first formally enumerated by Jean Baptiste Joseph Fourier in a series of manuscripts, presentations, and publications in the early nineteenth century.[3] Beginning in 1804, Fourier empirically came to understand conduction as a function of the heat transport, heat storage, and the boundary conditions of a material. By 1807 he had derived a partial differential equation that accurately and totally characterized the intricate problem of transient heat conduction. In 1822, Fourier formally presented and published his observations on the flow of heat in *Théorie analytique de la chaleur* (*The Analytic Theory of Heat,* as published in English in 1878).[4]

Fourier's equation, here represented in somewhat simplified form for transfer in a single direction, for the conduction of heat in solids was as follows:

$$q = -k \, (\Delta T / \Delta x)$$

The magnitude of heat flux (q) through a surface is proportional to the temperature (ΔT) across the thickness of the solid (Δx). In his terms, "the heat flux resulting from thermal conduction is proportional to the magnitude of the temperature gradient and opposite to it in sign."[5] Thus one of the salient contributions of Fourier's work is the identification of a coefficient of thermal conductivity (k).

Time plays a particular role in this equation.[6] In the steady-state form with the limit extended, time approaches infinity but the amount of proportional heat transmitted is increasingly small. Thus, as physicist Jennifer Coppersmith observes, for the first time in physics, an equation had arisen in which the solution is different if time is reversed (i.e. if $-t$ is substituted for t).[7] As an analytic of heat transfer phenomena, Fourier deduced the relative importance of the various phenomena involved. He simplified complex phenomena, but did not oversimplify them.

This famous Fourier equation had broad impact on a range of nineteenth and twentieth century scientific pursuits, well beyond the phenomena of heat transfer.[8] More than an equation that elegantly describes heat diffusion, it helped other scientists understand diffusive behavior in a range of contexts from electricity to molecular diffusion in, for instance, liquid bodies, gas diffusion, geology, and blood flow.[9]

What is thus at stake in Fourier's equation for the topic of energy in architecture *far* exceeds any immediate application to heat transfer. While heat transfer is important, the error habituated into the discipline of architecture is to privilege the short-term heat flux of a wall over the many forms of diffusion evident in the long list of diffusive phenomena characterized by the Fourier equation or its progeny. Rather than temperature flux alone, Fourier characterized diffusion more generally and this is applicable in the delirious range of flow structures that constitute a building: the bio-geophysical flows that yield raw matter, the flows of matter into material through production and transportation, the aggregation of knowledge and energy inherent to typology, the typical métier of operational energy flow and/or the flow of buildings into landfill. Each of these flows, and an understanding of how designers tap into each flow, is essential. While the insulation apparatus privileged one over the others, the matter of dissipation is not so easily isolated once a more totalizing view of the open, dissipative thermodynamics of buildings is in view. Like Fourier, certain phenomena must be isolated but, like Fourier, if the analytic is astute then the isolation serves to trigger insight about a range of phenomena, but does not serve to isolate them. This focus on diffusion and dissipation will continue throughout this book, as it is essential to a more complete understanding of the total thermodynamics of architecture.

The Heat Transfer Apparatus

Fourier's empirical observations about heat would not have been possible without means to measure heat. Thus, following work by Galileo Galilei and Santorio Santorio, Gabriel Daniel Fahrenheit's enclosed mercury thermometer (1714) and his subsequent scale of temperatures with determined freezing, melting, and boiling points (ca. 1724) were important preparatory, instrumental developments. One early conduction apparatus that aided in understanding conduction included the measurement of surface temperatures on a single metal bar at set distances from an isolated source. Although this early conduction apparatus was focused on conductivity, it eloquently illustrates the diffusion of heat in a homogenous material.

In the course of the emerging heat science, several thermometers were developed to study specific phenomena. For example, James Six's max-min thermometer was designed to record the maximum and minimum temperatures during a set period of time. Joule developed a thermometer that isolated thermal measurements from the influence of adjacent radiant bodies, an intricate apparatus involving a metal spiral suspended from a silk filament that inflects the position of a mirror. Abraham Bréguet's strain thermometer also involved a coil. Here the coil is a gold film sandwiched by platinum and silver ribbons. The thermal expansion difference of the platinum (lower) and silver (higher) turns a temperature indicator attached to the base.

In 1760, using Fahrenheit's instruments and scale, Joseph Black in turn observed, but did not enumerate, that ice absorbs heat energy without a change in temperature while melting. He thus coined the consequential terms "latent heat" and "specific heat."[10] Adding to this early history of instruments, Antoine Laurent de Lavoisier and Pierre-Simon Laplace built an ice calorimeter (1783) that afforded the measurement, for the first time, of latent heat and specific heat values for various materials.[11] Henri Victor Regnault, in turn, elaborated an apparatus for determining the specific heat of various material mixtures.

Simultaneous with Fourier's work, another Jean Baptiste — Jean-Baptiste Biot — made observations about the conduction of heat along a thin metal bar in 1804. Biot observed that heat was not only conducted but also emitted to its milieu. Biot would eventually contribute a numerical constant — now known as the Biot number — for transient heat flow calculations concerned with relative heat flow of interior and surface heat flow properties. As even this brief account affords, a larger apparatus of instruments, coefficients, and equations was essential to this early understanding of heat transfer.

These initial observations — on conduction, specific heat, diffusion, and emissivity — of heat transfer would prompt development throughout the nineteenth century. Maxwell, Ohm, Joule, and Lord Kelvin — amongst many others — extended early observations about heat into other scientific domains such as electricity and liquid diffusion. Other important foundations for the phenomena of heat

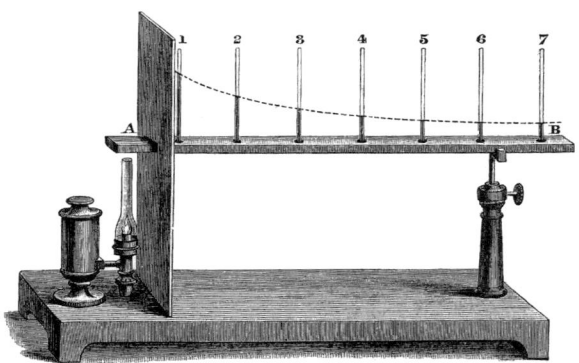
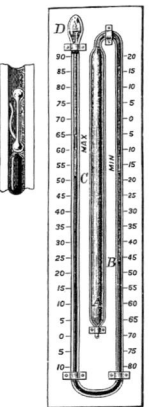
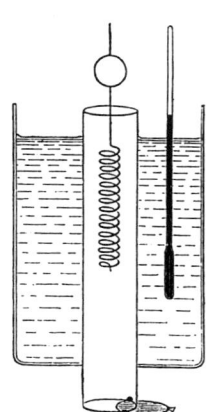
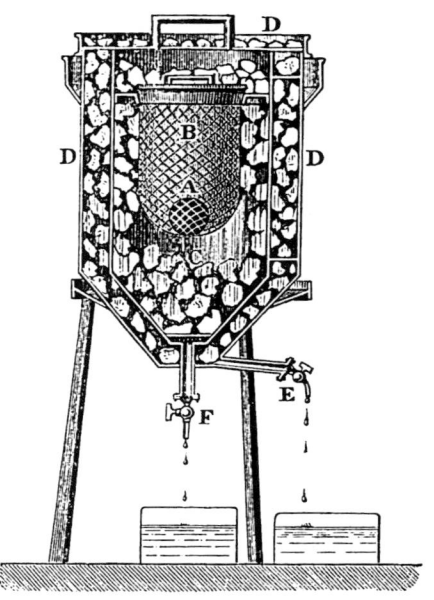
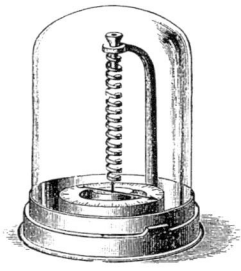

The Earliest Instruments of the Insulation Apparatus
Top: Early conductivity apparatus
Middle: Six's max/min thermometer; Joule's insulated thermometer; Lavoisier-Laplace's ice calorimeter
Bottom: Strain thermometer

transfer in buildings include work on radiation (by Stefan experimentally in 1879 and quantitatively in 1884, resulting in the Stefan-Boltzmann equation). These much more complicated and nuanced observations about heat are indicative of the complex behavior of heat flux and dissipation.

Maxwell, Ohm, Joule, and Lord Kelvin all play central roles in the development of thermodynamics at this time as well. While the development of heat transfer theories and practices in buildings engaged aspects of these early scientific developments, the profound implications of the intellectual transformations triggered by these early observations about heat did not always persist in architecture. In other words, the profundity of these observations about the dynamics of heat have — even now — yet to fully enter the minds and practices of architects, engineers, and building scientists.

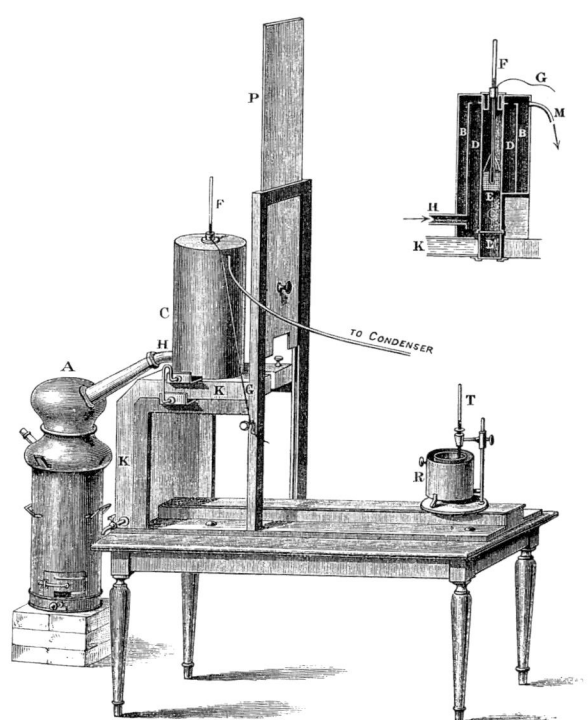

Fig. 60.—Regnault's Apparatus.

The Thermal Apparatus, Getting More Specific
Regnault's specific heat apparatus

Given the historical trajectory of heat transfer science in the insulation apparatus, our understanding of Fourier's equation is most clear for homogenous materials under certain circumstances. Many materials are, of course, generally homogenous. Most building assemblies, however, are not. Further, other materials, like wood, are anisotropic. While heat is transferred in building materials in conditions that are nearly steady-state, we know that buildings are often characterized by transient conditions. In particular, given the mass of buildings, the role of their density, heat capacity, and diffusivity can be very important, but only if conceived of as non-isolated, transient phenomena. The matter gets complicated quickly and the above concerns are still only regarding a single, isolated scale of dissipation.

As heat transfer theory was transferred to building science, engineering, and architecture, only partial heat transfer pedagogies and practices emerged. Conduction-based approaches to buildings dominate even now but are most efficacious only in certain circumstances. Under those circumstances, such equations and simulations are expedient and valid. But conduction does not alone provide a comprehensive understanding of the heat flux and dissipation of energy in buildings. The expeditious focus on conductivity has limited the thermodynamic understanding and imagination of architects. Partial descriptions of heat transfer in theory and in practice isolated insulation from the full reality and potential of heat flux in buildings. Whence this partiality?

The Transfer of Heat Transfer to Buildings

Collectively, these early scientific observations prepared the way engineers and, to a much lesser extent, architects conceptualized the questions of heat, warmth, materials, and systems in the context of buildings. How early scientific concepts were transferred to architecture is best documented in the treatises about heat in buildings published by early engineers, doctors, and the occasional architect. For example, the sequence of early British research and treatises that focused more directly on the "warming and ventilation" of buildings, as it was known, set the terms for a range of thermal phenomena in buildings and prefigured the apparatus of insulation.

One of the earliest British treatises — Thomas Tredgold's 1824 *Principles of warming and ventilating public buildings, dwelling houses, etc.* — dealt with a surprising range of heat transfer topics and systems for such an early consideration of the subject. He included basic matters of conduction and convection, and he also incorporated multiple observations about the roles of radiant transfer. While Tredgold surveyed a broad range of available techniques and described the various "advantages and disadvantages of distributing heat," he was taken by the uses of steam, revealing the close connections to the material culture that partially engendered thermodynamics in the first place.[12]

In 1845, Walter Bernan expanded on similar steam-based systems but included a historical consideration of the warming and ventilation of buildings in *On the History and Art of Warming and Ventilating Rooms and Buildings, etc.*[13] The focus of his first volume is a sweeping history of Egyptian, Roman, Palladian, and British approaches to heating. The second volume focuses primarily on stoves and contemporaneous heating strategies in the United Kingdom.

Joseph Gwilt's turn-of-the-century *Encyclopedia of Architecture: Historical, Theoretical, & Practical* follows a similar breadth of scope and organization. As with Bernan, Gwilt included expanded sections on the warming of buildings in the late-nineteenth century editions.[14] He articulated the respective roles of convection, conduction, and radiation as they pertain to various heating sources. Although the basic principles of heating a building are in place in this and similar contemporaneous volumes, insulation did not emerge as a topic of consideration in the practice of warming and ventilating buildings during this period.

Péclet's "Poor Conductors"

The French physicist Jean Claude Eugène Péclet included in his treatise a description of "poor conductors of heat," perhaps the earliest expression of a concern for insulation in a book on the heating of buildings.[15] Building on Fourier's work, Péclet reported as early as 1841 on an apparatus for testing the conductivities of various materials.[16] This was the first of such experiments. In his apparatus, homogenous plates of common building materials were placed in between a steam-heated chamber and a chamber of constant temperature on the opposite side. He thus experimentally deduced material conductivities.

This work culminated in the 1863 publication of *Traité de la chaleur*, the fourth edition of which includes work on the "poor conductors of heat" mentioned above.[17] This illustrated volume points to a range of insights about heat in the context of buildings and other structures. His early values for conductivity stood well into the twentieth century when more precise derivations of conduction emerged.

The conductivity of materials is a primary concern in Péclet's instruments and work. This focus on conduction would remain the basis of later nineteenth century and twentieth century work on insulation. For instance, the earliest architectural literature on insulation again focused on the compelling yet unreliable nomenclature of "non-conductors" of heat.[18] While conduction is a sound starting point for heat transfer research, this point of departure determined, if not over-determined, much insulation theory and practice.

The Transfer of Heat Transfer to Buildings from Refrigeration

In the North American context, the focus on thermal conductivity was in many ways a consequence of the fact that insulation theory emerged primarily from the refrigeration industry. There was mounting pressure from a burgeoning refrigeration industry to produce greater scientific knowledge about insulation.[19] Michael Osman has articulated some of the implications of refrigerators that were scaled up to the size of buildings in the form of cold storage warehouses.[20] This tendency to simply scale up refrigeration equipment to the scale and purpose of buildings, without much regard for fundamentally different purposes, possibilities, and system boundaries, severely constrains what buildings and architects can do with respect to heat transfer and powerful modes of energy dissipation.

Notably, heat transfer problems for refrigeration are fundamentally more of a steady-state condition: assumed stable interior temperatures and relatively stable exterior room temperatures provide an altogether different milieu for refrigeration insulation than that of buildings. Yet, the research means, methods, materials, and procedures of these refrigeration *and* building applications often remained the same, especially in the formative phases of insulation praxis.

Refrigeration approaches created procedures, and problems, for thinking about thermal insulation for buildings as well as for the human occupants of those buildings. The specificity of buildings (their climates and interior, human milieus) was not, of course, the impetus of research on "non-conductors" in the refrigeration industry. Much like the history of air-conditioning, the transfer of technology from the refrigeration industry to buildings did not always recognize the radical difference between these respective industrial and physiological contexts. This is evident in the procedures, protocols, and outcomes of early "heat insulator" research that was initiated in the refrigeration industry and then migrated to the building industry.

Refrigeration Research & Thermal Conductivity

A steady-state focus on conduction in refrigeration materials is problematic when transferred to the less steady-state circumstances of buildings, yet the principles and procedures were little changed for buildings. For instance, as an agent of the United States government, Hobart C. Dickinson was charged with the devel-

A History of Heat Transfer in Buildings   67

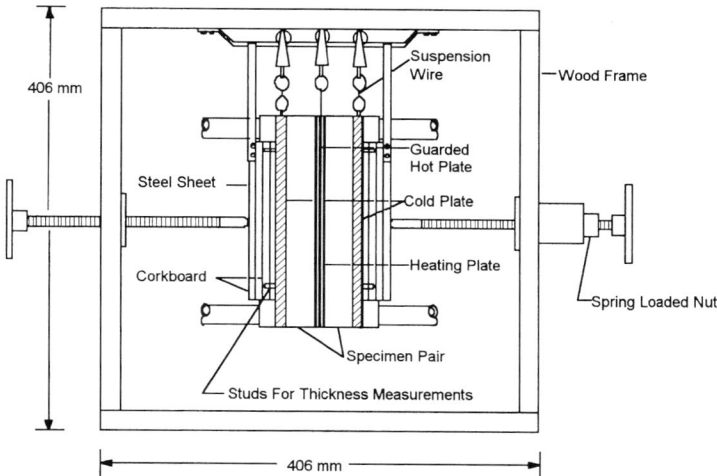

"A Troublesome Operation"
Early guarded hot-plate apparatus and schematic: 1929 NIST guarded hot-plate apparatus by M.S. Van Duesen. This apparatus consists of cold-plates that sandwich test samples and a central vertical hot-plate that separates test samples.

opment of standards for heat insulators. In that role, he published what became a standard text on the derivation of thermal conduction of materials in 1916, based on experiments beginning in 1912.[21] The funding for this research came from the United States Congress "at the request of the refrigeration industry."[22] Dickinson's approach, like many before and after, was based on a guarded hot-plate apparatus and process that determined the conductivity of flat, homogenous materials.[23] This apparatus was essentially a refinement of Péclet's plate apparatus.

Dickinson's hot-plate apparatus characterized a one-dimensional thermal flow through homogenous materials by limiting, if not eliminating, volumetric dissipation of heat. This suits the laboratory characterization of materials well, but not so readily the characterization of actual construction assemblies. Robert Zarr, a mechanical engineer for the National Institute of Standards and Technology, identified a key conflict in this approach:

> While the experimental physicist determined the thermal conductivity of homogenous solids, the engineer was interested in the practical problem of the transmission of heat through systems. The net result was two distinct groups of experimental investigations with disagreements within each approach (because of difficulties in measuring heat flow and temperature) and between each approach (because of erroneous application of results in heat transfer calculations).[24]

While scientifically sound on its own terms, the resulting steady-state conductivity values manifest only one mode of heat transfer involved in building assemblies. Yet this instrument, with its focus on conductivity, rapidly became the standard for the measurement of insulation material for refrigerators and buildings alike. Its technological momentum grew. By 1945, the American Society of Refrigeration Engineers, the American Society of Heating and Ventilating Engineers, the American Society for Testing and Materials (ASTM), and the National Research Council, had collectively produced the first standard regarding thermal conductivity based on this apparatus. As the momentum of this approach increased, the role of other important heat transfer parameters diminished. This is one example of the vexing tendency to isolate what in reality is non-isolated.

Conduction in Architecture

Of the 400 or so technical and architectural articles related to the phenomena of heat transfer in buildings published before 1925, only 22 had an overt focus on radiant transfer. Another 15 articles considered the relationship between radiation, conduction, or convection. Likewise, of the 400 texts related to the phenomena of heat transfer in buildings published before 1925, only 13 had an overt focus on convective transfer, with another 44 that considered convection in relationship with conduction or radiation. The clear majority of these articles focused on the problem of thermal conduction in building materials or, far less commonly, some combination of conduction, convection, or radiation; only about 30 articles considered the mixture of the three modes together.

This preoccupation with conduction in the literature reflects more an extension of the momentum of refrigeration research than the mixtures of heat transfer inherent in buildings. While these early texts exhibited clear knowledge about the various forms of heat transfer, few considered more than one and even fewer considered them in the transient simultaneity of their real conditions. At that time, it was clearly useful, if not necessary, to consider a building or building material as an isolated system for the purposes of study. However, the abstracted premises and outcomes of such work should be synthesized back into the non-isolated context of the building. This rarely happened in the modern apparatus of insulation. As a result, not all modes of heat transfer were as present in the historical development

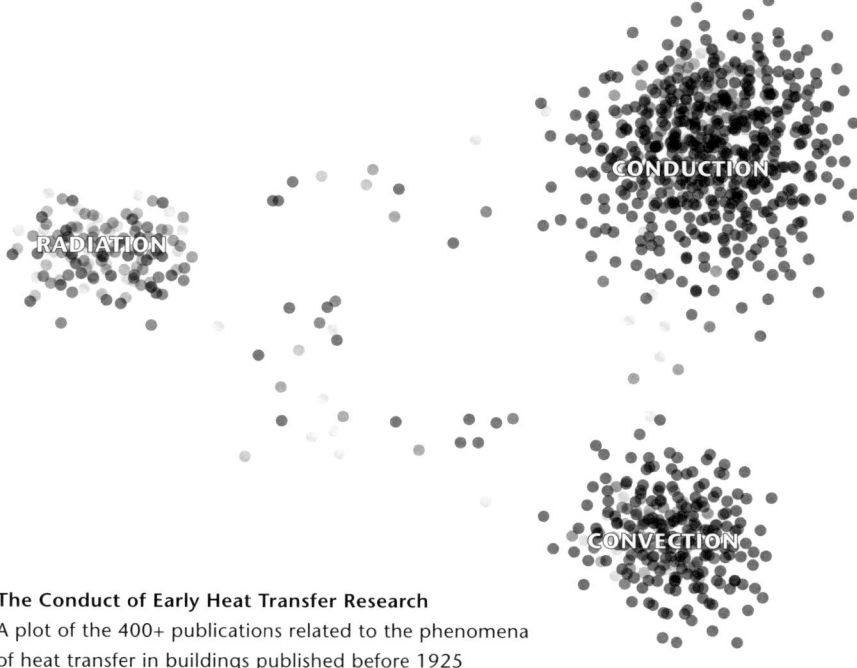

The Conduct of Early Heat Transfer Research
A plot of the 400+ publications related to the phenomena of heat transfer in buildings published before 1925

of building insulation theories and practices as they are in actual heat transfer conditions in building assemblies.

The preponderance of early literature that dealt with conduction reveals a focus grounded in reality: conduction is often the seemingly most problematic factor in terms of heat flux. It also proved somewhat difficult to derive as characterized and construed. But this preoccupation with conduction also suggests an unwarranted, disproportionate focus and momentum. A tendency for reductive study confounded the early work on thermal conductivity: the desire to simplify study occluded important, albeit difficult, aspects of heat transfer. So, again: whence this preoccupation?

"A Troublesome Operation"

As the topic of heat transfer shifted from scientific observation and premise to prescriptions for practice — especially in the context of building insulation as opposed to refrigeration insulation — certain recurrent problems and complications emerged. The opening line of one of the more scientific architectural journal articles on early (1938) heat-insulating building materials is very clear about what will become a theme in heat transfer theory for the rest of the twentieth century: that the "determination of the thermal conductivity of a material is a troublesome operation."[25]

This 1938 article also discusses an updated version of the hot-plate testing apparatus. In the resultant discussion of this apparatus in relation to insulating materials for buildings, the author acknowledges the difficulty of isolating one thermal phenomenon from the simultaneous roles of other modes of heat transfer, particularly in a discussion of increasingly common cavity wall construction:

> When the cavity is large enough to allow for convection currents, the advantage of the low thermal conductivity of air is to some extent nullified. Consequently it is advantageous to subdivide the space between the walls by the use of multiple partitions. Such partitions serve the additional purpose of acting as radiation shields.[26]

The presence of convection and radiation in the convention of modern layered construction complicates steady-state heat transfer logics based on thermal conductivity. Building assemblies are ultimately more unruly than the hot-plate derived values, a necessary aspect of the heat transfer apparatus but one that engenders many questions.

The steady-state functions of a hot-plate experiment mostly apply only for the conceptual conditions of construction assemblies. Actual construction assemblies are riddled with imperfections, poor installations, and, most importantly, shifting boundary conditions. The delta between the ideal assemblies assumed in steady-state calculations and the transient reality is filled with other thermal behaviors and mechanisms. The transfer of insulation research from refrigeration science to buildings was, despite appearances, not a simple task.

Early Heat Transfer Literature in Architecture

Given the uneven scientific treatment of heat transfer in the context of buildings in general, insulation theories and practices in architecture acquired a particular character. The understanding of heat transfer in buildings most certainly emerged from outside of architecture. For example, of the 400 or so publications related to the phenomena of heat transfer in buildings published before 1925, only three appeared in architectural journals.[27]

Further, of the nineteen articles about insulation published in architecture-related journals before World War II, only a few had any technical basis. Instead, in both professional and home journals, articles placed emphasis on the frugality of insulation, a social and economic rationale rather than a scientific basis. This persuasion of economic and energetic sensibility is evident in article titles such as: "The house that will save coal" (1920), "Prevention of heat losses" (1921), "Building to save coal" (1923), "Insulating the house" (1926), and "The value of insulation" (1928).[28] In this period, the very few scientific occurrences of conduction-based accounts of heat transfer are even further obscured by the frugality discourse in architectural journalism, as well as by the bias of marketing publications. From the beginning, practicing architects and architecture professors had few ready sources to help configure anything but an agenda based on steady-state assumptions, conductivity calculations, and claims about frugality.

More than a scientific question, these articles reflect a culturally based ethic about energy in architecture, one that persists even today. *Very importantly*, these articles reflect more about the convergence of what Max Weber called the "Protestant Ethic" with North American attitudes on energy than they say anything about heat flux itself.[29] In these articles, a Calvinist characterization of energy in buildings — a pious concern for frugality and economic gains/losses, a zealous avoidance of waste, an aversion to luxury if not comfort, and a preoccupation with "calling" — was coupled with the refrigeration-derived focus on conductivity.

These two habits of mind — the chilly Calvinist cultural assumptions coupled with refrigeration practices — proved excellent managerial bedfellows. As a set of cultural concepts about energy, however, this Calvinist perspective runs counter to a more Dionysian, non-modern perspective that better corresponds to the reality of open thermodynamic systems and that could otherwise shape an architectural agenda for energy.

Rather than a technical focus on the physical phenomena at hand, early architecture-related journals more commonly exhibited a qualitative analysis of heat. It is rare to encounter a technical account of heat transfer in architectural journals of this period. As such, the development of heat transfer theories and insulation practices occurred largely outside the discipline of architecture. Architects and consumers in this period were unequipped, and perhaps uninterested, in the thermodynamic basis of heat transfer in buildings in the resulting void. Calvinistic cultural postures about energy filled this void instead and persist even now.

As evidenced in the early insulation apparatus literature, heat transfer was not a central architectural concern. In this period, architects were largely isolated from the development of insulation theories, materials, and practices. They at best implemented the products and procedures of adjacent disciplines.[30] It is not surprising, then, that the theories and practices that ultimately emerged from this period of development are neither architecturally advantageous nor directly connected to the reality of buildings.

This left architects reaching for an agenda for energy, often with uneven results. Consider, for instance, one of the earliest publications related to heat transfer in a North American architectural publication. This 1896 article from the *American Architect and Building News* discusses the "[i]nfluence of temperature in our buildings." Lacking a more scientific approach, the author casts the topic of temperature largely in a realm of intuitive and relevant but unsubstantiated claims about the relative merits inherent in shifting habits of construction at the time: traditional versus modern construction, heavy versus lightweight modes of construction. The author claims that "the flimsiness of the modern wall is answerable largely for this rapid transmission of heat and cold."[31] He continues his attack, stating that "modern ideas of elegance and lightness have been prejudicial to the resistance of the structure to the weather, and of contributing to rob the modern house of comfort and equable temperature."[32] The article does eventually list some of the earlier ideas and palliative materials for insulating the cavities of these lighter-weight walls, such as slag wool.

More convincingly, this anonymous early building insulation article correctly placed emphasis on issues associated with what would be later described as bioclimatic design: orientation, size of windows, location of windows, as well as other architectural/climatic adaptations such as verandas and solar shading devices. This milieu-specific portrayal of the thermal behavior of buildings pointed to a more robust and totalizing agenda for the relationship between insulation, heat transfer, and building design, one less evident in the focus on an isolating perspective of "non-conductor" material properties. While its claims are unsubstantiated, this particular article is notable for its specificity about what architects can do in relation to heat transfer, as opposed to what an industry can impart from outside of the specificity of architecture. This was an example that later architectural publications and practices would rarely repeat.

Other early insulation documents, typically marketing reports from various manufacturers, instead focused more directly on insulating materials and products, as opposed to heat transfer theory or discipline-specific opportunities.[33] These seemingly innocuous observations from the early insulation literature and marketing reports set a particular tone for insulation discourse in architecture that, while not exclusive to insulation practices, determined multiple aspects of insulation's technological momentum. The interests of industry largely drove knowledge about insulation in buildings. Architects were to read, understand, and specify insulation products as prescribed, a matter of continuing education hosted by motivated sources.[34] This transactional approach to knowledge dissemination instituted a recurrent and acquiescent dynamic for the profession of architecture when it came to technological developments at large, most certainly including insulation.

Early architectural articles and reports only partially point to the scientific basis of heat transfer in buildings and more wholly to market-related solutions. By the end of nineteenth century, it was primarily manufacturer-based articles that articulated insulation problems and proprietary solutions, such as one on mineral wool written by the United States Mineral Wool Company that appeared in the journal *The Manufacturer and Builder*.[35]

Distinct from the development of building assemblies and structures that remained less proprietary in the same period, insulation from the beginning shifted from a material science problem to a proprietary product problem. No longer just a function of material properties alone, corporate properties and agency merged with material properties to shape insulation products and practices, a shift in material culture explored in the next chapter but that also pertains to this narrative of heat transfer.

Take, for instance, the National Institute of Standards and Technology engineer who observed that "early technical progress in cooling and heating for industrial processes and thermal comfort had a direct effect on the development of the methods for thermal insulation materials at the National Bureau of Standards.[36] As in Ruth Schwarz-Cohen's study on "How the Refrigerator got its Hum," the influence of corporate agendas on scientific purview is evident here.[37] The presence of the interests of pre-war refrigeration industry and of post-war insulation manufacturers plays a critical role in the generation and dissemination of insulation theories, standards, and practices.

By the early twentieth century, the terms and habits of mind regarding the general insulation apparatus had been established. A core of scientific observations, equations, and instruments, as parsed by refrigeration scientists and engineers, formed the technical basis of insulation. Managing this evolving body of knowledge, key governmental bodies began to set standards while various entrepreneurs and industries determined supply and value chains for proprietary products. Within this developmental apparatus of insulation, more than insulation was produced. However, the foremost role of the apparatus was to establish, professionalize, and monetize a *need:* in this case the need for insulation.

Need

As Foucault stated, "I understand by the term 'apparatus' a sort of — shall we say — formation which has as its major function at a given historical moment that of responding to an *urgent need*."[38] I would contend further that the existence of any apparatus in our given historical phase of capital and late-capital exists not only to *respond* to a need but often to more completely *create* need where none otherwise existed. The apparatus of heat transfer research and its resultant insulation practices professionalized a new need in buildings along with new expectations of comfort and performance.

Ivan Illich characterized the modern period as a novel one in which "people had 'problems,' experts had 'solutions,' and scientists measured such imponderables as 'abilities' and 'needs'."[39] This is key to the dynamics of technological momentum in general, but to those of insulation specifically. The apparatuses of nascent needs, experts, and solutions appeared throughout modernity with great purpose and motivation, often with a targeted monopoly as a goal. As Illich stated,

> The new organized specialists must, though, be carefully distinguished from racketeers... By establishing this kind of monopoly... they initially seem to fit the dictionary definition of gangsters. But gangsters, for their own profit, corner a basic necessity by controlling supplies. Today, [specialists] gain legal power to create the need that, by law, they alone

will be allowed to satisfy… They turn the modern state into a holding corporation of enterprises which facilitate the operation of their self-certified competencies: equal needs are laid on the citizen/client, only to be fulfilled in a zero-sum game.[40]

In such a process, professional responsibility changes dramatically. In this way, the insulation apparatus transformed not only buildings but architects and the practice of architecture as well. Take as but one example, the posture adopted in this quotation from *Ice & Refrigeration* in 1920:

Insulation can be thought about in two ways. You can either follow your own theories or the theory of some architect, who cannot possibly know about insulation unless he is making a specialty of the refrigerating trade; or you can buy manufactured insulation from a reputable manufacturer who will guarantee that it will permit only a certain rate of heat transmission…[41]

This transformation of building practices from guilds, unions, and other syndicates to professions and corporations was obviously not ideologically or physically neutral. Architects and buildings were fundamentally transformed in this engenderment and professionalization of need, and insulation is a very emblematic example of this dynamic. The enabling acquiescence of many parties is essential to professionalization processes of this type. "The public acceptance of domineering professions," Illich notes,

is essentially a political event. Each new establishment of professional legitimacy means that the political tasks of law-making, judicial review and executive power lose some of their proper character and independence. Public affairs pass from the layperson's elected peers into the hands of a self-accrediting elite.[42]

The professionalization of heat transfer in buildings introduced new subjects and altered extant subjects: *it was most certainly anything but a benign political event*. The guarantees of product-peddling often supplanted fiduciary responsibilities in the new needs. Significantly, in this arrangement, architects were not expected to know the intricacies of heat transfer or insulation, just to be able to incorporate the demands and imposed systems. Yet architects, even now, curiously retreat towards this capitulating arrangement, an extremely problematic failure of praxis *and* politics. It is difficult to claim that buildings, cities, or practices benefit from such non-architectural demands, architectural acquiescence, and failed politics.

The early marketing reports in industrial journals presented the primary way through which heat transfer and insulation were professionalized in architecture. Taking only what they needed to know about their products, entrepreneurs marketed their wares to architects. Architects in turn were educated on the nascent need and the means to resolve it. These demands were then passed onto clients first through market differentiation and later through market expectations, national standards, and building codes. Illich, again: "professionals tell you what you need and claim the power to prescribe."[43] This was certainly the case with insulation.

In many cases, a fiscal argument drove the professionalized need and the resultant managerial posture of the Calvinist ethos of efficiency resonated with the bottom line of clients. This resonance served the larger apparatus well enough, but it did not necessarily serve architects, buildings, or rudimentary building science. In contrast, other perspectives were possible and other perspectives were occasionally pursued, and together these alternative perspectives form the minor histories and studies of insulation development in modernity. While a reductive approach to insulation — driven by industrial ambitions more than scientific efficacy — dominated insulation theory and practices, it was not the only perspective.

Non-steady-state Characterizations

The one or two-dimensional, steady-state characterizations of heat transfer in building envelopes that constitute what architects gleaned from the insulation apparatus, while expedient, have limitations. Building scientists John Straube and Eric Burnett are very clear about these limitations: "It is very important to recognize that these approximate two-dimensional methods do not always provide an accurate representation of heat flow within the enclosure."[44] They explain further, "heat flow is rarely a steady-state process. Although the air temperature usually does not change very quickly, the effect of the sun on heat flow through envelopes is usually more important than air temperatures."[45]

Further, the steady-state assumptions of conduction calculations neglect other important parameters. The roles of convection and radiation transfer, while often present in discussions of insulation, played a far more minor role regardless of the thermal circumstances under consideration. For instance, one early experiment showed that radiation "is responsible for roughly two-thirds of the heat transferred across a vertical air space, bounded by wood or plaster, such as found in the walls of typical frame construction."[46] This uneven treatment of heat transfer phenomena persists on account of the technological momentum accrued through parallel research momentum, market momentum, and pedagogical momentum.

"So erroneous as to be absolutely useless"

Despite great momentum in aspects of research and standard setting in the first half of the twentieth century based on conduction, the validity of the steady-state focus on conductivity was frequently questioned on scientific grounds. One such observation occurred at a 1939 symposium on insulating materials:

> While our technical knowledge and understanding of the conductivity of building materials, methods for determining the same, and the resulting effect on heat transfer through walls built up of such materials for steady-state conditions were being perfected, other factors such as condensation of moisture within the wall, the effect of heat capacity of the structure, and the wide departures from steady-state flow encountered in applications to summer cooling and air-conditioning have confronted the industry.[47]

These known factors presented multiple problems for the steady-state approach to heat transfer in buildings, as these same authors continue:

> Only for steady-state or equilibrium conditions is the heat flow through a wall per unit time equal to the product of the transmittance coefficient, U, the area and the temperature difference between the air on the warm and cold sides. That is, this relationship is only valid when the temperature difference between the two sides of the wall, made up of its various materials, remains constant for a sufficient length of time so that the rate of flow at all points throughout the thickness of the wall is the same, or when equilibrium exists between the rate of heat entering the warm side of the wall and the rate of heat flowing from the cold side. Of course, this ideal condition never exists even under most favorable heating practice. Granted that the inside temperature may be kept constant throughout the 24 hour period, still there is a daily cyclic change in the outside temperature, which may be increased by solar radiation impinging directly against the structure.[48]

The authors conclude, "under these conditions the assumption of steady-state flow is so erroneous as to be absolutely useless to the designing engineer."[49] The persistent presence of steady-state flow assumptions in the context of buildings is thus difficult to resolve as a matter of building physics alone. Other factors must contribute to its technological momentum and thewse will be explored in the second and third chapters. What will remain the focus in the balance of this chapter are the exceptions to the reductive characterizations of heat transfer that occupy the periphery of the modern insulation apparatus. Given the physical and fiscal focus of the insulation apparatus in North America, the development and reception of more dynamic and periodic understandings of heat transfer before World War II occurred outside of North America.

The Diffusion of Heat Flux

It is revealing to understand the relative heat transfer approaches taken throughout North America, Europe, and Australia in the early to mid-twentieth century. Each country had related agendas for the characterization of heat transfer but deployed varying methods and agents, resulting in varied outcomes. If anything, this brief survey indicates that what may be most familiar in any country about heat transfer today is often the product of both historic contingencies and, later, more universal assumptions about heat transfer.

Norway: 1920–1927

A minor but poignant exception to the pre-World War II series of North American insulation research and publications was a 1924 article — one of few with a technical orientation to appear in an American architectural journal — titled "Architectural Engineering: Heat transmission through dwelling house walls."[50] This article reports on research conducted not in North America but at the Norwegian University of Science and Technology in Trondheim. The article is exceptional in the early record of insulation publication for its more technical orientation as well as for how its experimental apparatus was a precedent for many non-North American research premises, protocols, and outcomes.

The article reports on the construction, instrumentation, and behavior evaluation of 27 huts built in northern Norway (1919–27). In 1919, Norwegian architect Andreas Fredrik Bugge undertook a series of experiments related to the thermal behavior of wall assemblies. Bugge was an architect already involved with the Norges Tekniske Høiskole as construction manager, with Bredo Greve as architect, of the main building on the campus in 1905–06. The campus is located on the Gløshaugen plateau southeast of central Trondheim.

In an approach that contrasted with the concurrent isolated, analytical preoccupations with conductivity in North America, this early Norwegian in-situ, in-climate empirical test of thermal behavior is interesting to consider in the context of the non-isolated, dissipative reality of buildings. The ambition of the investigators for the test houses is well summarized in the subtitle of their subsequent publication, "Result of Tests with Wall-Constructions and Materials for Building Warm and Cheap Dwelling Houses." In a two-year evaluation, the thermal behavior of twenty-seven serially distributed 2 x 2 meter huts of common construction types was tested for a few primary criteria, quoted verbatim here:

1. Measurement of the quantity of heat energy which must be supplied to each individual house in order that the temperature of the air in the house be maintained at a constant level (usually 20 degrees C).
2. Tests to find how quickly the air in the houses cools when the supply of heat is cut off and how quickly the air, on the other hand, is heated by a constant heat supply.
3. Measurements of the humidity of the air in the houses and in the wall cavities.
4. Measurements of the air temperature in the hollows of the walls, especially in connection with the action of the sun's rays.
5. Ordinary meteorological observations.[51]

Each of the test huts housed little more than an electric heater and thermostat. A physics professor, Sem Sæland, designed the heating and measurement apparatus in each test house, testing for the "heat-isolating power" of the wall constructions in "the rough and very shiftly climatic conditions" of Trondheim.[52] In Bugge's words,

In each house an electrical heater is placed, by which the room is heated. The consumption of energy in each separate house is measured by means of a watt-hour-register. The damp conditions have also been observed in the houses as well as the hollow spaces in the walls, and also the outside metrological conditions. These latter are exactly identical for all the houses and the said measurements will thus give relative values of the heat-isolating capacity of the individual types of walls, etc., under conditions which answer as nearly as possible to the conditions in an ordinary house.[53]

The huts were constructed in a single straight row atop the Gløshaugen plateau, along what is now Sem Sælands Vei on the Norwegian Technical University campus. Bugge claimed that winds and insolation struck the buildings equally: "All the houses," as lead Andr. Bugge noted, "received, therefore, the same weather effects."[54] However, the twenty-seven houses were lined northwest to southeast, facing southwest, so the insolation on the first would have been greater than the other twenty-six. The houses were tested as transient, but as equally transient as possible for the purpose of comparison. Certain variations were inevitable, such as varying interior surface wall area due to thickness of construction or slightly dif-

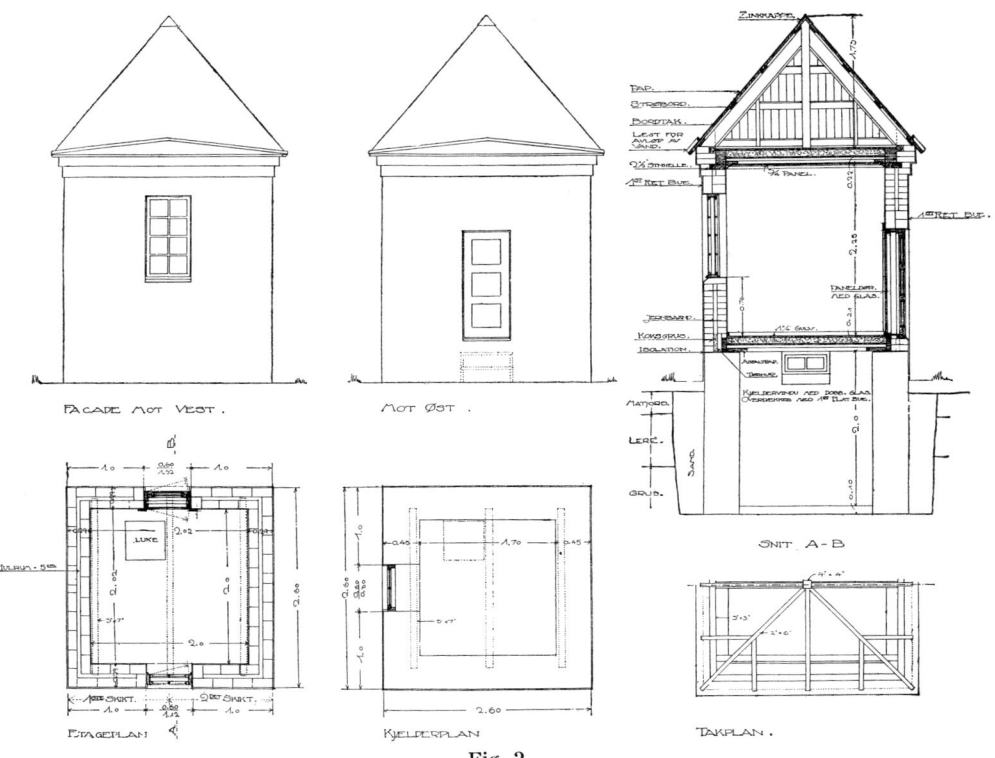

Fig. 2.

Non-isolated Testing: Bugge's Early In-situ Testing of "Warm and Cheap" Construction
Above: The schematic of the primitive thermal testing hut
Opposite top: The seriality of huts testing the seriality of heat transfer
Opposite bottom: Hut number 1

ferent floor and roof assemblies depending on the construction type. These variants, however, were compensated for in the team's final results.[55]

While not exactly an ideally structured experiment, what is unique and refreshing here, from an architectural and thermodynamic point view, is testing the fundamentally transient behavior of full assemblies over longer periods of time, rather than steady-state study of individual materials over very short periods and in very isolated laboratory conditions. This reflects a core methodological difference between isolated and non-isolated approaches. Bugge's interest in the dissipation of heat, the influence of insolation, shifting humidity, shifting climate, and

82   Insulating Modernism

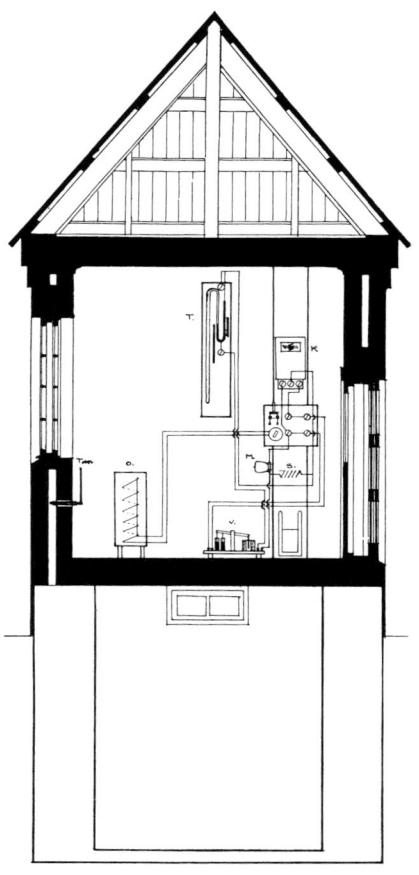

Mounting af the Instruments.

K = kilowatt-hour-indicator.   M = resistance-lamp.
O = electrical stove.   T = thermo-regulator.
S = shunt.   V = current-switch.
Tm = Thermometer.

Fig. 30.

Sustained Testing
The mounting of the instruments and the test huts, as seen in the upper right quadrant of aerial photographs of the Trondheim campus, ca. 1930, and below, ca. 1960.

other boundary conditions all place construction assemblies in a very different milieu than the analytical, laboratory-based approach.

Some of the more unique types of construction assemblies tested in this early experiment are pertinent to this book and deserve specific attention. Bugge tested both masonry and wood construction types. Of the masonry types, House I was the most straightforward: a 1–1/2 wythe brick load-bearing masonry wall. Like all the other masonry types tested, House I was plastered on the inside and outside. House II introduced air pockets into the same wall thickness by means of its masonry pattern. This wall was tested first with just plaster and then with a thin wood panel. House IV was thicker (46 centimeters) with a half brick on the outside to create an air cavity and was thinly plastered on the inside. House V was similar, but with the cavity on the inside. House VI used iron ties to tie double- and single-wythe walls together with a cavity on the interior side. House VII used iron ties to tie two single layer walls together. During the testing, screened pieces of coke were added as insulation in the House VII cavity. House VIII tested "Leanstone," a three-cavity cement masonry block product developed in Stockholm. House IX used the "Rexstone" masonry unit, an interlocking three-cavity cement masonry block. House X used a 10 centimeter layer of in-situ cast reinforced concrete wall. A "molerstone" layer lined the inside. House XI consisted of a hard-burnt outer layer of masonry and a "molerstone" inner layer, bound with iron ties. The final masonry type, House XXV, tested an interesting mixture of cement block units. Inner and outer blocks were mixed with different types of aggregate: the interior layer had coke additives to lighten the block and increase its insulating capacity.

Of the wood specimens, House XII most closely followed the construction norms of wood house construction in Trondheim at the time. It consisted of three layers of wood in alternating directions, separated by 1–5/8 inch air cavities. House XIII was the most robust of the wood houses tested. It also served as the baseline of behavior evaluation as it represented the local building regulations for house construction in Trondheim. Its assembly consisted of a 5 x 5 inch frame with vertical 3 inch tongue-and-groove planks as the primary structure. On the outside of the primary structure, a thin layer of impregnated pasteboard lined the wood. A sheet of sheet wool pasteboard lined the inside face of the primary wood. An outer 2 inch air cavity separated the outer wood cladding and a 3/4 inch cavity separated the interior wood finish material. This wall maintained its quite low energy input very well. Houses XIV–XIX tested varied assemblies of wood frames, layers of tongue-and-groove surfaces, and cavity number and dimensions. House XX was conventionally framed but plastered on the exterior with 3 centimeters of the "Bacula" plaster system. House XXI, also conventionally framed, used sawdust for 4 inch of cavity

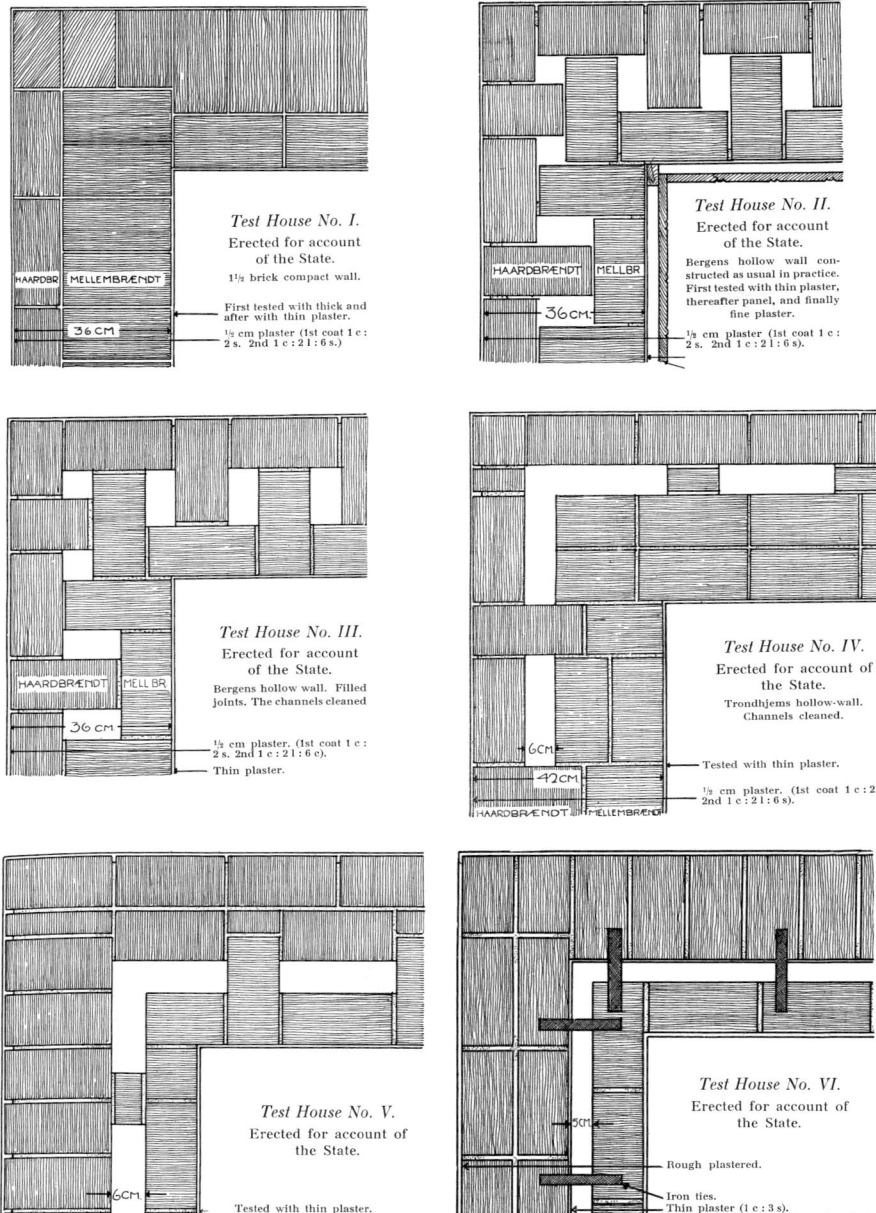

"Warm and Cheap" Wall Types Tested
Above and following pages: Test Houses I–XXVII

A History of Heat Transfer in Buildings   85

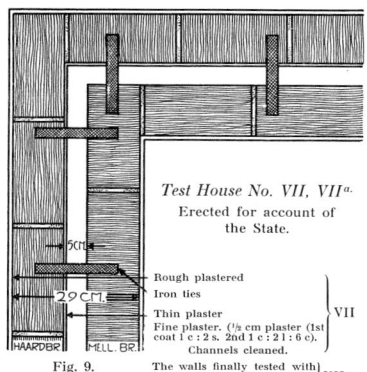

Fig. 9.

Test House No. VII, VII^a*.*
Erected for account of the State.

Rough plastered
Iron ties
Thin plaster
Fine plaster. (½ cm plaster (1st coat 1 c : 2 s. 2nd 1 c : 2 1 : 6 c). Channels cleaned.
The walls finally tested with wholly filled-up hollow space with coke of hazel nut size. } VII^a

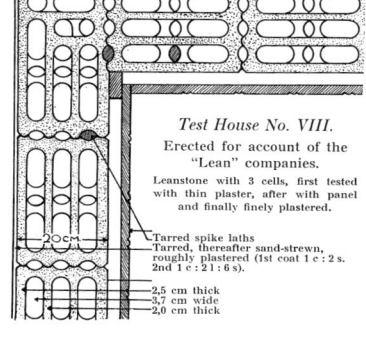

Test House No. VIII.
Erected for account of the "Lean" companies.

Leanstone with 3 cells, first tested with thin plaster, after with panel and finally finely plastered.

Tarred spike laths
Tarred, thereafter sand-strewn, roughly plastered (1st coat 1 c : 2 s. 2nd 1 c : 2 1 : 6 s).

2,5 cm thick
3,7 cm wide
2,0 cm thick

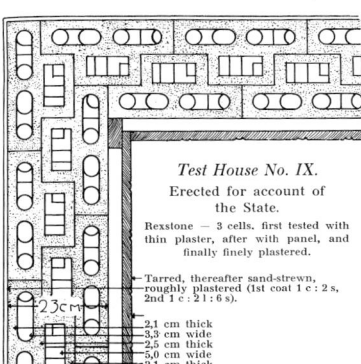

Test House No. IX.
Erected for account of the State.

Rexstone — 3 cells, first tested with thin plaster, after with panel, and finally finely plastered.

Tarred, thereafter sand-strewn, roughly plastered (1st coat 1 c : 2 s, 2nd 1 c : 2 1 : 6 s).

2,1 cm thick
3,3 cm wide
2,5 cm thick
5,0 cm wide
2,1 cm thick

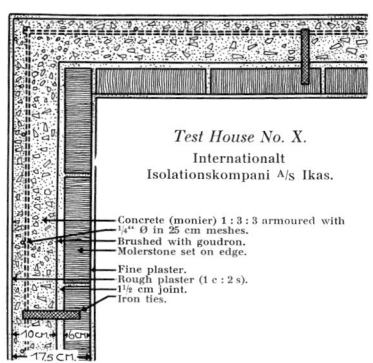

Test House No. X.
Internationalt Isolationskompani A/s Ikas.

Concrete (monier) 1 : 3 : 3 armoured with ¼" ∅ in 25 cm meshes.
Brushed with goudron.
Molerstone set on edge.
Fine plaster.
Rough plaster (1 c : 2 s).
1½ cm joint.
Iron ties.

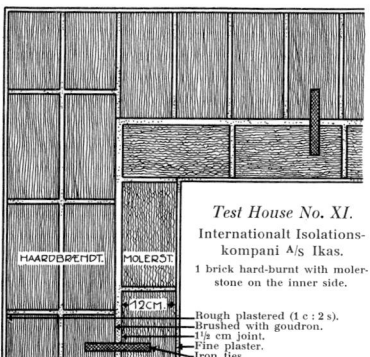

Test House No. XI.
Internationalt Isolationskompani A/s Ikas.
1 brick hard-burnt with molerstone on the inner side.

Rough plastered (1 c : 2 s).
Brushed with goudron.
1½ cm joint.
Fine plaster.
Iron ties.

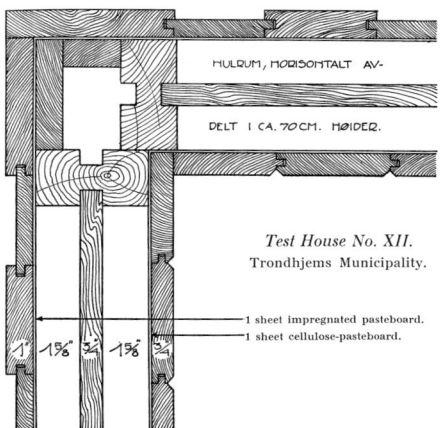

Test House No. XII.
Trondhjems Municipality.

1 sheet impregnated pasteboard.
1 sheet cellulose-pasteboard.

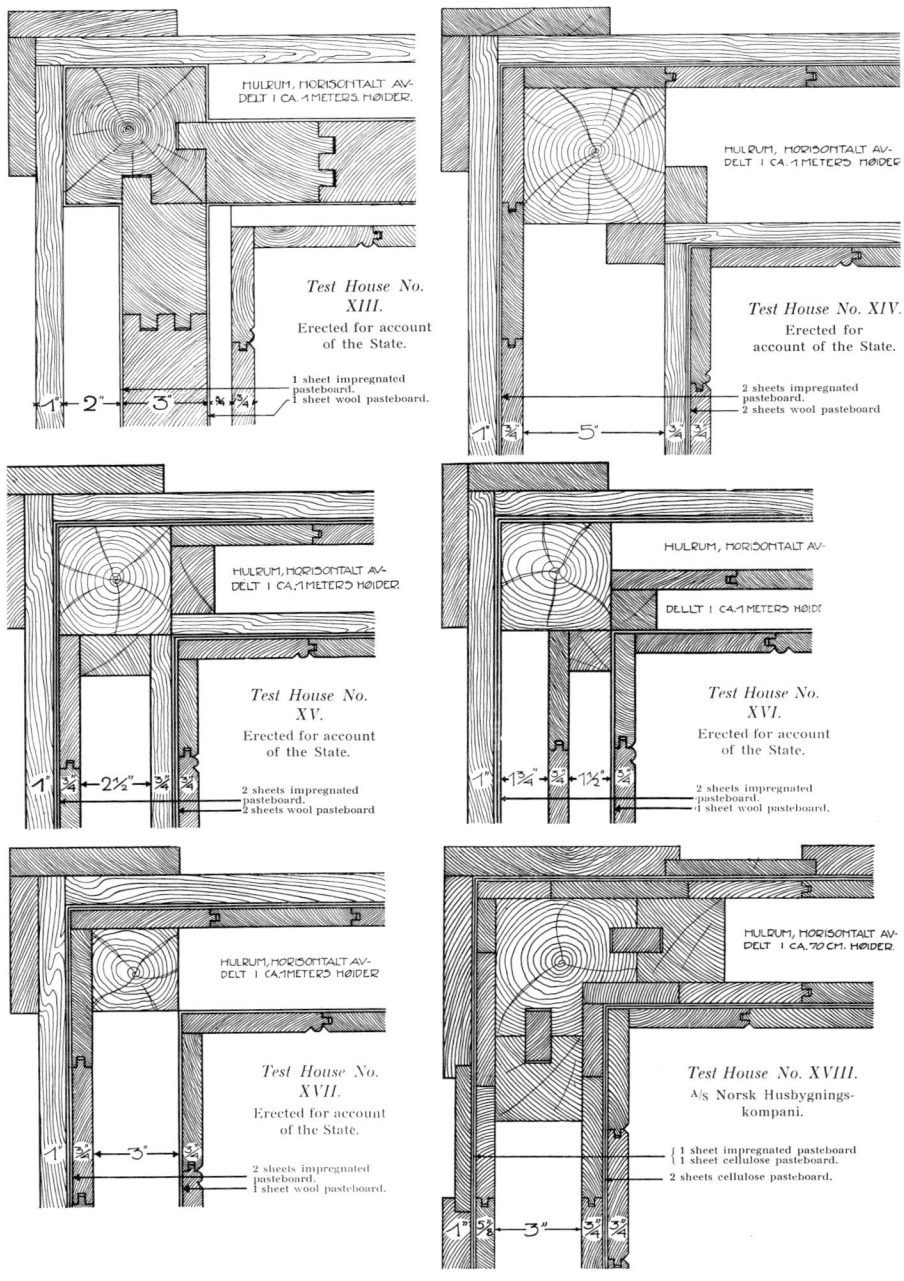

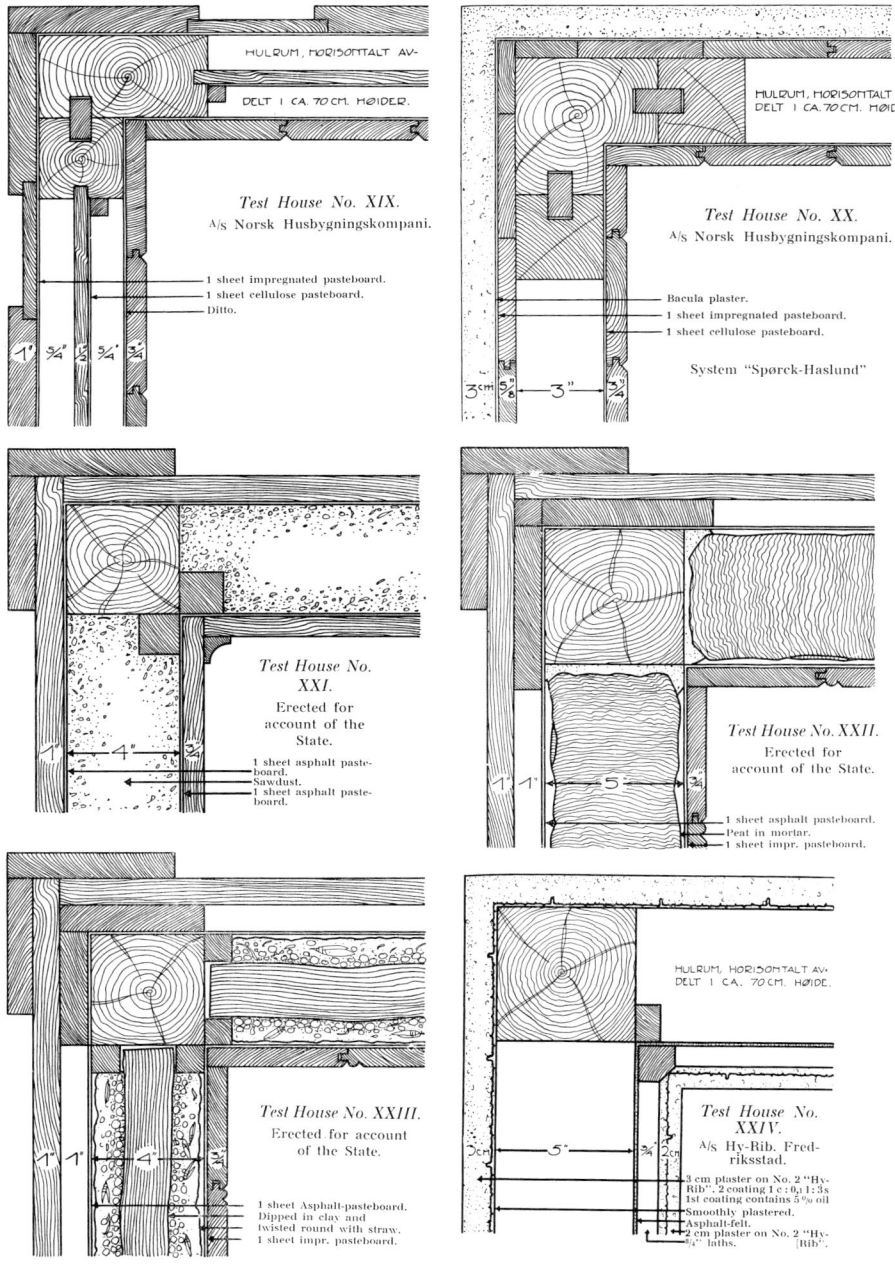

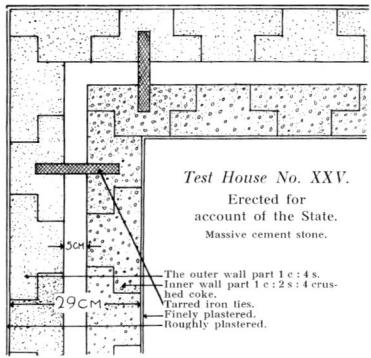

Test House No. XXV.
Erected for
account of the State.
Massive cement stone.

The outer wall part 1 c : 4 s.
Inner wall part 1 c : 2 s : 4 crushed coke.
Tarred iron ties.
Finely plastered.
Roughly plastered.

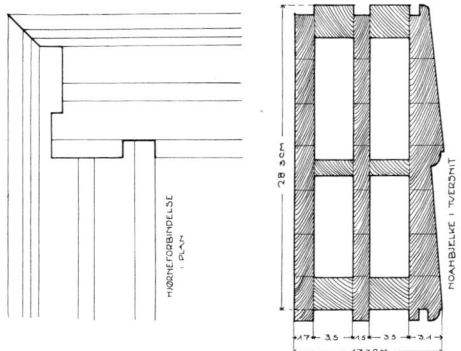

Test House No. XXVI. Fig. 28.
A/S Trækonstruktion, Christiania.
Block walls with "Noah"-joists.

insulation sandwiched between asphalt-faced planks of wood. It included provisions to add sawdust over the time it settled. House XXII, similarly framed (5 x 5 inches), used 5 inch peat bricks as insulation and House XXIII used a system of lath wrapped in rope, dipped in clay. House XXIV tested the "Hy-Rib" system: a conventional wood frame clad on both faces with a ribbed metal panel that is a substrate to 3 centimeters exterior plaster and 2 centimeters interior plaster. Two wood construction types were added late in the evaluation period on the foundations of other test huts. House XXVI consisted of built-up timbers ("Noah" beams) with built-in air spaces and a special corner joint cut. The small pieces of wood used in the fabrication of the beam were held together with adhesive. In the context of contemporary wood construction, House XXVII is interesting to consider. This assembly consists of two single-ply layers of wood panels and a central, single layer of two-ply wood panel, laminated with adhesive (an early form of cross-laminated wood panels). The cavity face of each outer layer of wood was lined with sheep wool. The following table assembles the results of the different types in the experiment.

As the report table illustrates, the rate of dissipation was measured at 5, 10, and 15 hours after the heat source was shut off. A composite thermal resistance value for the construction assemblies was deduced based on the amount of energy input required. Professor Bugge's recommendation after observing the twenty-seven constructions and their results for two years was straightforward: "It appears, therefore, from these tests that to obtain a really good and warm house, which shall be comparatively cheap in construction and maintenance, it should be built of wood."[56]

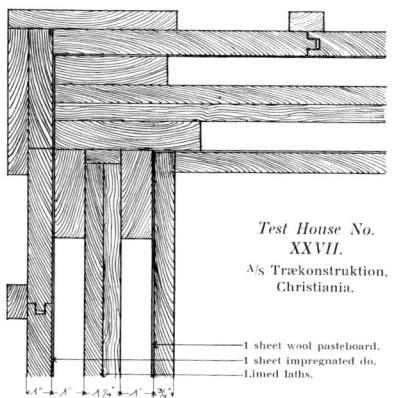

Test House No. XXVII.
A/s Trækonstruktion, Christiania.

—1 sheet wool pasteboard.
—1 sheet impregnated do.
—Limed laths.

House Number	Construction	Relative Heat Consumption As Shown	Temperature After Cutting off heat at 20°C			input to 5 hour retention	Ranked according to input to 15 hour retention
			5 Hours	10 Hours	15 Hours		
XXII	Wood	105	9.2	7.1	5.5	11.41	19.09
XIII	Wood	100	8.8	6.1	4.4	11.36	22.73
IV	Brick	159	10.2	8.4	6.9	15.59	23.04
XXI	Wood	96.5	8.5	5.5	3.9	11.35	24.74
VI	Brick	157	9.8	7.5	5.9	16.02	26.61
V	Brick	164	10.0	7.8	6.1	16.40	26.89
XXIII	Wood	119.5	7.8	5.6	4.1	15.32	29.15
II	Brick	175	9.2	7.1	5.5	19.02	31.82
III	Brick	179	9.2	7.1	5.5	19.46	32.55
VII	Brick	178	9.6	7.1	5.3	18.54	33.58
XI	Brick	156	8.2	5.9	4.6	19.02	33.91
XV	Wood	111	7.2	4.4	3	15.42	37.00
XIV	Wood	116.5	7.2	4.4	3	16.18	38.83
IX	Cement Block	181.5	8.8	6.2	4.5	20.63	40.33
XVI	Wood	108.5	6.5	3.6	2.3	16.69	47.17
XII	Wood	109	6.5	3.6	2.3	16.77	47.39
XXIV	Hy-Rib	176	8.6	5.2	3.7	20.47	47.57
VIII	Cement Block	200	8.0	5.2	3.4	25.00	58.82
XIX	Wood	115	5.5	2.8	1.7	20.91	67.65
XVII	Wood	128	5.5	2.8	1.7	23.27	75.29
XVIII	Wood	129	6.3	3.1	1.7	20.48	75.88
XX	Wood	145	5.5	2.7	1.7	26.36	85.29
X	Reinforced Concrete	221	4.6	2.6	1.6	48.04	138.13

"Warm and Cheap" Wall Types Tested
Bugge's published results

As a rough indication of behavior, it is useful to rank the huts retroactively comparing the energy input to energy retained. Additionally, the temperature of the retained air is worth noting. In this ranking, the local-code-driven wood frame system does the best. The more massive House XIII, with 3 inch tongue-and-groove infill, intakes energy rapidly and maintains it through the diffusivity of the wood. House IV was the best of the masonry types. This is likely because most of the masonry mass is on the interior of the assembly.

Conduction, of course, also played a fundamental role, but in their study it is only one of many composite factors that ultimately determined the behavior of these small buildings. As Sæland noted, if "an even temperature is maintained in

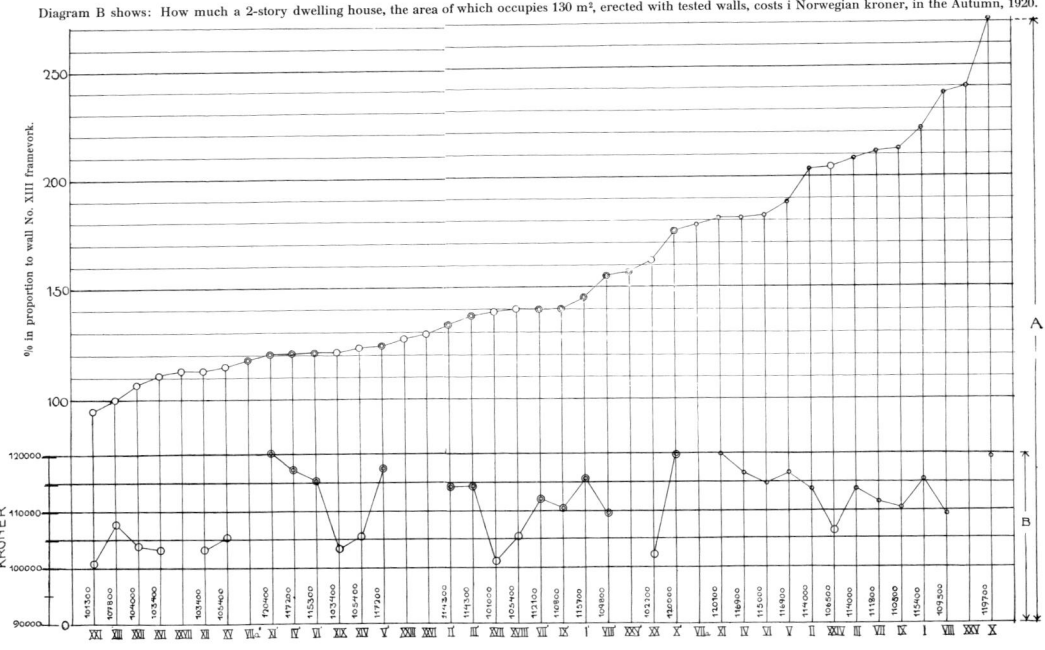

"Warm and Cheap" Wall Types Tested
Bugge's chart of "Medium heat-consumption of the walls of the test houses, (in percentage), in proportion to No. XIII (=100)"

a house through a lengthened period, the heat consumption of the house will depend exclusively on the co-efficient of the heat transmission for the external roofs, floors and walls (especially the walls)."[57] In other words, in a near-steady-state condition, conductivity matters most. However, as Sæland notes, "When, on the other hand, as is generally the case in practical life, the supply of heat is interrupted for a shorter or longer interval, and the inside temperature, as a result thereof, varies, the heat capacity of the said construction has to be reckoned with."[58] In this early transient test of thermal dissipation, the more transient role of the thermal diffusivity, for instance, played a role in their results and the retroactive ranking included here.

TABELL 5. *Tabell over resultaterne ved Norges Tekniske Høiskoles forsøkshus. 1920—1927.*

Hus-nummer	Veggkonstruksjon	Forholdstall for vegger alene	Varme-forbrukstall kcal m² °C time	Varme-motstand m² °C time kcal	Anmerkning	Hus-nummer	Veggkonstruksjon	Forholdstall for vegger alene	Varme-forbrukstall kcal m² °C time	Varme-motstand m² °C time kcal	Anmerkning
XXXIII	2P K.4cm Puss	93,0	0,63	1,59	λ av kork=0,052	XVI	2P 1P	111,0	0,760	1,32	
XXI	1P S 1P	95,5	0,65	1,54		XII	1P 1P	112,5	0,77	1,30	
XXX	2P K.3cm	95,5	0,65	1,54	λ av kork=0,048	XXVII	1P 1P	112,8	0,77	1,30	
XXXVIa	Tm. "Arki" 2P	99,5	0,676	1,48	Varmemotstand av «Arki»=0,365	XV	2P 2P	115,0	0,79	1,27	
XIII	1P 3" 1P	100,0	0,68	1,47		VI'	1½ st. eng hulmur	118,0	0,80	1,25	
XXXII	Arm. Slag-beton. K.4cm	102,7	0,70	1,43	λ av kork=0,051	XI'	1 st. hardbr ½ st. moler	120,0	0,814	1,23	
XXII	1P T 1P	107,0	0,73	1,37		XIV	2P 2P	120,5	0,82	1,22	
XXXVIb	Tm. "Arki" 1P	111,0	0,755	1,32		IV'	1¾ st. Thjems-hulmur	121,0	0,82	1,22	
						XIX	1P 1P 1P	122,0	0,83	1,20	
						V'	1¾ st. Thjems-hulmur	123,0	0,836	1,196	

"Table of results for the Norwegian Institute of Technology Experimental Houses"
Above and next page: The column headings translated: "house number," "wall construction," "ratio to reference wall (House XIII)," "heat consumption figures," "thermal resistance," and "note."

Hus-nummer	Veggkonstruksjon	Forholdstall for vegger alene	Varme-forbrukstall kcal m² °C time	Varme-motstand m² °C time kcal	Anmerkning
VII'a	1steng hulmur Koks	123,0	0,838	1,193	
XXXVIII	2P Celotex	124,5	0,85	1,175	λ av Celotex = 0,05
XXIII	1P L 1P	127,0	0,87	1,15	
XXVI	«Noah» bjelke	129,0	0,88	1,135	
XXXIX	Tm. «Arki» Hy-R.	131,5	0,89	1,12	
II'	1½ st. Bergens hulmur	133,0	0,90	1,11	
III'	1½ st. Bergens hulmur Fylite fuger	134,0	0,91	1,10	
I'	1½ st. mur.	136,0	0,926	1,08	
VII'	1steng hulmur	138,0	0,94	1,065	
XVII	2P 1P	138,5	0,95	1,050	

Hus-nummer	Veggkonstruksjon	Forholdstall for vegger alene	Varme-forbrukstall kcal m² °C time	Varme-motstand m² °C time kcal	Anmerkning
XVIII	2P 2P 2.stk	140,0	0,96	1,04	
XXVIII'	Celle-beton	142,0	0,963	1,038	
IX'	«Rex-sten»	149,0	1,01	0,99	
VI	1½ st. eng. hulmur	154,5	1,05	0,95	λ av mur 0,63
XXXI'	Arm. slag-beton	157,0	1,063	0,94	
XXV'	1st.eng hulmur Cement-sten	157,0	1,067	0,94	
XI	1st. hårdbr. ½ st. moler.	158,0	1,075	0,93	
VIII'	Lean-sten	160,0	1,09	0,92	
IV	1¾ st. Thjems. hulmur	161,0	1,095	0,92	λ av mur 0,64
XXIX	Arm. beton. K. 3 cm.	161,5	1,097	0,91	

Hus-nummer	Veggkonstruksjon	Forholdstall for vegger alene	Varme-forbrukstall kcal m² °C time	Varme-motstand m² °C time kcal	Anmerkning
XX	Bacula-puss 1P 1P	162,0	1,11	0,90	
VIIa	1st.eng. hulmur Koks	163,0	1,112	0,893	
V	1¾ st. Thjems-hulmur	164,0	1,115	0,896	
X'	Arm. beton. ¼ st. moler	174,0	1,185	0,844	
II	1½ st. Bergens hulmur	180,0	1,23	0,81	
III	1½ st. Bergens hulmur	182,0	1,24	0,80	
I	1½ st. mur.	188,0	1,28	0,78	λ av mur 0,63
VII	1st.eng. hulmur.	192,0	1,307	0,765	
XXVIII	Celle-beton	199,5	1,357	0,738	
XXIV	Hy-R. 1P Hy-R.	204,0	1,40	0,715	

Hus-nummer	Veggkonstruksjon	Forholdstall for vegger alene	Varme-forbrukstall kcal m² °C time	Varme-motstand m² °C time kcal	Anmerkning
IX	«Rex-sten»	214,0	1,45	0,69	
XXXI	Arm. slag-beton	229,0	1,56	0,641	λ av slagbetong 0,465
XXV	1st.eng. hulmur Cement-sten	231,0	1,57	0,637	λ av betong 0,88
VIII	Lean-sten	239,0	1,62	0,617	
X	Arm. beton. ¼ st. moler.	271,0	1,84	0,544	

As part of the transient consideration of the test houses, Sæland also considered the effect of insolation on the wall assembly behavior, noting that one would have to "reckon also with a supply of energy from without, e.g. radiation from the sun."[59] In the test, Sæland observed that,

> the measurements which have been made of the temperature in the hollows of the walls show clearly that the effect of the sun's radiation is quite considerable and that the radiating heat which a brickwall absorbs in the course of a clear sun-shiny day is a dimension which must be reckoned with in exact heat-economical calculations.[60]

Ultimately, though, he closed his contribution to the report with a resignation about the complexity of such factors: "the effect of the radiation of the sun, like the influence of the heat capacity of the walls, in this connection, is at any rate so complicated that it would require to be subjected to more particular examinations."[61]

The published results of this experiment were a series of dissipation curves and a table. The heat capacity of the masonry walls, of course, performed rather well for this "cool-off" test. The wood houses XIII (Trondheim building regulation standard) and XXII (wood panel with peat infill) both performed well. By combining the conductivity, diffusivity, and cost evaluations of the houses, the solid wood assembly (XIII) and the peat assembly (XXII) were the most consistently high-performing. The peat used on the test houses was 5 inch fuel peat bricks.[62] Bugge recommended, however, that 6 to 7 inch broad peat planks should be used in any future constructions and that the peat must be allowed to dry fully.

Bugge and Sæland took a decidedly empirical, in-situ approach to the thermal behavior of wall assemblies. This approach reflects the composite action of multiple, simultaneous factors and forces on walls that are difficult to abstract and synthesize in the laboratory model adopted in North America. This isolation-based North American habit of mind can work well. However, it underestimates the difficult nuances and complexities of heat transfer that Sæland indicated in his explication of the Trondheim test house experiments. Much was (and can be) learned from conductivity studies alone; but if left isolated, as it so frequently is in the practices and pedagogies of the topic, such studies offer only limited insight into the difficult whole of heat transfer in buildings.

These early Norwegian test huts took a different tack: a fundamentally non-isolated approach to what were then abstract and unknown phenomena. Rather than isolating individual phenomena as the basis for building a theory and prac-

Cooling off test.
Curves for houses Nos. IV, V, VI, VII, VIII, IX.

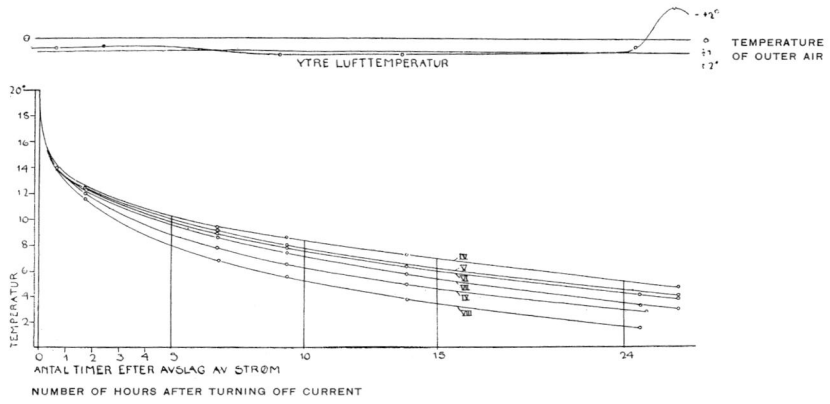

Curves for brick houses Nos. X, XI.
— » wood — » XII, XIII, XIV, XV, XVI.

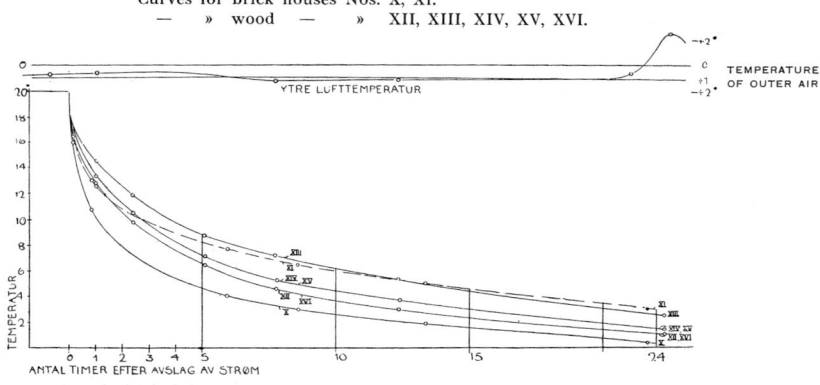

Curves for brick houses Nos. II, III.
— » wood — » XVII, XVIII, XIX, XX, XXI, XXII, XXIII XXIV.

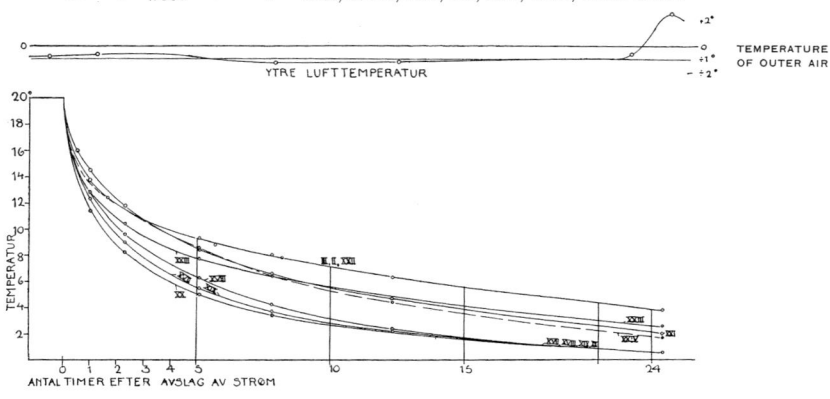

tice of heat transfer, these researchers choose to start with the difficult whole, evaluate the results and then begin to identify problems, capacities, and opportunities. Perhaps the most profound difference between the two approaches, however, is the degree to which the role of time and the non-steady-state complexity that it introduces are allowed to pervade our characterization of heat transfer in buildings.

It is interesting to note that, after the initial in-situ wall assembly tests, Bugge conducted a research trip to the United States (1926) that resulted in an extension of the in-situ experiment. His focus was the North American construction assemblies for houses, as counter-points to European examples.[63] He was interested in,

House Number	Construction	Table of Test Data — House XIII is the basis of consumption			Temperature After Cutting off heat at 20°C				
		Relative Heat Consumption							
		As Shown	With Added Wood Panel	With Added Coat Plaster	5 Hours	10 Hours	15 Hours		
I	Brick	188.5		185.5					
II	" "	175	124	172	9.2	7.1	5.5	19.02	31.82
III	" "	179			9.2	7.1	5.5	19.46	32.55
IV	" "	159			10.2	8.4	6.9	15.59	23.04
V	" "	164			10.0	7.8	6.1	16.40	26.89
VI	" "	157			9.8	7.5	5.9	16.02	26.61
VII	" "	178			9.6	7.1	5.3	18.54	33.58
VIII	Cement Block	200	140	194.5	8.0	5.2	3.4	25.00	58.82
IX	" "	181.5	129	176.2	8.8	6.2	4.5	20.63	40.33
X	Reinforced Concrete	221			4.6	2.6	1.6	48.04	138.13
XI	Brick	156			8.2	5.9	4.6	19.02	33.91
XII	Wood	109			6.5	3.6	2.3	16.77	47.39
XIII	" "	100			8.8	6.1	4.4	11.36	22.73
XIV	" "	116.5			7.2	4.4	3	16.18	38.83
XV	" "	111			7.2	4.4	3	15.42	37.00
XVI	" "	108.5			6.5	3.6	2.3	16.69	47.17
XVII	" "	128			5.5	2.8	1.7	23.27	75.29
XVIII	" "	129			6.3	3.1	1.7	20.48	75.88
XIX	" "	115			5.5	2.8	1.7	20.91	67.65
XX	" "	145			5.5	2.7	1.7	26.36	85.29
XXI	" "	96.5			8.5	5.5	3.9	11.35	24.74
XXII	" "	105			9.2	7.1	5.5	11.41	19.09
XXIII	" "	119.5			7.8	5.6	4.1	15.32	29.15
XXIV	Hy-Rib	176			8.6	5.2	3.7	20.47	47.57
XXV	Cement Block	198							
XXVI	Wood	121							
XXVII	" "	109							

Bugge's Expanded Test Results
Opposite: Results of the "cooling off test"
Above: All construction types compared

for example, the Craftsman houses of the early twentieth century.[64] As a result of the trip, Bugge added twenty-six more huts to his experiments, for fifty-three in total by the end of the experiments.[65] It is this second group of test houses that can be seen in aerial photographs of the Trondheim Gløshaugen campus in the 1930s and that remained into the 1950s before modern campus buildings were built on their site.

From 1925–1927, Bugge field-tested a number of these North American examples and materials. The second round of tests included proprietary products such as Celotex, cork insulation boards, and newer aerated concrete block types.[66] The result of these additional experiments, as assembled in a beautiful chart, is a startlingly comprehensive field-based set of data on trans-Atlantic examples. It should be noted, however, that Sæland's compelling dissipation "cool-off" tests

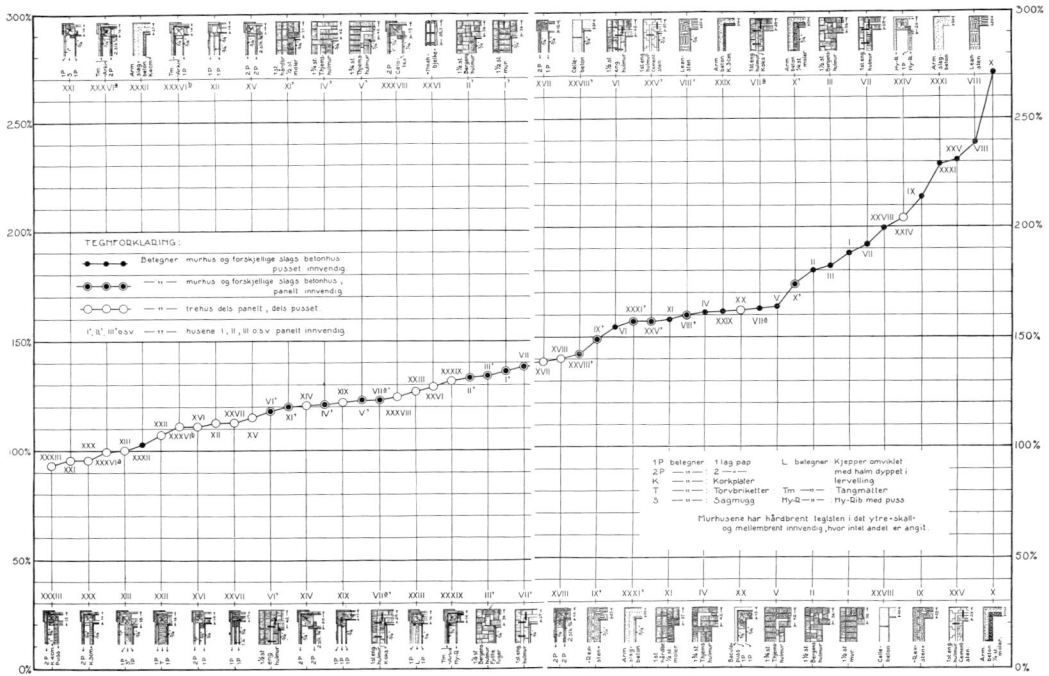

Bugge's Expanded Test Results
All construction types plotted

were not conducted on this second round of examples. So, whereas some of the North American insulation products faired rather well in the conductivity tests, their reduced mass and testing (i.e. no recognition of relative heat-capacity and dissipation tested in the first examples), should temper one's reading of the results. The North American test houses did not benefit from Sæland's full insight and observations.

More generally, Bugge and Sæland's early in-situ experiment — the first according to their documentation — forms a cogent and compelling challenge to the contemporaneous preoccupation with a laboratory-based approach centered on abstraction, analysis, and synthesis. The inherently synthetic and composite reality of the in-situ tests allowed Bugge and Sæland to observe and document thermal phenomena beyond the bounded domain of conductivity that over-determined other research trajectories regarding insulation. This reflects a radically different epistemology and subsequent practices no less scientific, for the dissipation of energy in buildings.

In this way, the various approaches to understanding, and then designing, thermal phenomena are highly dependent on their experimental apparatus, as indicated by the reference to Peter Galison in the introduction. The choice of instrument and apparatus is not neutral and conditions not only the experiment but also the outcomes, as well as the practices those outcomes co-determine. The contrast struck between the Norwegian and North American experimental apparatuses at this time provides insight into the respective trajectories the history of insulation took in subsequent decades.

The frank optimism, as opposed to the resistance, of the inevitable dissipation of heat in the Norwegian example exhibits not just a divergent testing apparatus but an extremely refreshing engagement with composite thermodynamic effects that, if not completely lost, were at least difficult to simulate in the North American laboratory culture and its refrigeration antecedents. Perhaps paradoxically, it was Norway's lack of necessary refrigeration that allowed it to learn about heat dissipation more directly and more quickly than its North American counterparts.

Germany: 1918–1931

This Norwegian experiment was not the only early example of full-scale, longer-term transient thermal evaluations of buildings. In this period, Germany had a building research arm, the Reichsforschungsgesellschaft für Wirtschaftlichkeit im Bau- und Wohnungswesen (translation: The Reich Research Society for Economy in Construction and Housing). The RFG was established in 1926 to address the German housing shortage and to rationalize the means of housing production. Its members included Walter Gropius, Ernst May, and other architects, researchers, and politicians. The RFG analyzed housing in a totalizing scope and this included the role of heat transfer in building components. It was dissolved in 1931.

Ernst May's efforts on the Praunheim housing settlement in Frankfurt serve as one example of the RFG work.[67] The design of the building system for this housing system included extensive testing of the thermal conductivity of floor slabs, wall slabs, and their joints. In 1926, as part of this larger effort, the Technical University in Munich tested different weights of concrete and different concrete admixtures — such as Ernst May's pumice for lightweight concrete panels — for their effect on thermal behavior.[68] One material test focused on a mix of one part cement, one part sand, four parts slag, and five parts pumice. The particular mixture of cement, slag, and pumice occupied much attention in the research. Importantly, they also evaluated the behavior of the buildings once they were constructed of the various material mixes. The relative density and thus weight of construction was one concern, the relative conductivity of wall and panel types was another. In their tests, masonry had a thermal conductivity of .75 while the value for pumice concrete panels was .33 and for aerated concrete was .845 or .567 depending on the panel thickness.[69] In the case of Praunheim, conductivity tests were used as an indicator of thermal behavior. In short, rigorous forms of thermal testing were seen as essential to a larger form of housing research and housing delivery.

These German full-scale in-situ field experiments were an extension of a robust period of laboratory research on insulation. After World War I, fuel costs in Germany provoked formal insulation research located in the Technische Universität München.[70] This laboratory — the Forschungsheim für Wärmewirtschaft — was charged with standardizing how German insulation manufacturers could evaluate and label insulation products. The laboratory was led by Dr. Oskar Knoblauch and included a hot-plate apparatus for the evaluation of materials. By the mid-1920s, conductivity values derived in the Forschungsheim für Wärmewirtschaft were published and part of practice, including the RFG discussed above.

Whereas the early German insulation apparatus was laboratory-based, later, a post-war German building testing station was established in Holzkirchen near Munich, in Upper Bavaria. The extreme weather of Holzkirchen was the primary reason for the site selection. The post-war years included the full-scale in-situ testing of identical housing of different building assemblies, not unlike the Norwegian example above.[71] This testing center became the basis of the now well-known Fraunhofer Institute.

Poland: 1926–1939

Poland had no formalized, government-funded institutes or experiments related to heat transfer or building science in general. Any activity in the area was the result of individual and group initiative. There were two key building science protagonists in Poland in the middle third of the twentieth century. The architect couple Szymon Syrkus and Helena Syrkus became members of CIAM and worked to incorporate building science as part of Polish modernism.[72] The introduction of new materials for the exterior wall of buildings was of particular interest to the couple.

Other Polish architects shared this interest. Some, such as architect Lech Tomaszewski, made a series of astute observations about the properties of multi-layer exterior walls, such as desirable amounts of aeration of concrete mixtures, distinguishing between materials that capture and channel heat flow (interior re-radiation of accumulated heat) and materials that protect against heat flow (more familiar applications of insulation).[73] Tomaszewski's characterization of wall material properties is compelling to consider today because it is not based on a false dichotomy of insulating materials and non-insulating materials but rather more fundamentally characterizes a spectrum of relative velocities of dissipation and how that might be productive in various circumstances.

Netherlands: 1925–1940

Like Poland, Dutch research activity related to heat transfer largely stemmed from ambitious architects advancing a modernist ethos and who saw building science as a constitutive aspect of such architecture. The key Dutch protagonists were architects Jan Duiker and Johannes Bernardus van Loghem along with the engineer Jan Jacobus De Ridder. Duiker was a leader of progressive architects and served as editor of the journal *De 8 en Opbouw,* which hosted advanced technical research for Dutch architects at the time.

Following aspects of the Norwegian and German model, the Netherlands did have one set of full-scale, in-situ experiments. The Betondorp experiment, built outside Amsterdam, tested various lightweight concrete mixtures between 1923 and 1928.[74] Based on this work, the Dutch protagonists focused on the development of multi-layered walls as one aspect of heat transfer.[75] An outline by polish architect Szymon Syrkus at CIAM IV in Athens in 1933 articulated the notion of the "Functionally Differentiated Wall." This outline was subsequently published in *De 8 en Opbouw* in 1939 and motivated Dutch research on the topic of heat trans-

fragebogen : „funktionell differenzierte aussenwand"
beilage no. 2: beispiel einer entwicklungstabelle der schichtenwand

1. ausgangspunkt:
vollziegelmauer dicke verschieden je nach dem klima.

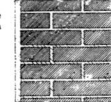

aussenputz 2 cm
vollziegel 38 cm resp. 52
innenputz 1,5 cm

8. warschau.
dasselbe prinzip in backsteinmaterial: isolationsschicht aus dünnwandigen hohlziegelblöcken an der aussenseite, akkumulationsschicht aus dickwandigentragenden hohlziegelblöcken von gleichem format fugen versetzt.

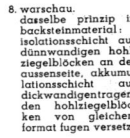

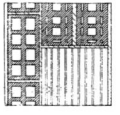

aussenputz 2 cm
dünnwandige ziegelhohlblöcke 27 x 13 x 27 cm
mörtelfuge 1 cm
dickwandige ziegelhohlblöcke 27 x 13 x 27 cm
innenputz 1,5 cm

2. warschau
43 cm dicke vollziegelwand, in der die wärmebrücken der mörtelfugen mit pappe unterbrochen sind, wirkt wärmetechnisch gleich wie eine 55 cm dicke vollziegelwand.

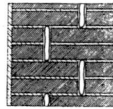

aussenputz 2 cm
vollziegel 13 en 27 cm
pappfuge 3 cm
innenputz 1,5 cm

9. zürich - komposit skelett.
das prinzip der wände 6, 7 u. 8 ist in diesem beispiel durch anwendung stark differenzierter materialien weitergeführt: als isolation dient kork, als akkumulation kiesbeton, als aussenverkleidung kunststeinplatten.

kunststein 6 cm
korkplatten 4 cm
betonbrüstung 10 cm
innenputz 1,5 cm

3. basel - stahlskelett bimsbetonwand; wärme-Isolierung -und-accumulierung nicht genau differenziert.

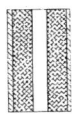

aussenputz 2 cm
bimsplatten 8 cm
luftraum 6 cm
bimsplatten 8 cm
innenputz 1,5 cm

10. warschau
die luftschicht wirkt als körperschallisolierung, gleichzeitig trennt sie die organische isolierplatte von der event. feuchtigkeit der zementmörtelplatten und des mauerwerks.

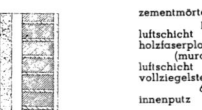

zementmörtelplatten 4 cm
luftschicht 4 cm
holzfaserplatten (muroblock) 5 cm
luftschicht 4 cm
vollziegelsteine 6 x 13 x 27 cm
innenputz 1,5 cm

4. warschau - eisenbetonskelett
wand aus ziegeln u. zellenbetonplatten: isolierungsschicht nach innen, akkumulierungsschicht nach aussen verlegt.

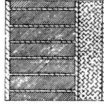

aussenputz 2 cm
vollziegel 27 cm
zellenbeton 10 cm
innenputz 1,5 cm

11. berlin - holz skelett.
beispiel einer serienweise nergestellten und trocken montierten aussenwand.

kupferwellblech 0,05 cm
1 lage asbestbitumenpappe
1 lage aluminiumfolie
2 lagen aluminiumfolie
2 lagen asbestbitumenpappe mit
1 lage aluminiumfolie dazwischen
2 lagen asbestbitumenpappe mit
1 lage aluminiumfolie dazwischen
1 lage aluminiumfolie
1 lage asbestbitumenpappe
aluminium innenblech

5. paris - stahlskelett
dasselbe prinzip in bezug auf die anordnung der akkumulierungsschicht (backstein) und der isolierungsschicht (heraklith), aussen-u. innenputz durch verkleidungsplatten ersetzt.

kunststein 5 cm
mörtelfuge 1 cm
hohlsteine 11 cm
luftraum 5 cm
heraklith 3 cm
verkleidung 0,5 cm

6. warschau - stahlskelett
weitere durchbildung der wand 4: zellenbetonplatten als wärmeisolierung nach aussen, wärmeakkumulierungsschicht aus backstein nach innen versetzt, (die zellenbetonplatten werden in der fabrik mit ihrer verkleidung aus natursandstein versehen).
auf der windanfallseite celotex auf latten.

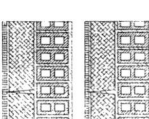

natursandsteinplatten 2,5 cm
cementmörtel mit dichtungszusatz 1 cm
zellenbeton 10 cm
mörtelfuge 1 cm
hohlziegel 6 x 13 x 27 cm
innenputz 1,5 cm
auf windanfallseite celotex auf lattung 3,6 cm

12. zürich.
differenzierung d. wände je nach der lage im hause: südbrüstung in zusammenhang mit der fensterkonstruktion gebracht. (auf wärmeakkumulierung wird bewusst verzichtet) nordwand normale artübliche hohlziegelwand.

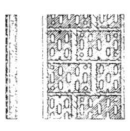

südbrüstung:
eternit 0,3 cm
luft 0,7 cm
kork 0,8 cm
sperrplatte 0,3 cm
nordwand:
aussenputz 2 cm
ziegelhohlblöcke 30 cm
innenputz 1,5 cm

7. warschau-stahlskelett.
zur verminderung der fugenzahl und möglichkeit der anordnung versetzter fugen wurden hohlziegelblöcke von vierfachem normalformat verwendet.
das prinzip der anordnung der schichten bleibt dasselbe.

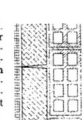

natursandsteinplatten 2,5 cm
zementmörtel mit dichtungszusatz 1 cm
zellenbeton 10 cm
mörtelfuge 1 cm
hohlziegelblock 27 x 13 x 27 cm

13. basel.
dasselbe prinzip in anderer ausführung.

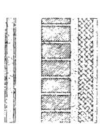

südbrüstung:
pressinsulite 0,6 cm
korkplatte 3 cm
pressinsulite 0,6 cm
nordwand:
ziegelsteine 6 x 12 x 25 cm
kork 6 cm
tuffplatten 6 cm
innenputz 1,5 cm

"Functionally Differentiated Walls"
Wall assembly conductivity values from *De 8 en Opbouw*

fragebogen: „funktionell differenzierte aussenwand"
beilage no. 1: anzuwendende zeichen
questionnaire: „le mur extérieur differencié fonctionnellement"
annexe no. 1: signes conventionnels

backstein / brique de terre cuite / brick		kunststeine / pierres artificielles / artificie stone
		natursteine / pierres naturelles / natural stone
holz: weichholz / hartholz / bretter / sperrholzplatten		
bois: bois tendres / bois durs / planches / bois contreplaqués		eternit und schiefer / eternit et ardoise / eternit and slates
wood: soft wood / hard wood / planks / plywood		
		lockere füllstoffe / materiaux de remplage / lose filling in materials
metalle / métaux / metal		kork und leichtbauplatten / liège et dalles de construction légère / cork and lightbuildingslabs
kiesbeton / beton de gravier / gravel concrete		filz / feutre / felt
beton légers / leichtbeton / bims etc. concrete		bitumen gewebe / falzplatte / paste
		bitumes jutes / plaques / pâtes
		bitumen sheet, paste
		glas / verre / glass
zementmörtel u. estriche / enduit de ciment / cementplaster		rohrmatte / planches de roseaux / schilfmatting
putze / enduits / plaster		terrain / ground

fer in relation to issues of construction.[76] There was also significant interest in the interaction of building envelopes and heating systems. The two systems were seen as more highly intertwined than in other countries.

For instance, Wessel de Jonge has documented one such debate between more conservative engineers and Jan Duiker.[77] The engineers argued that Duiker's large panes of sheet glass were, when considered in isolation, significant sources of heat loss. Duiker argued, however, that large sheet glass, properly considered, was not as significant of a heat loss penalty when considered as a functional aspect of heat gains, the specific heat and absorptivity of adjacent material and the novel radiant-based heating systems developed with De Ridder. De Ridder would later advance the argument that large sheet glass improves the efficiency of systems on account of less mass and faster system reaction times.[78]

De Ridder, likewise, took issue with more isolated considerations of conductivity. Against the use of dry, laboratory conductivity values, De Ridder was aware

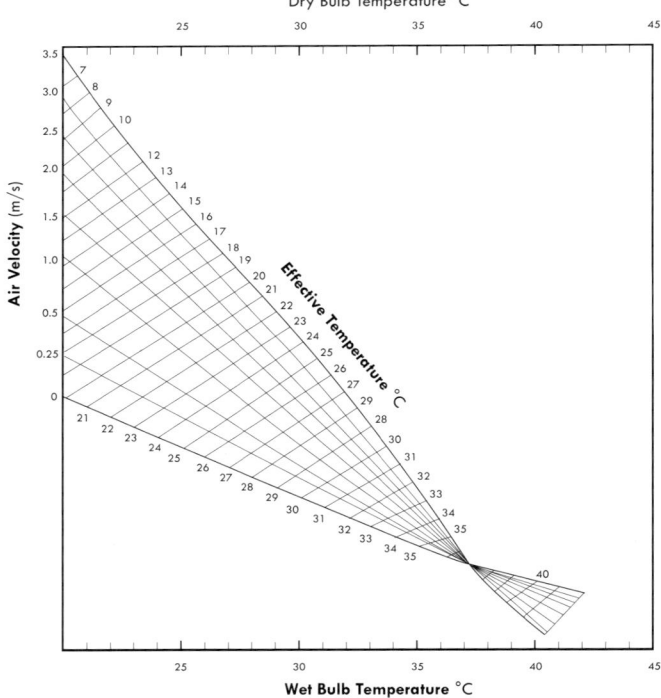

Dufton's "Effective Temperature" Chart
Temperature, become more intricate

that humidity and moisture detrimentally affected the insulative capacity of materials and argued for humidity-based adjustments and for much greater attention to detailing for dry insulation and the correct placement of air cavities.[79] In each of these cases, these designers considered heat transfer dynamics as fundamental aspects of their architecture. The results, such as Duiker and Bijvoet's Open Air School (1930) and Zonnestraal Sanatorium in Hilversum (1928–31) and Duiker and De Ridder's Gooiland Hotel (1935), are the canonical Dutch examples of thermal innovations in this period and are indicative of a latent and untapped architectural agenda for energy.

These arguments and experiments are best documented in *De 8 en Opbouw*, re-published in the 1980s in the Netherlands. Eventually, a pair of Dutch books related to heat transfer in buildings were published, but not until 1936, long after many of the modern architecture building experiments were already established.[80] One of these authors, Van Loghem, had unambiguous advice for the architects seeking to disrupt the longer tradition of load-bearing masonry, "the elimination of the load-bearing function true eliminated one problem but the requirements of 'het Nieuwe bouwen' on the other hand created at least ten new problems."[81]

Great Britain: 1921–1948

The British Research Establishment in the United Kingdom was also quite active on thermal topics and with related approaches in the period between the World Wars. The British Research Establishment was established in the early 1920s to address a rapidly evolving building industry. By 1921, initial tests on various materials were underway.[82] By the mid-1930s, the Building Research Station (BRS) employed over two hundred researchers engaged in different aspects of research.

In the context of heat flux in buildings, several notable advancements occurred at the BRS in the years leading to and during World War II. Most significant to the present historical account, one of the earliest (1934) technical treatments of the transient thermal behaviors of buildings was developed there by A. F. Dufton.[83] His article, "The Warming of Walls," established some fundamental terms that extended far beyond the steady-state analysis of heat transfer in walls. The article is in many ways a watershed point in this history of insulation. Dufton pushed the topic of heat flux away from steady-state concerns and towards the sinusoidal oscillations, periodic states, and mixed-mode oscillations that are farther and farther from equilibrium in their behavior. But, as Dufton notes, "a complete analysis of the warming of a room is not easy."[84]

Dufton continued this line of thought by developing a new heating laboratory at the Building Research Station in Garston. He observed that "the introduction of panel-heating, for example, has not only made necessary a revision of ideas as to what constitutes a comfortably warmed environment but has brought out also the important fact that the received methods of computing the heat losses of a building are only directly applicable when the building is heated in the 'traditional' manner."[85] To this end, the BRS constructed a building around a single room with controlled surfaces and conditions that enabled empirical comparisons of different heating methods and installations. So as to maintain consistent conditions from test to test, the room was isolated as much as possible within the new building. Great technical effort focused on maintaining equivalent conditions for the tests.[86]

Aiming to move beyond the limitations of laboratory-derived conductivity values of dry materials, the BRS eventually moved heat transmission research outside to actual climate-exposed conditions through both the construction, in 1926, of an experimental house in Garston and, in 1936, a series of smaller chambers with easily replaced envelope assemblies.[87] Similar exterior tests of roof assemblies began in 1931 in two small buildings.[88]

A 1938 volume, *Principles of Modern Building*, brought many aspects of the BRS research to the broader profession of practicing British architects.[89] This rich resource considered walls in a comprehensive manner — structural, thermal, moisture level, durability, etc. — but its engagement with thermal concerns stopped with conductivity. The author's summary of insulation for architects — titled "The practical treatment of insulation" — ends with the statement that "the rate of heat flow through a wall is directly related to the temperature difference between the two faces."[90] While such observations brought principles of heat transfer to practicing architects, they did not reflect Dufton's earlier insights nor comport with the greater intricacies of heat transfer available at the time. Once again, architects were presented with an under-complex perspective on heat transfer.

Another important British publication from this period was Thomas Bedford's 1948 volume on *Basic Principles of Ventilation and Heating*.[91] This book, rich in both historical and technical accounts of heat transfer, represents the most comprehensive consolidation and distillation to date of many aspects related to heat transfer. Its clarity and reach are unmatched. One chapter focused on the thermal properties of buildings.[92] In this chapter both the role of thermal capacity and the role of radiant transfer are considered in the context of insulation, pointing towards a more nuanced post-war characterization of heat transfer.

Australia: 1945–1955

As in the Norwegian, German, and British examples, the Commonwealth Experimental Building Station in Australia was similarly engaged in the analysis of over a hundred structures of various materials and construction types.[93] The Australian Building Station initially focused on construction issues, but shifted to issues of climate, physiology, and heat flux after World War II. This shift in research agenda is reflected in several research bulletins with titles such as *Climate and House Design, Sunshine and Shade in Australia* and *Natural Ventilation, Ceiling Height, and Room Size*.[94] Collectively, these research bulletins reflect a much more dynamic, non-isolated characterization of heat transfer in the realm of buildings.

The research program on issues related to temperature, physiology, and climate emerged from a concern about the imported British concept of effective temperature. Australian researchers stated that "the opinion was formed that existing comfort scales were not applicable to the problem of climate and house design when comfort at elevated temperatures was considered."[95] The combination of specific climate issues in Australia, coupled with the heavier research focus on construction technique, eventually yielded a sustained research that more directly merged issues of construction, climate, and comfort. In that context, this took research into various types of earthen construction.

In 1949, at the Commonwealth Experimental Building Station in Sidney, an English architect and engineer, George Middleton, constructed rammed-earth buildings.[96] After testing, Middleton was confident about the role of earthen construction, as the title of his subsequent book introduction reveals: *The establishment of a well-tried technique for the identification of suitable earths, and a standard of practice for the methods of construction, should place earth wall construction, which has so much to commend it, high among the accepted building methods.*[97]

The United States, Canada, & France

By World War II, the focus on conductivity as the central concern of heat flux in buildings was questioned in the United States. The Texas Engineering Research Station was engaged in this question in the early 1940s.[98] Elmer Smith completed, for the North American context, a refreshingly idiosyncratic dissertation at the University of Texas that developed methods for evaluating transient conditions in buildings.[99] He also accounted for the interior walls and furniture for buildings in transient conditions.[100]

His line of research is boldly exemplified by the report "Summer Comfort Factors as Influenced by Thermal Properties of Building Materials."[101] This report begins by taking exception to the tendency for steady-state, conductivity-based heat transfer research in buildings in the North American literature. As a less reductive alternative, the Texas Engineering Research Station aimed to characterize far more transient problems that connect heat transfer to climate, orientation, diurnal flux, and building use.

Elsewhere in the United States, Charles Osborn Mackey and Lawrence T. Wright Jr., were unequivocal about their exasperation with the preoccupation with steady-state characterizations of walls and insulating materials:

> if the wall has any heat storage capacity, and wall materials in any finite thickness do, the steady-state heat flow can exist only when (1) the temperature of the indoor air and the motion of that air remain constant with respect to time, (2) the amount of radiation received or emitted by the inside surface of the wall remains constant, (3) the temperature and motion of the outdoor air remain constant, and (4) the amount of radiation received or emitted by the outside surface of the wall remains constant. These conditions are seldom, if ever, exactly constant through a one-hour period and are never constant throughout the day. It is basically incorrect to select a material for a building wall on the basis of its steady flow thermal properties; engineers are aware of this fallacy, but the laws of unsteady heat flow are very complex and departures from the steady flow laws are not easy to derive in any case and may not be important in some cases.[102]

By the end of World War II, there was considerable work on the non-steady-state heat transfer in buildings. In the United States, Mackey and Wright Jr. advanced their work, publishing two important articles at the end of the war on the periodic heat flow of homogeneous and composite building envelopes.[103] This work assumed sinusoidal variation in outside air and thus developed methods for calculating heat flux to the interior accordingly. This produced methods for introducing shifting boundary conditions in the equations of heat transfer.[104] Another American, J. P. Stewart, was likewise engaged in questions of shifting boundary conditions triggered by solar gains and their effects on cooling load calculations.[105]

Not to be left out, the French were occupied with related work. In 1947 A. Nessi and L. Nisolle published a non-steady-state account of heat flow in building envelopes.[106] This article significantly advanced the topic. In relation to this, by 1957 in Canada, the overlooked role of transient conditions in the thermal evaluation of building envelopes was being addressed. D. G. Stephenson's report on "Periodic Heat Flow in Walls and Roofs" begins with a telling abstract:

> The Division of Building Research is currently reexamining and appraising the methods employed in predicting the thermal energy exchange between the indoor and the outdoor environment through the enclosing walls and roofs of buildings. Summer conditions which give rise to periodic flow are of special interest, and heat exchange between the ground surface and the outdoor environment is also of concern.
> Although the path of the thermal energy exchange between indoors and outdoors may be broken into parts for special study, these are actually interrelated and in a rigorous treatment must be considered together.[107]

Building on Mackey and Wright Jr.'s work, British research on non-steady-state heat flux continued in the context of the Institution of Heating and Ventilation Engineers. Among the publications to emerge from this institution in the 1960s was E. Danter's work on the "Periodic heat flow characteristics of simple walls and roofs."[108] Part of Danter's contribution in this period was "admittance method." This concerned the sinusoidal fluctuation of interior and exterior temperatures, what he then referred to as temperature "swing." The amount of swing was determined by the "admittance" of the room, the relative density of the room construction assemblies. Commenting on the admittance method, building scientist Neville Billington observed that "admittance is clearly related in some way to the diffusivity and thickness of the materials of construction" and "is defined as the range of variation of heat output required to produce unit swing of the room temperature."[109]

By the late 1960s, these non-steady-state accounts of heat flux were well incorporated into building science books. Notably, Billington published a 1967 volume on heat that provided comprehensive accounts of both "The Steady Transfer of Heat," the first chapter, and "The Variable Flow of Heat," the second chapter. By this point, numerical models for non-steady-state heat flux were well established.

Heat Transfer, Fiscal Transfer

Despite the international recognition of the nuanced and dynamic thermal behavior of buildings, a reductive characterization of heat transfer nonetheless prevailed in architecture. Given the paucity of knowledge about heat transfer in the practices of architects (as evident in the dialogue on heat transfer in architectural journals and textbooks at the time), it is ultimately apparent that concern about the dynamics of heat transfer alone did not motivate this reductive characterization of the matter.

There was significant pressure after World War II, especially in North America, to determine a nationally accepted, non-complicated rating of building insulation materials to regulate the claims of the insulation industry. The interests of the insulation industry were quite evident in a post-war definition of insulating materials found in a common textbook on the topic:

> A thermal building insulation is defined as a proprietary (owned) manufactured building material having a heat flow (conductivity) not to exceed 0.50 B.t.u. per hour per square foot per degree Fahrenheit per inch thickness as tested by an authoritative laboratory and which, when installed, shall have a total heat resistance of not less than 1.0.[110]

If the early literature on thermal conductivity suggests that it was troublesome to derive, the above statement suggests that it was also cumbersome to define. Further, the focus on proprietary materials is difficult to discern in this definition given that, at the time, so much of a building was constructed from largely non-proprietary materials such as masonry, wood, and concrete. This shift from materials to products, from material science to product development, corresponds to an important corollary shift in this historical narrative and the building industry at large. The history of insulation is not only a history of astute scientific interests; it also reflects the influence of astute commercial interests.

R-value: Thermal Resistance & Electrical Metaphors

The call for a nationally accepted parameter — necessarily reductive — was met by the conceptualization of the R-value in North America. R-value refers to the capacity of a material to resist heat flow: what was characterized as thermal resistance. Under conceptually constant (but actually evasive) conditions, thermal resistance is a measure of the temperature difference between two surfaces of a material relative to the total heat flux through it. In architectural literature and training, thermal resistance is neatly but mistakenly presented as the reciprocal of conductivity. This is not always the case and it is excellent evidence of the will for a reductive, easily conveyed and understood parameter of heat transfer as a primary conceptual and physical unit, even if it does not correspond to reality.

Everett Shuman, an engineer, is frequently credited with the development of the R-value concept in 1945. In 1961 he joined the Building Research Institute at Pennsylvania State University as its director and an associate professor. In this role, he continued to advocate the function of R-values in simplifying and reducing thermal insulation to this single parameter.[111] Despite the common attribution of the R-value concept to Shuman, the notion of thermal resistance was an established aspect of the heat transfer discourse prior to 1945. Several articles from as far back as 1902 focus on the subject.

An early and persistent advocate for a simplified measure of thermal resistance was Carl Hering, a mechanical engineer. In a 1911 article, Hering makes a direct analog between electrical flow and thermal flow, going so far as to name it the "thermal Ohm's law."[112] He observes: "When heat or electricity flows through several bodies in series it is far simpler to use resistances in the calculations, instead of conductances." He thus derives the term "thermal resistance," as represented by the letter R, praising it as a "great convenience and time-saving device."[113] Hering's work built on the first statement about thermal resistance, dubbed "heat resistance," by Kent in 1902.[114]

Very importantly, this electrical metaphor for heat transfer is not only problematic for heat transfer but even more problematic when similar metaphors were used in the post-war years to characterize the metabolism of large-scale energy systems. Notably, Jay Forrester of MIT extended an electrical network metaphor to large-scale energy systems, resulting in a number of compelling but errant characterizations of energy systems.[115] In particular, by the time that Forrester's associates scaled up his metaphor, which led to the publication of *The Limits to Growth*, significant problematics had emerged with this metaphor.[116]

Again, as a steady-state concept, R-value is an inadequate proxy of heat transfer in building assemblies. Yet it is a habitually cited and taught platitude that claims it is as the reciprocal of conductivity. But this reciprocity is only valid under certain conditions, as John Straube notes:

> If all heat flow was by conduction, and if all materials where homogenous and exhibited no temperature sensitivities, then it would be appropriate to assume that the R-value was equal to the inverse of thermal conductivity divided by thickness (R = 1/k). However, heat transfer through most materials and assemblies is a combination of heat flow by the modes of conduction, radiation, and convection.[117]

Despite these basic errors and problems, the standardization of thermal conductivity, and R-values in due course, became an increasingly important issue for insulating materials in the second half of the twentieth century. This was especially the case in the 1970s as energy emerged as a topic of consideration in new ways. Thus, for the first time, in 1979, the United States Federal Trade Commission (FTC) stipulated that claims about the thermal resistance of insulation materials must henceforth be tested following ASTM standards and reported in a consistent manner.[118] The impetus for such standards reflects the degree of market saturation of conductivity as the proxy of heat transfer. These standards specified the laboratory procedure for the determination of building product conductivities.[119] In so doing, they also reinforced and popularized the momentum of steady-state characterizations of heat transfer.

The R-value concept, broadcast in the mid-twentieth century but developed decades before, foremost expresses market frustration with and exhaustion by the complexities of heat transfer that are evident in the early publications on the topic. The R-value concept ultimately manifested the intellectual and fiscal pressure to portray the unavoidable complexities of heat transfer in simple terms. The higher the R-value, the better insulated a building would be, ostensibly, according to the marketing literature and associated pedagogies. Consequently, however, the more this proxy serves as an index of an architect's knowledge of heat transfer, the more reduced and isolated that architect's knowledge of heat transfer is, an outcome that befits the apparatus of insulation more than it befits buildings or architects.

After decades of development of insulation theory, the industry needed, and so developed, a reductive and instrumentalized metric. But in no way did the reductive R-value concept resolve fundamental questions about heat transfer in buildings and in no way does it reflect the nuanced reality of heat transfer in buildings. In other words, as the twentieth century progressed, the insulation of modern architecture became increasingly isolated from the full realities of heat transfer. As the insulation apparatus progressed in this direction, architectural engagement and agency regressed.

Resistance to Thermal Resistance

As the deficiencies of the R-value approach have become more apparent, newer standards, such as ASTM 1326-05, use full-scale buildings assemblies to assess the actual behavior of assemblies.[120] This shift in testing procedure references early non-North American experiments such as the mid-1920s work in Trondheim, Norway. The newer standards acknowledge that there are multiple phenomena present in any building heat transfer scenario. Curiously, the original FTC definition of insulation from the 1970s was less focused on thermal resistance and offered a broader perspective than the early ASTM standards did: "Insulation is any material mainly used to slow down heat flow." This characterization synchronizes much better with reality. It suggests that thermal diffusivity and specific heat — important transient parameters of a material — would be of importance, yet the impetus of the standard focuses on steady-state R-values. It suggests, quite cogently, that the problem of heat transfer in buildings is first and foremost a design problem of modulating velocities. The FTC shift to test full-scale assemblies finally began to fulfill the terms of its own definition.

The persistent presence of steady-state theories, practices, and pedagogies of insulation, however, remains in contemporary practice. Whether in energy codes, energy simulations, building science courses or architectural journals, steady-state characterizations of heat transfer (U-values, R-values) dominate architecture and the building industry. While thermal conductivity and the concept of thermal resistance are important factors in the behavior of buildings in many circumstances, they are by no means the only parameters active in the thermodynamic reality of a building operating with diurnal and seasonal flux, not to mention in larger temporal and spatial exchanges of energy.

These parallel but often overlooked parameters point more towards a non-steady-state, transient grasp of heat transfer that more directly corresponds to the reality of building material assemblies. How these parameters operate in a solid wood building, for example, challenges many assumptions that are embedded in contemporary energy codes and green building certification checklists.

Given the realities of buildings and their transient conditions, the role of other thermal parameters — such as specific heat, thermal diffusivity, and thermal effusivity — warrants more attention in an architectural agenda for energy. Such an approach would not emerge from the dominant, steady-state "non-conductor" theories of insulation but indicate an approach that manages heat transfer in more robust and nuanced ways. The role of thermal mass in the diffusion, dispersion, and delay of heat transfer is but one trajectory. How such an approach — especially in the case of solid wood construction — could relate to human bodies, climate, ecologies, and other building-related energy flows extends the concern for heat transfer beyond the envelope of a building into a much broader, more ecological realm of interest. This is distinct from the abstractions of a hot-plate test for thermal conductivity. This non-modern agenda would reflect a much more ambitious and accurate understanding of the thermodynamic depth of buildings. It would also provide greater agency to the realm of design and less to the realm of industrial interests.

Thermal Mass & R-Values: the Thermal Mass Program, 1979–1985

Despite persistent reductive accounts of heat transfer, there was a minor resurgence in the topic of other, more transient thermal parameters in North America in the final quarter of the century. In the late 1970s and early 1980s, there was a spate of interest in non-steady-state characterizations of building envelopes that would include the role of mass in assemblies.

This can be seen as a consequence of the publication of Billington's book, as well as J. F. van Straaten's 1967 volume on the *Thermal Performance of Buildings*.[121] In this book, Van Straaten discusses a number of factors that point towards mass effects. In the first page of the book, the South African researcher notes that common conductivity equations refer

> to the ideal case where temperatures remain constant with time, so that all the heat entering on surface also leaves the opposite surface. As far as buildings are concerned, however, this state of affairs is seldom realized. Because of the ever-changing outdoor climate, steady-state temperature conditions can only be approximated and heat transfer through building elements is mostly unsteady or transient.[122]

Unlike other building science researchers, Van Straaten was not so ready to limit his understanding of heat transfer to this ideal case. He clarified when this ideal care is relevant:

> Strictly speaking, steady-state heat transfer is only approximated in practice because of the fact that outdoor climate never remains constant over long periods. Nevertheless, unilateral heat transfer theory can be applied with sufficient accuracy under the following conditions.
>
> (i) When the temperature difference between inner and outer air is large with respect to the short-term fluctuation in outdoor air temperature. This is more or less the case in cold climates where buildings have to be heated in winter.
> (ii) When the thermal or heat-storing capacity of the element concerned is small in comparison with the total heat flow. All light-weight structural elements such as curtain walls and thin roofing materials fall into this category.[123]

Outside of these circumstances, Van Straaten stated that "the U-value is not by itself an accurate indication of the thermal performance of the element under conditions of periodic or transient heat flow, i.e., under conditions of daily fluctuations in dry-bulb temperature, or solar radiation intensity, or both."[124] As such, before addressing insulation as a chapter, Van Straaten focused on "Heat Transfer Through Opaque Elements under Periodically Fluctuating Conditions."[125] In this chapter, Van Straaten reports on full-scale, multi-year experiments in South Africa that aimed to account for the thermal mass effects on some houses in various climates. The book featured several now familiar charts illustrating the thermal swing profiles of various construction assemblies compared to the outside air temperature swing profile, as measured in the test houses. The documentation of this fundamentally more transient treatment of heat transfer revealed much about more nuanced transfer mechanisms: both the capacity of more mass in building assemblies and also the complexity of characterizing that transfer analytically. Van Straaten noted:

> The actual mathematical treatment of this equation is beyond the scope of this book... All that can be said at this stage is that analytical approaches to the problem of thermal behavior of buildings under unsteady heat flow conditions become unwieldy without introducing simplifications. On the other hand, field studies of the problem, though apparently straightforward in principle, are exceedingly difficult to carry out.[126]

With a revealing mixture of utility and futility, several attempts have been made to account for some of the effects of aperiodic, volumetric heat diffusion as an amendment to typical R-Values. These studies, often called "mass correction" studies, try to combine the roles of steady-state resistance values with transient, dynamic effects of mass that occur on account of density, heat capacity, and diffusivity. But this is throwing good knowledge after bad and does not at all resolve the basic R-value problematic.

Some eighty years after the initial speculation about the respective roles of heavy- and lightweight construction in the thermal behavior of buildings in the first architectural article on insulation, a study on the mass effects of heavy constructions appeared in the United States in 1977.[127] With the resurgence of energy interest in the 1970s came, once again, attention to the power of mass. Around 1982, the National Bureau of Standards (NBS), the Department of Housing and Urban Development (HUD) and the Department of Energy (DOE) conducted a "thermal mass" study program, and the National Institute for Standards and Technology compared the thermal behavior of lightweight frame, masonry, and solid wood constructions outside of Washington, DC.[128] For this study, six test buildings of various construction assemblies were monitored and evaluated to determine the energy consumption required to maintain a constant interior temperature over a twenty-eight week period. Recalling the Norwegian, British, German, and Australian studies a half-century prior, this study used full-scale construction exposed to climate to determine behavior. The approach thus took stock of a far broader range of factors.

In this same period, the New Mexico Energy Research and Development Institute (NM-ERDI) conducted a similar and related DOE-sponsored thermal mass study, now over a multi-year period.[129] The NM-ERDI constructed lumber-framed cavity walls, solid wood walls, concrete block, and a range of adobe brick walls with thicknesses up to 15 inches. Otherwise, in terms of orientation, fenestration, roof and floor assemblies and sizes, the houses were identical. By 1984, ASHRAE papers documenting the thermal work of massive assemblies were published.[130] The conclusions were largely directed to the evolving energy codes.

Mass Effects on Energy Codes

Despite the great thermal inertia of more massive constructions, there remained within energy codes a resistance to the inclusion of non-steady-state characterizations of heat flux. At one point an "M-factor" (mass factor) was proposed as adjustment to the burgeoning ubiquity of the R-value.[131] The recurrent ambition in these studies, however, was to combine the transient state of the mass effects with the steady-state assumptions of the R-value approach: an odd pairing to be sure.

Some of this work persists in the form of the International Energy Conservation Code. Section 502.2.1.1.2 of the 2003 IECC code focuses on the effects of mass walls. Based on various specific heat capacities of a wall and a range of heating degree days, this section of the energy code makes provisions for the effects of non-standard construction strategies. This was adapted in the 2009 version as a "mass wall r-value," based on climate zone in the 2006 version and subsequent versions of the IECC. Despite this brief interest, the role of mass effects is certainly not as substantive as it should be in the minds of architects, building scientists, or engineers, given the great weight of buildings.

Conclusion: 1973

The historical development of heat transfer theories and practices in the North America insulation apparatus placed persistent emphasis on certain principles and behaviors of heat at the expense of others. These historical preoccupations demand explication, not only as one way to better understand the thermal behavior of buildings but also as an account of the only intermittently cogent development of technology and practices in a domain as complicated as buildings.

The futility of insulation's very name was perhaps a poorly selected, inadequate, and misleading starting point. Any theory of heat transfer in buildings should more fundamentally acknowledge the full dissipative reality of the thermodynamics of buildings: from their very construction through their operation and the emergy and bound energy involved therein. When this is the case, the problem of heat transfer becomes a question of heat flow modulation: a problem of designed velocities and deliberate dissipations of various quantities of various qualities of energy, not just quantities of resistance. It is also becomes far more productive than a Calvinist atonement for the inevitability of energy dissipation.

By avoiding the more nuanced non-steady-state diffusion of heat over time in the thermal storage capacity of matter, the theory of heat transfer, especially in North America has been reduced to the diffusion of reductive memes about thermal conductivity. This dissemination of oversimplified understandings of heat transfer in insulation theory and practice, especially as motivated by fiscal transfer, has hindered more architecturally and ecologically powerful practices of heat flux. Only when the actual thermal parameters of building materials — time, specific heat, diffusion, emissivity, etc. — become central concerns of insulation theory will heat transfer begin to have more correspondence with reality and be realized as a more efficacious, ecological basis for practice.

The historical emergence, and thus dependence, of insulation theories on refrigeration research in North America in particular was perhaps as inevitable as it was necessary. In this history, then, it is of little surprise that the first building with a double-skin glass façade — Occidental Chemical Corporation, 1982 — was built in Buffalo, NY; the fount of the air-conditioning logics that conditioned how so many buildings were designed and built throughout the twentieth century. In many ways, this building — and so many double-envelope buildings like it — comes closest to fulfilling the refrigeration-driven ambition of creating an isolated building; a building inside a building. When coupled with obviously necessary mechanical ventilation equipment, the building and refrigerator share unequivocally related operational logics. This persists as a managerial response to the perceived energy shortages and appeals for energy efficiencies that emerged in architecture after 1973.

However, any thermodynamically inclined architect should not be content to diminish the energy flux in buildings to the reductionist logics of "energy," es-

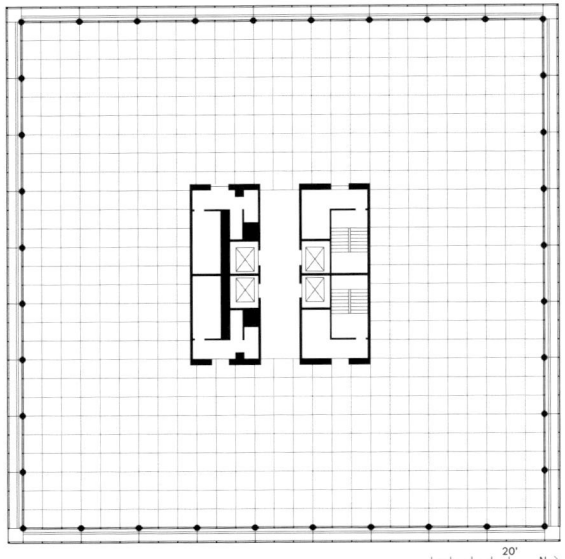
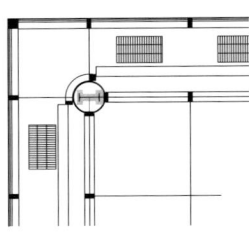
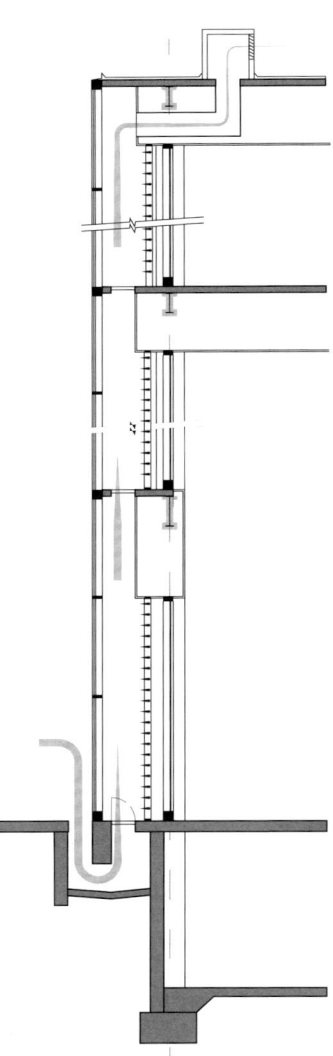

Buffalo's Propensity for Refrigerated Boxes
Top: Occidental Chemical plan
Right: Occidental Chemical envelope

pecially as manifest in the refrigeration-insulation apparatus of building science. Buildings have radically different obligations and opportunities which refrigeration alone cannot support. There is far more energy flux inherent to architecture than the periodic or steady-state heat flux in a wall. Especially in the case of double-glazed buildings, the preponderance of required emergy for this strategy, that achieves relatively similar "work" of a less redundant approach, reveals the cascade of compensations required for the over-illuminated and under-insulated conceit of a fully-glazed building. Architects today need a less isolated grasp of heat and energy than the insulation apparatus provided in modernity.

When a non-isolated perspective on heat flux guides an architect's engagement with energy, it becomes very difficult to be concerned only with building envelope heat flux. A much broader system boundary is absolutely necessary to make any claim about energy. The subsequent chapters expand on these system boundaries and their implications while also considering other fundamental phenomena often absent in the architectural consideration of heat flux and insulation.

Rather than a body of knowledge built around the thermodynamically absurd assumption of thermal isolation and quantitative neutrality (first-law thinking), perhaps the dissipative structure inherent to universal energy exchange could today more clearly shape architects' assumptions about heat diffusion and qualitative flux (second-law thinking).

What was established in the preceding historical narrative was the origins of the dominant praxis of heat transfer and thus insulation praxis in architecture, both of which heavily focused on conductivity over many other, more complex forms of energy exchange. This historical account also evinces a more minor, but very present paradigm that, by questioning the validity of the dominant conductivity and R-value paradigm, helps push our thermodynamic understanding of buildings closer to the open, non-isolated reality of architecture.

Notes

1. John May, "In the Aftermath of Modernity's Nature," interview, *The New City Reader*, ed. Sevin Yildiz, no. 7, 2012.
2. Paul Dunham Close, *Building Insulation: A Treatise on the Principles and Application of Heat and Sound Insulation for Buildings*, Chicago: American Technical Society, 1945. p. 1.
3. A concise summary of these reports and presentations is contained in: I. Grattan-Guinness, *Landmark Writings in Western Mathematics 1640–1940*, London: Elsevier Science, 2005. pp. 355–356.

4   J. B. J. Fourier, *Théorie analytique de la chaleur*, Paris: F. Didot, 1822.
5   Ibid.
6   For a more elaborate account of this equation, see Jennifer Coopersmith, *Energy: the Subtle Concept*, Oxford: Oxford University Press, 2010. pp. 203–205.
7   Ibid. p. 204.
8   T. N. Narasimhan, "Fourier's Heat Conduction Equation: History, Influence, and Connections," *Earth Planetary Science*, vol. 108 (3), September 1999. pp. 117–148.
9   Ibid. p. 123
10 As quoted from Black's lectures published from manuscripts by John Robinson in 1803, in W. F. Magie, ed., *A Source Book in Physics*, New York: McGraw Hill, 1935.
11 Antoine-Laurent Lavoisier and Pierre Simon Laplace, "Mémoire sur la chaleur," Paris, Gauthier-Villars et cie, 1920. Reprinted from *Mémoires de l'Académie des sciences, année 1780*, pp. 355–408.
12 Thomas Tredgold, *Principles of warming and ventilating public buildings, dwelling houses, manufactories, hospitals, hot-houses, conservatories, etc.; and of constructing fire-places, boilers, steam apparatus, grates, and drying rooms; with illustrations experimental, scientific, and practical…*, London: J. Taylor, 1824.
13 Walter Bernan, *On the History and Art of Warming and Ventilating Rooms and Buildings, etc., Vols I and II*, London: G. Bell, 1845.
14 Joseph Gwilt, *Encyclopedia of Architecture: Historical, Theoretical & Practical*, London: Longmans and Green, 1903.
15 J. C. E. Péclet, *Traité de la chaleur: considérée dans ses applications*, Vol. 1, 4th ed., Paris: G. Massor, 1878. pp. 542–555.
16 J. C. E. Péclet, "Experiments on Conductivity of Various Substances and the Coefficient of Conductivity." *Annales de chimie et de physique*, vol. 2, 1846. p. 107.
17 J. C. E. Péclet, *Traité de la chaleur considérée dans ses applications*, Paris: G. Masson, 1861.
18 The earliest applied publication regarding heat transfer to building materials is C. E. Emery, "Experiments on Non-conductors of Heat," *Transactions of the American Society of Mechanical Engineers*, vol. 2, 1881. p. 34.
19 Robert Zarr, "A History of Testing Heat Insulators at the National Institute of Standards and Technology," *ASHRAE Transactions*, vol. 107(2), 2001. p. 1.
20 Michael Osman, "Preserved Assets," in Aggregate (Architectural History Collaborative), ed., *Governing by Design: Architecture, Economy, and Politics in the Twentieth Century*, Pittsburgh: University of Pittsburgh Press. pp. 1–20.
21 H. C. Dickinson and M. S. Van Duesen, "The Testing of Thermal Insulators," *American Society of Refrigeration Engineers Journal*. vol. 3 (2), 1916. pp. 5–25.
22 Zarr, "A History of Testing," p. 2.
23 F. G. Hechler and A. J. Wood, "Review of methods of testing insulating and building materials for heat transmission," *The American Institute of Refrigeration*, vol. 17, 1928. pp. 151–178.

24 Zarr, "A History of Testing." p. 2.
25 E. Griffiths, "Heat insulating materials," *Journal of Scientific Instruments*, vol. **15** (4), **1938**. pp. 117–121.
26 Ibid. p. **120**.
27 These three articles are: "Heat Insulation," *Inland Architect & News Record*, vol. **26, 1895**. p. 52; "Influence of temperature in our buildings," *American Architect & Building News*, vol. **53**, 29 Aug. **1896**; and "Insulation," *Western Architect*, vol. **20, 1914**. pp. 16–20.
28 "The house that will save coal," *Country Life*, vol. **37**, Apr. **1920**. pp. 70–71; "Prevention of heat losses," *Architectural Forum*, vol. **35**, Sep. **1921**. pp. 93–98; R. B. Whitman, "Building to save coal," *Country Life (New York)*, vol. **45**, Dec. **1923**. pp. 61–63; "Insulating the house," *House & Garden*, vol. **49**, June **1926**. p. 86; TC. S. Taylor, "The value of insulation," *Country Life (New York)*, vol. **64**, Oct. **1928**. pp. 41–43.
29 Max Weber, *The Protestant Ethic and the Spirit of Capitalism*, trans. Stephen Kalberg, New York: Oxford University Press, **2011**.
30 Chapter **2** of this book expands on this observation.
31 "Influence of temperature in our buildings," *American Architect & Building News*, vol. **53**, Aug. **1896**. p. 68.
32 Ibid. p. **68**.
33 Consider, as one of the earliest examples in a trade journal: "Concerning Mineral Wool," *The Manufacturer and Builder*, vol. **21** (7), July **1889**. p. **152**.
34 The "material history" chapter of this book expands on this observation.
35 "Concerning Mineral Wool." p. **152**.
36 Zarr, "A History of Testing." p. **1**.
37 Ruth Schwartz Cowen, "How the Refrigerator got its Hum," in Donald MacKenzie and Judy Wacjman, eds., *The Social Shaping of Technology*, Philadelphia: Open University Press, **1985**. pp. 214–216.
38 Michel Foucault, "The Confession of the Flesh" (**1977**) interview, in Colin Gordon, ed., *Power/Knowledge: Selected Interviews and Other Writings 1972–1977*, New York: Pantheon Books, **1980**. p. **195**.
39 Ivan Illich, *Disabling Professions*, New York: M. Boyars, **2000**. p. 11.
40 Ibid. p. **16**.
41 J. H. Bracken, "The Storage of Ice," *Ice & Refrigeration*, vol. **38** (4), April **1910**. p. **211**.
42 Illich. p. **20**.
43 Ibid. p. **17**.
44 John Straube and Eric Burnett, *Building Science for Building Enclosures*, Somerville, MA: Building Science Press, Inc., **2005**. p. **214**.
45 Ibid. p. **214**
46 G. B. Wilkes, *Heat Insulation*, New York: John Wiley & Sons, Inc., **1950**. p. 16.
47 F. C. Houghten, C. Gutberlet, and A. Rosenberg, "The effect of solar radiation on the heat transmission through walls," in *Symposium on Thermal Insulating Materials: Columbus Regional Meeting, American Society for Testing Materials*, March **8, 1939**. p. **106**.
48 Ibid. p. **106**.

49 Ibid. pp. **107**–**108**.
50 "Architectural Engineering: Heat transmission through dwelling house walls," *American Architect and the Architectural Review*, vol. **126**, Sept. **1924**. pp. **299**–**306**.
51 Ibid. p. **300**.
52 Sem Sæland, "The Heat-Transmission Measurements **1919**–**1922**," in Andreas Bugge, ed., *Test Houses: Erected by Norges Tekniske Hoiskole (The Norwegian Technical University), Trondheim; Results of Tests with Wall-Constructions and Materials for Building Warm and Cheap Dwelling-Houses*, trans. J. Craig, Trondheim: F. Bruns Bokhandels Forlag, **1924**. p. **25**.
53 Andreas Bugge, "The Plan for the Test Houses," *Test Houses: Erected by Norges Tekniske Hoiskole (The Norwegian Technical University), Trondheim; Results of Tests with Wall Constructions and Materials for Building Warm and Cheap Dwelling-Houses*, trans. J. Craig, Trondheim: F. Bruns Bokhandels Forlag, **1924**. pp. **6**–**7**.
54 Ibid. p. **7**.
55 Ibid. p. **7**.
56 "Architectural Engineering: Heat transmission through dwelling house walls." p. **301**
57 Sæland. p. **58**.
58 Ibid. p. **58**.
59 Ibid. p. **61**.
60 Ibid. p. **61**.
61 Ibid. p. **62**.
62 Bugge. p. **65**.
63 Jos Tomlow, "Building Science as Reflected in Modern Movement Literature," in Jos Tomlow, ed., *Climate and Building Physics in the Modern Movement: Proceedings of the 9th International DOCOMOMO Technology Seminar*, Paris: Docomomo International, **2006**. p. **10**.
64 David Martinsen, *Erfaringen fra Amerika: Norske arkitekter i USA 1880–1930 – et glemt kapittel? Eksempler fra remigrantenes arbeider i Oslo*, University of Oslo, **2008**. p. **15**.
65 Andreas Bugge, *Amerikas små hjem, deres planlegning, konstruksjon og utførelse*, Oslo: Grøndahl & Søn, **1927**.
66 Ibid. pp. **79**–**85**.
67 See Ivan Rupnik, "Building Systems/Building Territories: Industrialized Housing Delivery and the Role of the Architect," in Kiel Moe and Ryan Smith, eds., *Building Systems*, London: Routledge, **2012**. p. **92**.
68 "4. Versuche über Wärmeleitung von Bimsbetonplatten und Leichtbetonplatten," *Forschungsbeim für Wärmeschutz*, München, **14**. Juli **1926**. p. **122**.
69 Ibid. p. **132**. The report does not specify any units.
70 Roland Gellert and Horst Zehender, "First Steps in Establishing the Discipline of Building Science: The Research Institute of Thermal Insulation in Munich," in Jos Tomlow, ed., *Climate and Building Physics in the Modern Movement: Proceedings of the 9th International DOCOMOMO Technology Seminar*, Paris: Docomomo International, **2006**. pp. **7**–**8**.

71 Victor Olgyay, "Appendix B," in *Design with Climate: Bioclimatic Approach to Architectural Regionalism*, Princeton: Princeton University Press, **1963**. p. **180**.
72 Jadwiga Urbanik, "Szymon Syrkus: CIAM Representative of Poland and Pioneer in Integrating Building Science in Modern Movement Architecture," in Jos Tomlow, ed., *Climate and Building Physics in the Modern Movement: Proceedings of the 9th International DOCOMOMO Technology Seminar*, Paris: Docomomo International, **2006**. pp. **53–60**.
73 Ibid. pp. **56–57**.
74 Wessel de Jonge, "The Unbearbale Lightness of Building: the 'Functionally Differentiated Outer Wall' and the Preservation of Modern Movement Buildings," in Jos Tomlow, ed., *Climate and Building Physics in the Modern Movement: Proceedings of the 9th International DOCOMOMO Technology Seminar*, Paris: Docomomo International, **2006**. p. **29**.
75 Ibid. p. **31**.
76 *De 8 Opbouw*, no. **17/18**, 2 Sep. **1939**. pp. **190–191**.
77 de Jonge. p. **32**.
78 Ibid. p. **32**.
79 Ibid. p. **32**.
80 Koen Limperg, *Naar Warmer Woningen: overzicht van de warmtetechnische eigenschappen van bouwmaterialen en bouwconstructies met bespreking van de wijze waaop hiermede in de praktijk kan worden gerekend*, Amsterdam: Van Holkema & Warendorf, **1936**; Johannes Bernardus van Loghem, *Acoutisch en Thermisch Bouwen voor de Praktijk*, Amsterdam: Veen, **1936**.
81 Johannes Bernardus van Loghem, *Acoustisch en Thermisch Bouwen voor de Praktijk*, Amsterdam: Veen, **1936**. pp. **136–137**, as quoted in de Jonge. p. **33**.
82 Roger Courtney, "Building Research Establishment: Past, Present and Future," *Building Research & Information*, vol. **25 (5)**, **1997**.
83 A. F. Dufton, "The warming of walls," *Journal of the Institution of Heating and Ventilating Engineers*, vol. **2**, **1934**. pp. **416–417**.
84 Ibid. p. **416**.
85 A. F. Dufton, "New Heating Laboratory at the Building Research Station," *Nature*, 22 Aug. **1936**. p. **335**.
86 Ibid. p. **335**.
87 F. M. Lea, *Science and Building: A History of the Building Research Station*, London: Her Majesty's Stationary Office, **1971**. p. **63**.
88 Ibid. p. **64**.
89 Robert Fitzmaurice, *Principles of Modern Building, Vol. 1: Walls, Partitions, and Chimneys*, London: His Majesty's Stationary Office, **1938**.
90 Ibid. p. **55**.
91 Thomas Bedford, *Basic Principles of Ventilation and Heating*, London: H. K. Lewis & Co. Ltd., **1948**.
92 Ibid. p. **216–235**.
93 J. W. Drysdale, *Thermal Characteristics of Dwellings: Technical Study 27*, Sydney: Commonwealth Experimental Building Station, **1948**.

94 For example: J. W. Drysdale, *Climate and House Design*, Sydney: Department of Works and Housing, Australia Commonwealth Experimental Building Station, Bulletin No. **3**, June, **1947**. J. W. Drysdale, *Climate and House Design, Summary of Investigation 1945–1947*, Sydney: Department of Works and Housing, Australia Commonwealth Experimental Building Station, June, **1947**. *Duplicated Document No. 21*. J. W. Drysdale, *Natural Ventilation, Ceiling Height and Room Size : Notes Regarding Minimum Provisions in Dwellings with Respect to Australian Conditions*, Sydney: Department of Works and Housing, Australia Commonwealth Experimental Building Station, **1947**. *Duplicated Document No. 22*. J. W. Drysdale, *Climate and House Design: Physiological Considerations*, Sydney: Department of Works and Housing, Australia Commonwealth Experimental Building Station, March, **1948**. *Duplicated Document No. 25*. J. W. Drysdale, *Thermal Characteristics of Model Structures*, Sydney: Department of Works and Housing, Australia Commonwealth Experimental Building Station, **1948**. *Duplicated Document No. 26*. J. W. Drysdale, *Climate and House Design: Thermal Characteristics of Dwellings*, Sydney: Department of Works and Housing, Australia Commonwealth Experimental Building Station, June, **1948**. *Duplicated Document No. 27*. J. W. Drysdale, *Natural Ventilation: Study of Air Movement in Buildings by use of Models*, Sydney: Department of Works and Housing, Australia Commonwealth Experimental Building Station, **1949**.

95 J. W. Drysdale, *Climate and House Design*, Sydney: Department of Works and Housing, Australia Commonwealth Experimental Building Station, Bulletin No. **3**, June, **1947**.

96 There are related documents as well: G. F. Middleton, *Earth Wall Construction 1: Pisé or Rammed Earth*, Sydney: Department of Works and Housing, Australia Commonwealth Experimental Building Station, Bulletin No. **17**, **1947**. G. F. Middleton, *Earth Wall Construction 2: Adobe or Puddled Earth*, Sydney: Department of Works and Housing, Australia Commonwealth Experimental Building Station, Bulletin No. **18**, **1948**. G. F. Middleton, *Earth Wall Construction 3: Stabilized*, Sydney: Department of Works and Housing, Australia Commonwealth Experimental Building Station, Bulletin No. **19**, **1948**.

97 George Frederick Middleton, *Build Your Own House of Earth: Pisé and Adobe Construction,* Melbourne: Compendium, **1953**.

98 Elmer Gilliam Smith, "A Simple and Rigorous Method for the Determination of the Heat Requirements of Simple Intermittently Heated Exterior Walls," *Journal of Applied Physics*, vol. **12**, **1941**. pp. **638–642**.

99 Elmer G. Smith, *The Heat Requirements of Intermittently Heated Buildings*, PhD thesis, The University of Texas at Austin, **1941**. See also, Elmer Gilliam Smith, "The Heat Requirements of Intermittently Heated Buildings," *Texas Engineering Research Station Bulletin*, no. **62**, **1941**.

100 Elmer Gilliam Smith, "The Heat Requirements of Simple Intermittently Heated Interior Walls and Furniture," *Journal of Applied Physics*, vol. **12**, **1941**. pp. **642–644**.

101 Charles Osborne Mackey and Lawrence T. Wright, Jr., *Summer Comfort Factors as Influenced by Thermal Properties of Building Materials*, New York: The John B. Pierce Foundation, **1943**.
102 Ibid. p. **7**.
103 Charles Osborn Mackey and Lawrence T. Wright, Jr., "Periodic heat flow – homogeneous walls or roofs," *Transactions of the American Society of Heating & Ventilating Engineers*, vol. **50**, **1944**. p. **293**; Charles Osborn Mackey and Lawrence T. Wright, Jr., "Periodic heat flow – composite walls or roofs," *Transactions of the American Society of Heating & Ventilating Engineers*, vol. **52**, **1946**. p. **283**.
104 Simon Rees, Jeffrey Spitler, Morris Davies, and Philip Haves, "Qualitative Comparison of North American and U.K. Cooling Load Calculation Methods," *HVAC&R Research*, vol. **6**, **2000**.
105 J. P. Stewart, "Solar Heat Gain through Walls and Roofs for Cooling Load Calculations," *Transactions of the American Society of Heating & Ventilating Engineers*, vol. **54**, **1948**. pp. **361–388**.
106 A. Nessi and L. Nisolle, "Fonctions d'influence de flux de chaleur des parois de construction," *Rapport, Comité Tech. Indust. Chauffage*, Paris, **1947**.
107 D. G. Stephenson, *Periodic Heat Flow in Walls and Roofs, Report # DBR-R-132, Division of Building Research*, National Research Council Canada, September, **1957**.
108 E. Danter, "Periodic heat flow characteristics of simple walls and roofs," *Journal of the Institution of Heating and Ventilation Engineers*, vol. **28**, **1960**. pp. **136–146**.
109 Neville S. Billington, *Building Physics: Heat*, Oxford: Pergamon Press, **1967**. p. **83**.
110 Close, *Building Insulation*. p. **9**.
111 The best biographic information on Everett Shuman is available on a Penn State University website associated with his papers in the Penn State University archives special collections: http://psu.edu/dept/findingaids/ead/**1338**.xml (consulted June **19**, **2012**). "He traveled extensively to professional meetings, achieving nearly two million miles of air travel. His first flight occurred in **1905** at the age of **3** when he was strapped to his father who was a wing passenger on an aircraft flown by Orville Wright."
112 C. Hering, "Thermal Resistance and Conductance; The Thermal Ohm," *Metallurgical and Chemical Engineering*, vol. **9** (**1**), **1911**. p. **13**.
113 C. Hering, "Simplifying Some Thermal Calculations by the Use of the Thermal Ohm," *Journal of Franklin Institute*, **1911**. p. **576**.
114 W. Kent, "Heat Resistance, the Reciprocal of Heat Conductivity," *Transactions of the American Society of Mechanical Engineers*, vol. **24**, **1902**. p. **278**.
115 See, for instance, Jay Wright Forrester, *Urban Dynamics*, Cambridge, MA: The MIT Press, **1969**. Whether industrial, urban, or world dynamics, the electrical metaphor was constant through this work.

116 Donella H. Meadows, et al., *The Limits to Growth: A Report for the Club of Rome's Project on the Predicament of Mankind*, New York: Universe Books, 1972.
117 J. Straube, "Thermal Metrics for High Performance Enclosure Walls: The Limitations of R-Value," *Building Science Press, Research Report – 0901*, 2007. p. 2. <http://buildingscience.com/documents/reports/rr-0901-thermal-metrics-high-performance-walls-limitations-r-value> (consulted October 13, 2012).
118 *FTC Title 16: Commercial Practices, Part 460: Labeling and Advertising of Home Insulation.*
119 ASTM C 177–85 (Reapproved 1993), "Standard Test Method for Steady-State Heat Flux Measurements and Thermal Transmission Properties by Means of the Guarded-Hot-Plate Apparatus;" ASTM C 518–91, "Standard Test Method for Steady-State Heat Flux Measurements and Thermal Transmission Properties by Means of the Heat Flow Meter Apparatus;" ASTM C 1114–95, "Standard Test Method for Steady-State Thermal Transmission Properties by Means of the Thin-Heater Apparatus;" ASTM C 236–89 (Reapproved 1993), "Standard Test Method for Steady-State Thermal Performance of Building Assemblies by Means of a Guarded Hot Box;" ASTM C 976-90, "Standard Test Method for Thermal Performance of Building Assemblies by Means of a Calibrated Hot Box."
120 ASTM 1326–05 "Standard Test Method for Thermal Performance of Building Materials and Envelope Assemblies by Means of a Hot Box Apparatus."
121 J. F. van Straaten, *Thermal Performance of Buildings*, Amsterdam, New York: Elsevier Pub. Co., 1967.
122 Ibid. p. 6.
123 Ibid. p. 55.
124 Ibid. p. 81.
125 Ibid. pp. 81–103.
126 Ibid. p. 99.
127 Mario J. Cantani and Stanley E. Goodwin, "Heavy Building Envelopes and Dynamic Thermal Response," *Journal of the American Concrete Institute*, Feb. 1976.
128 D. M. Burch, W. E. Remmert, D. F. Krintz, and C. S. Barnes, "A Field Study of the Effect on Wall Mass on the Heating and Cooling Loads of Residential Buildings," in *Proceedings of the Thermal Mass Effects in Buildings Conference*, Oakridge National Laboratory, Oakridge, TN, 2–3 June, 1982.
129 David K. Robertson, "Observation and Prediction of the Heating Season Thermal Mass Effect for Eight Test Buildings with and without Windows," *US DOE Thermal Mass Program Phase 2*, NM-ERDI, University of New Mexico Albuquerque, 1984.
130 D. M. Burch, D. F. Krintz, and R. S. Spain, "The Effect of Wall Mass on Winter Heating Loads and Indoor Comfort-An Experimental Study," *ASHRAE Transactions*, vol. 90 (1), 1984; D. M. Burch, K. L. Davis, and S. A. Malcolm, "The Effect of Wall Mass on the Summer Space Cooling of Six Test Buildings,"

ASHRAE Transactions, vol. **90 (2)**, **1984**.; D. M. Burch, W. L. Johns, T. Jacobsen, G. N. Walton, and C. P. Reeve, "The Effect of Thermal Mass on Night Temperature Setback Savings," *ASHRAE Transactions*, vol. **90 (2)**, **1984**.

131 Mario J. Catani and Stanley E. Goodwin, "The Effect of Mass on Heating and Cooling Loads and on Insulation Requirements of Buildings in Different Climates," *ASHRAE Transactions*, vol. **85 (1)**.

"In every instance considered, natural selection will operate so as to increase the total mass of the organic system, to increase the rate of circulation of matter through the system, and to increase the total energy flux through the system, so long as there is present an unutilized residue of matter and available energy."[1] **Alfred J. Lotka**

A Material History of Insulation in Modernity

In general, the history of building materials in the twentieth century is a narrative in which *materials* increasingly became proprietary *products*. Previously throughout human history, raw *matter* — as extracted, harvested, transported, processed, and manipulated by humans — became *materials* in the process of human-world interactions. At each stage of this process, humans captured and channeled the affordances of energy in matter into all the stuff of life. They did so with matter from different locations, formats, and purposes; constantly aiming to maximize the power of the system. For thousands of years, buildings were assemblies of materials selected for various behaviors in self-organized selection processes of trial and error.

In modernity, another stage was introduced to this processing of matter in the human history of material culture. The succession from bio-geophysical source matter, to material, *to products* reflects increasing degrees of human handling, energy transformities, accumulation, and modes of capital accumulation.

The important twentieth century shift from materials to proprietary products took materials and endowed them with other properties, including unique research and development programs, idiosyncratic characteristics, standardization, and regulatory protocols and the vagaries of marketing prowess. In short, building products became imbued with as many corporate properties as they had innate structural or thermal properties. In turn, buildings of the twentieth century became assemblages of products and, accordingly, architects increasingly became a hybrid of building design professionals and professional consumers of products. This is decidedly an important epistemological and energetic shift in the history of architecture's practices.

This chapter charts a material history of insulation in particular. Its historical arch spans from archaic modalities to contemporary practices. Prior to the late nineteenth century, these materials were not named as such. Insulation materials emerged alongside the industrialization of building materials into insulation products. In the periods and paradigms before insulation emerged in name, certain materials and material properties were, of course, selected for thermal modulation capacities, but did not yet suffer the conceit of an isolated theory of insulation. These early builders simply aimed to amplify the power of their constructions.

It was not until the late nineteenth century that material approaches to the modulation of heat flow through building materials began to be routinized into a limited set of parameters, industrial products and interests, regulatory protocols, and codes. This was a significant shift in habits of mind and practice. These early modern dynamics in the material world routinely advanced certain practices while constraining others. The result was an experimental and evolving mix of results that did not always align with the claims of manufacturer marketing, scientific methods, or the assumptions of early modern progress ideologies.

An important aspect of this pre-modern material history is that older heat exchange modulation materials were often cunning redirections of a dissipation structure: the materials were the "waste" and by-products of other processes. As this material history evolved, increasingly high emergy materials were developed and employed to do relatively low-quality but important thermal work, a complete inversion of the prevailing emergy-to-exergy ratio of the prior material regime.

This transformation was justified, perhaps, because the sources of heat for the building also became inappropriately high-quality sources such as electricity and the various combustion-based fuels, again used for a low-quality function. In this way, high-emergy heating and "cooling" sources, when coupled with insulation, co-escalated through modernism, always beyond the realm of more powerful exergy and emergy design on account of fewer feedbacks and a higher environmental loading ratio. A key aspect of this book suggests that rather than a habit of mind based on resistance to heat transfer, what is needed today is a habit of mind focused on the productive and deft exchange of heat and non-isolated dissipation through design in service of maximum power. This material history thus documents a shift from low-emergy materials with complex feedback for thermal modulation to higher-emergy materials with less complex feedback and suggests a better exergy/emergy design for heat exchange in the future.

The Scientism of Insulation

The science of insulation and the industrial capacity of insulation emerged together. This industrial paradigm is still a self-organizing trial and error system, however, industrial entrepreneurship and marketing, professionalization, and code compliances all complicate and inflect the process of trial and error of design and construction. For instance, as in the medical profession, architects occasionally iatrogenically design and specify counterproductive materials and processes.

One early example discussed in this chapter includes carcinogenic asbestos, hailed as a wonder insulation product at its introduction. Directly related to the insulation apparatus, sick building syndrome cases resulted from an expert-driven demand for hermetic, super-tight construction and inadequate ventilation, especially as coupled with numerous nefarious fumes from a plethora of toxic products specified in late twentieth century buildings. The routines of practice in different periods professionalized these counterproductive, iatrogenic outcomes that complicate our current modalities of self-organization. To be clear, these counterproductive outcomes were, in fact, designed and specified by architects and experts.

As suddenly as some insulating materials appeared, many disappeared, but often not until some damage was done and some lessons were learned. When materials so rapidly become products, complicated human-material dynamics take shape. As the complicatedness increases, the many potential intersections and feedbacks are difficult, if not impossible, to anticipate. Thus, with respect to modern insulation products and practices, this material paradigm is characterized as much by what architects know about these dynamics as what they do not know. This "inability-to-know" is a fundamental characteristic of what Ulrich Beck calls the "risk society."[2] Insulation serves as an excellent case study about how risk conditions nearly all aspects of contemporary life, including the radically contingent practices of architecture.

Progress narratives in modern material culture tend to focus on the escalating production, accumulation, and distribution of goods. These narratives — found in everything from architects' manifestos to advertisements in Popular Mechanics — invoke both material goods (e.g., refrigeration or medication) and social goods (e.g., civil liberties and access to education). Progress narratives rarely acknowledge the progressive production, accumulation, and distribution of "bads" — material and social — that also constitute modernity. These constitutive bads include pollution, externalities, and tyranny (including the tyranny of naive progress ideologies). The history of insulation practices shares this form of occlusion. While insulation produced material and social good, its constitutive bads cannot be overlooked or isolated. The productive and counterproductive aspects of the internalities and externalities of architecture's material practices must be part of its historical narratives, especially in as much as historical narratives temper contemporary practices.

The ostensible aim of insulating materials was the thermo-regulation of buildings. However, with iatrogenics and risk in mind, we must adopt a broader, more interrogative historical account of material properties and capacities than that which a simple material taxonomy of insulation alone would provide. In this history, the adoption of certain scientific protocols discussed in the first chapter quickly regulated more than just thermal flow. The evolution of insulation's material culture also regulated the flow of capital and the flux of architectural pedagogies, and thereby ultimately co-determined systemic aspects of contemporary buildings and their assembly in productive and counterproductive ways. To be sure, this technological momentum definitely selected certain materials and processes over others.

This material history, like any material history, was never neutral. Materials, products, techniques, and technologies are never technically, socially, or ecologically neutral agents. As with other aspects of the insulation apparatus, the missions and omissions of insulating products are of central concern to this history. Conspicuously present are certain habits of mind and conspicuously absent are alternative modes of thinking about heat transfer. These less isolated accounts of materials are not only possible, but ultimately prove very compelling from an architectural point of view. So, to the physical and corporate properties inherent to insulation materials, one ambition of this history of insulating materials is to in due course identify certain architectural properties inherent in the insulating matters and to identify certain architectural capacities latent in the insulation apparatus.

The practice of insulation in modern architecture is largely a history of imposed products and procedures generally incorporated as deftly and intelligently as possible given the demands of the apparatus. Yet, from human-world interactions to the immense material reality of a building, there is great architectural potential in an expanded understanding of material/energy dissipations in architecture. A grasp of the materials and properties of various approaches to heat flux are essential to an architectural agenda for energy exchange and transition. In what follows, I track the use of various insulating strategies and materials from archaic periods through contemporary methods. As with the history of concepts and derivations of insulation in the first chapter, this chapter generally focuses on the developmental phases of insulating practices in modernity, and then follows a few consequential materials and strategies through to contemporary practices.

Clay, Seaweed, Hair, & Asbestos

My taxonomy of insulating matter and materials begins with the primary bio-geophysical categories of mineral, vegetal, animal, and synthetic sources. Organizing insulating materials into these source categories helps create a hierarchy of generally increasing scales of required energy transformations. Contemporary synthetic spray-foam insulations, for instance, have radically different emergy values and transformations when contrasted with a mineral source like adobe, even if the

A Material History of Insulation in Modernity   131

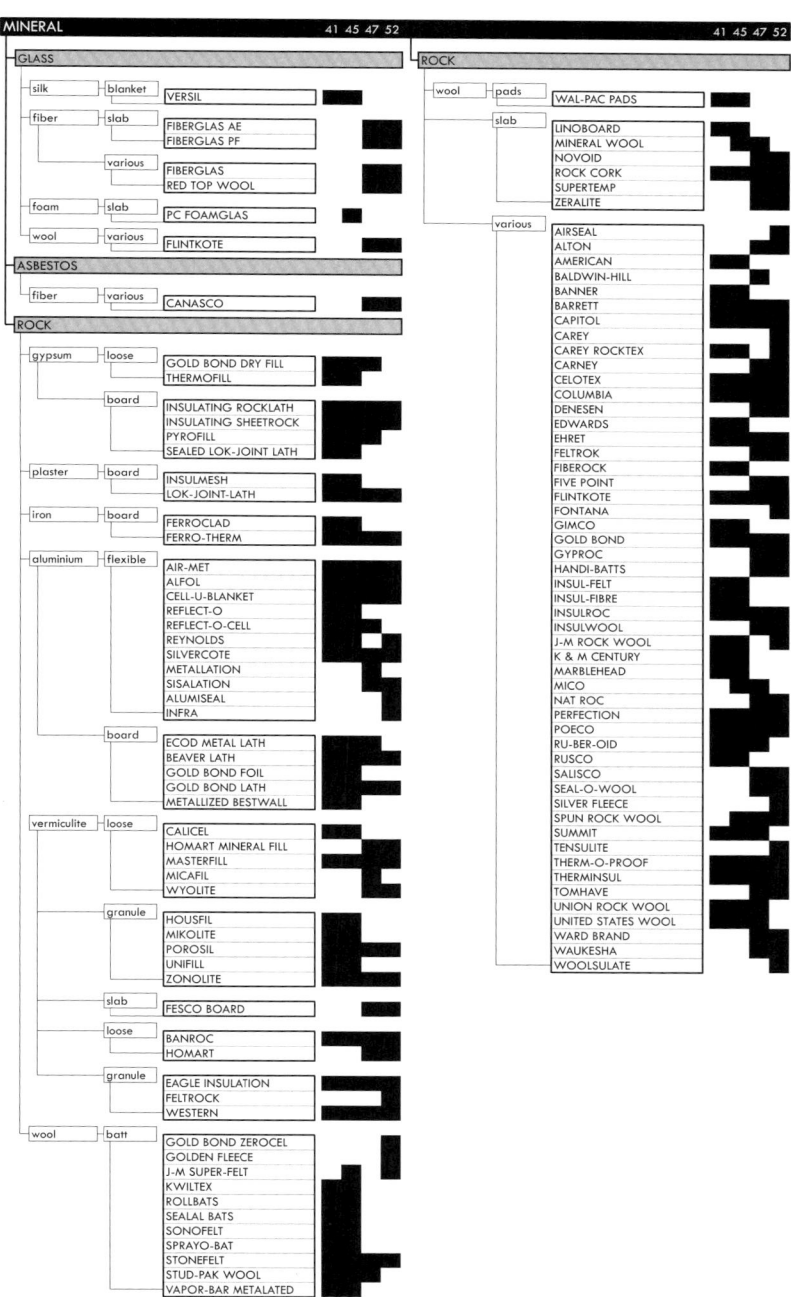

Clay, Seaweed, Hair, and Asbestos
A taxonomy of insulating material types

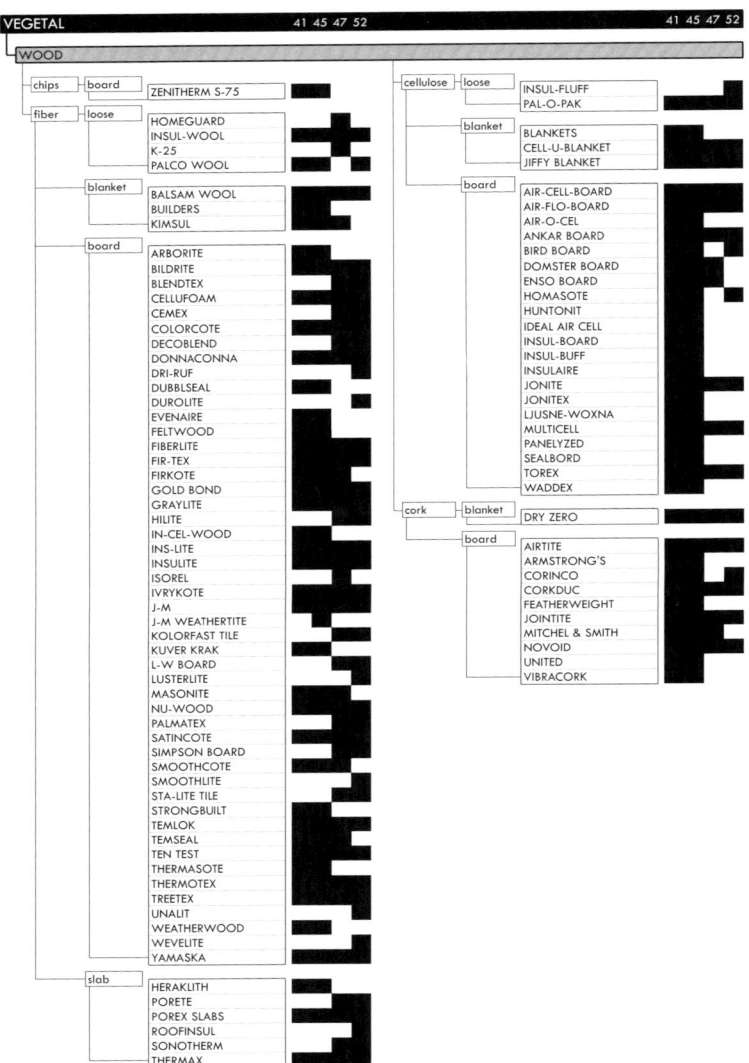

Clay, Seaweed, Hair, and Asbestos
A taxonomy of insulating material types

A Material History of Insulation in Modernity

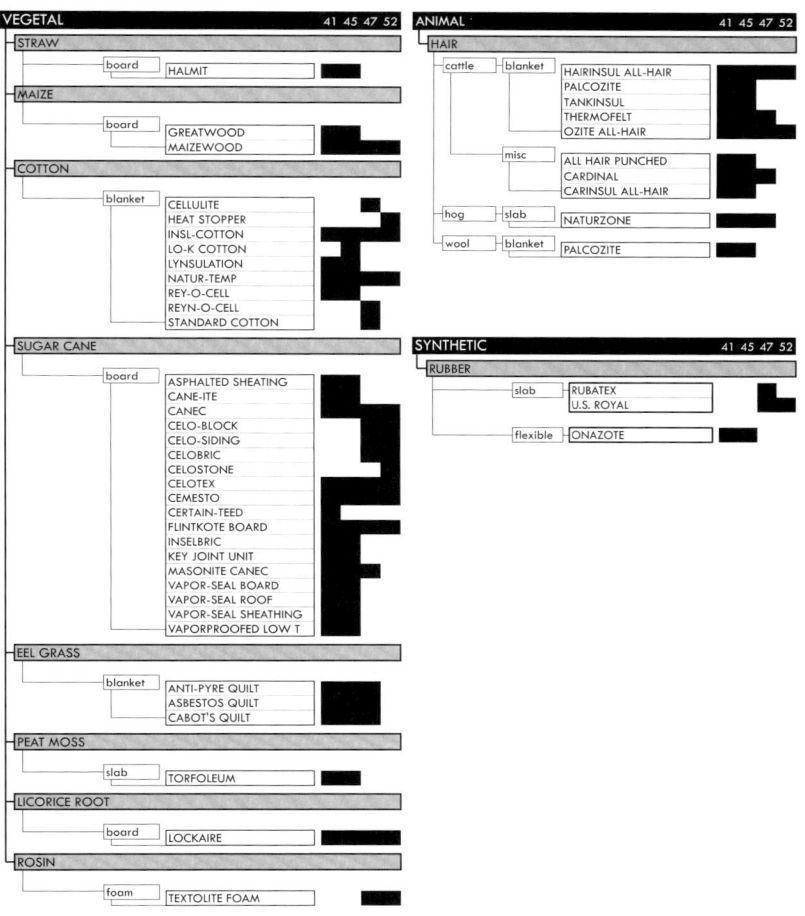

Clay, Seaweed, Hair, and Asbestos
A taxonomy of insulating material types / Bottom: Specific emergy values of various insulation materials

materials might have similar thermal modulation capacities. This reflects their equally varied supply and value chains, as considered in economic and ecological registers. I focus on the transformities of this rubric in order to not isolate long-term energy cycles from our consideration of short-term insulating materials.

In this book's insulation material taxonomy presented in the preceding pages, insulating materials are organized vertically according to rough energy transformities — from the mineral, vegetal, and animal to the synthetic: roughly from low to high transformity. Horizontally, the materials are also organized according to time. It is useful to see when various materials and products were introduced and their relative transformity. This also visualizes which materials have benefited from centuries, and occasionally millennia, of self-organized use and evaluation from products which benefit from more robust research, testing, and marketing in a shorter period. Finally, this taxonomy also helps identify materials — archaic or modern — that could benefit from further research and use in a non-modern, non-isolated agenda for energy in architecture.

The structure of this chapter follows that of this taxonomy. It moves diagonally across these charts, from archaic and low-transformity materials to the industrial mode of high-transformity products. The purpose of this organization, again, is to keep the short-term energy transitions in architecture — such as the diurnal thermal flux through walls, for instance — in a framework that situates short-term exergy functions in a context of longer-term cycles of emergy dynamics and maximal power design.

Furs, Pelts, Felts, & the Origin of Architecture

Preindustrial milieux were replete with compelling, sometimes surprising, appropriations of matter for the purpose of thermal modulation. The persistent struggle for homeorhesis in human development provoked the earliest attempts to modulate human comfort with what remains, even now, the most efficacious means: directly tempering the capacities of the body before that of its milieu. The principle of tempering the body directly has great efficacy and is often underestimated in the thermodynamic imagination of contemporary buildings.

For centuries, humans have captured and channeled the boundary layers of various flora and fauna species to augment and amplify their own boundary dynamics. Whether it was vegetal headwear for shading or animal skins for thermal modulation, when early humans appropriated these animal and vegetal assemblages, they also appropriated the biologically adapted evolutions of adjacent species in various milieux. These were also co-products of metabolically related and necessary harvesting and hunting practices. In this fundamental way, humans tapped deeply into the thermodynamics and metabolism of any region. But these early exosomatic devices would not cling solely to the body for long.

Performance-enhancing Rugs

As early humans became more settled, the furs, pelts, and felts often left the body and were hung on a wall or other scaffold. The Mezhyrich huts in central Ukraine provide one of the more direct examples. Part shelter, part trophy, the 15,000-year-old dwellings were constructed from stacked mammoth bones and likely clad with mammoth pelts. This was one archaic paradigm: kill it and then reassign every body part to thermodynamic advantage. Similar structures were found in what is now South Dakota.

More common, perhaps, would have been the modulation of caves with a similar material palette. Imagine, as one archaic scenario, an early hunter, returning on a stormy night from his metabolic quest to his temperate Mediterranean cave. Wanting to more directly heat his body while cooking, a dampened pelt is removed on entry and slung in on the fireside of a rock surface to dry, joining the pelts of others. In each step, one can trace a series of thermodynamic and physiological adaptations that occur with respect to various modes of heat transfer in this rudimentary milieu: exposure of the body to radiant heat sources and modification of the thermal effusivity of radiant sinks by the drying pelts. Likewise, that same early human might sleep on a mat of reeds or other flora to at once stay drier and warmer than the ground might allow, another composite thermodynamic "design."

In this hypothetical scenario — as well as in contemporary life — the modulation of heat flux, material practices, and eating habits are all essential to a broader understanding of metabolism. As such, the fire was more than just a heat source: it also pre-digested food and thus aided metabolic intake. Harvard anthropologist Richard Wrangham, for instance, claims that "cooking substantially increases the amount of energy we obtain from food."[3] This builds on other related work that aims to explain why humans have big brains and small guts. Leslie Aiello and Peter Wheeler's "Expensive Tissue Hypothesis" articulates various physiological adaptations that co-evolved with higher-quality, higher-yield diets enabled by cooking.[4] The point here is simply to acknowledge that fire is essential to making the human metabolism (of both food and buildings) more powerful. Fire is not only a primary source of thermal modulation in early human development, but, in deep and systemic ways, it also is a primary source of what makes humans human. Grasping the thermodynamics of short and long-term metabolisms of human life is essential, for it helps structure the relative roles of human comfort, insulation, and construction energies.

As such, it is not at all surprising that fire precedes the first of *any* references to insulating materials in architecture. In the first chapter of his second book on architecture, Vitruvius provides his account of the origin of the dwelling house. For Vitruvius, architecture began with the fire and the assembly of early humans around the fire. His narrative account of this origin constitutes the first two paragraphs of his second chapter. In the third paragraph is the first mention of thermal insulation in architecture, included here in full because of his concise and poignant account of human metabolism and development that resonates so well with contemporary accounts of human evolution:

The Origin of the Dwelling House

1. The men of old were born like the wild beasts, in woods, caves, and groves, and lived on savage fare. As time went on, the thickly crowded trees in a certain place, tossed by storms and winds, and rubbing their branches against one another, caught fire, and so the inhabitants of the place were put to flight, being terrified by the furious flame. After it subsided, they drew near, and observing that they were comfortable standing before the warm fire, they put on logs, and while thus keeping it alive, brought up other people to it, showing them by signs how much comfort they got from it. In that gathering of men, at a time when utterance of sound was purely individual, from daily habits they fixed upon articulate words just as these had happened to come; then, from indicating by name things in common use, the result was that in this chance way they began to talk, and thus originated conversation with one another.

2. Therefore it was the discovery of fire that originally gave rise to the coming together of men, to the deliberative assembly, and to social intercourse. And so, as they kept coming together in greater numbers into one place, finding themselves naturally gifted beyond the other animals in not being obliged to walk with faces to the ground, but upright and gazing upon the splendor of the starry firmament, and also being able to do these with ease whatever they chose with their hands and fingers, they began in that first assembly to construct shelters. Some made them of green boughs, others dug caves on mountain sides, and some, in imitation of the nests of swallows and the way

they built, made places of refuge out of mud and twigs. Next, by observing the shelters of others and adding new details to their own inceptions, they constructed better and better kinds of huts as time went on.

3. And since they were of an imitative and teachable nature, they would daily point out to each other the results of their building, boasting of the novelties in it; and thus, with their natural gifts sharpened by emulation, their standards improved daily. At first they set up forked stakes connected by twigs and covered these walls with mud. Others made walls of lumps of dried mud, covering them with reeds and leaves to keep out the rain and the heat.[5]

Taken together, these first three paragraphs of Book II form a compelling program for architecture: *first fire; then the means to modulate it*. This was a motivating impulse of architecture for Vitruvius: a description of the means to capture and channel energy for the metabolic prosperity of humans and the amplification of social life. In this regard, Vitruvius was essentially a thermodynamic chronicler, summarizing thousands of years of self-organized construction and the thermodynamic emergence of language, society, buildings, and cities. The remainder of his *Ten Books*, and ostensibly the history of architecture, is the means to elaborate on this thermodynamic and metabolic origin/purpose of construction.

His last sentence in the above excerpt is the first architectural articulation of thermal modulation in buildings. But, to grasp the thermodynamics of architecture, it is essential not to isolate the dissipation of energy discussed in his last sentence from the thermodynamic evolution chronicled in his discussion of the origin of the dwelling house. To reduce the thermodynamics of architecture to the passage of heat in a wall severs it from the vitality and plenitude of the larger collectivity of human settlement and evolution. There is no better introduction to the emergence of human thermodynamic purpose than Vitruvius, especially when coupled with a contemporary bibliography on non-equilibrium thermodynamics.

My earlier allegorical transfer of body cladding to cave wall lining connects Vitruvius's observations to those of Adolf Loos. For Loos, building as adjunct skin is at the origin of architecture. Loos, too, describes the origin of architecture in explicit thermodynamic and metabolic terms:

> Even if all materials are of equal value to the artist, they are not equally suited to all his purposes. The requisite durability, the necessary construction often demands materials that are not in harmony with the true

> purpose of the building. The architect's general task is to provide a warm and livable space. Carpets are warm and livable. He decides for this reason to spread out one carpet on the floor and to hang up four to form the four walls. But you cannot build a house out of carpets. Both the carpet on the floor and the tapestry on the wall require a structural frame to hold them in the correct place. To invent this frame is the architect's second task.
>
> **This is the correct and logical path to be followed in architecture. It was in this sequence that mankind learned how to build. In the beginning was cladding. Man sought shelter from inclement weather and protection and warmth while he slept. He sought to cover himself. The covering is the oldest architectural detail. Originally it was made out of animal skins or textile products.**[6]

For Loos, the architect's first task is overtly thermodynamic: the disposition of performance-enhancing rugs and other thermally deliberate or active surfaces. As such, Loos detected a problematic divergence from what was deemed to be the "correct and logical path to be followed in architecture" in early modernity. Within the emergence of the insulation apparatus,

> There are architects who do things differently. Their imaginations create not spaces but sections of walls. That which is left over around the walls then forms the rooms. And for these rooms some kind of cladding is subsequently chosen, whatever seems fitting to the architect.[7]

Insulation, as but one product focused solely in the section of the wall, is implicated in Loos's critique of methodologies that do not achieve the architectural implications of providing "a warm and liveable space." For Loos, the architect's other obligations should precede the architectural and thermal concerns of insulation as construed in the early insulation apparatus. This helps articulate an architectural agenda for thermal energy and thermal flux: first determine the inevitability of thermal flux, and then discern an appropriate surface for that thermal dissipation.

Be it Vitruvius, Loos, or contemporary anthropological work on the role of fire in human development, we are reminded that humans are animals, seeking thermodynamic and metabolic affordance in as direct and powerful ways as possible. As such, we humans are anything but autonomous or self-sustaining. We are in fact highly dependent — in the most rudimentary *and* high-technology milieux — on the affordances of our environments. It is no mistake that Vitruvius and Loos see the convergence of building and fire in the thermodynamics and metabolism of early humans. While such convergence and resourcefulness is rarely matched in contemporary design, sustained reflection on these raw thermodynamic and metabolic roots of architecture can profoundly enrich and amplify any architectural agenda for energy.

De re mineralibus

Moving from the fire to the inevitability of baked and fired materials, the thermodynamic, metabolic, and social affordance of robust constructions motivated many of the earliest buildings. The means to capture and channel the affordances of mineral flows is enlivened by Manuel De Landa's observations about this phase in human development. He first describes the mineralization of various life forms over a long period:

> In the organic world, for instance, soft tissues (gels and aerosols, muscle and nerve) reigned supreme until 500 million years ago. At this point, some of the conglomerations of fleshy matter-energy that made up life underwent a sudden mineralization, and a new material for constructing living creatures emerged: bone... A stiff, calcified central rod that would later become the vertebral column, made new forms of movement possible among animals, freeing them from many constraints and literally setting them into motion to conquer every available niche in the air, in the water, and on land.[8]

De Landa then moves on to the mineralizing actions of humans with an important insight:

> The human endoskeleton was one of the many products of that ancient mineralization. Yet that is not the only geological infiltration that the human species has undergone. About eight thousand years ago, human populations began mineralizing again when they developed an urban exoskeleton: bricks of sun-dried clay became the building materials for their homes, which in turn surrounded and were surrounded by stone monuments and defensive walls. This exoskeleton served a purpose similar to its internal counterpart: to control the movement of human flesh in

and out of a town's walls. The urban exoskeleton also regulates the motion of many other things: luxury objects, news, and food, for example.[9]

To add to De Landa's account we must note that this mineral exoskeleton of course modulated thermal flows as well. As settlement thickened, so did the corpus of buildings. Whether we consider the piling and daubing of mud as part of the first mention of heat insulation by Vitruvius, or De Landa's perspective on the convection of language, information, genetics, food, and tectonic plates, the exoskeleton of architecture modulates the flow of energy in both short and long-terms. In these self-organized designs, form follows the thermodynamics of metabolistic flow. Heat fluxes, metabolisms are modulated, and information feedbacks are all captured and channeled by the means in which humans compose their mineral walls.

In other words, from the beginning, thermal phenomena were not isolated from other flows in the most archaic and profound sense possible. It is in fact a grave digression to begin to isolate thermal flux from the larger dissipations of energy and life. Throughout preindustrial milieux, insulating materials are but one aspect of the multiple energy/matter flows of powerfully linked ecological and architectural metabolisms. Thus, it is instructive to ask for what mineralization processes and systems are thermal modulating materials merely a by-product?

I include De Landa in this material history of insulation to help restate one of the primary claims of this book: that buildings are non-isolated, open, dissipative structures and that thermal flows cannot be considered in isolation of other energy flows of matter and information. Only in what Bruno Latour describes as modern age of constitutive separations would such an isolated view of insulating materials appear rational.[10] Thus, in this material history of insulation, I aim to focus on insulating materials, but only within larger contexts and the apparatus of insulation.

As one more observation on the short and long-term thermodynamics of insulating materials — about how these materials modulate short and long fluxes of energy — it is important to consider an operating principle that appears to guide the spontaneous configuration of flow patterns, whether of minerals or solar energy, as well as what appears to guide human interactions with flows in general. This principle is what Adrian Bejan describes as the Constructal Law:

> For a finite-size system to persist in time (to live), it must evolve in such a way that it provides easier access to the imposed (global) currents that flow through it.[11]

Much like Howard T. Odum's observations about how energy systems that select and reinforce powerful configurations will prevail, Bejan claims that the most powerful configurations of flow will prevail. In both Odum and Bejan, systems acquire greatest power when processes of feedback amplify the flow of matter as captured energy. With regard to insulating materials, the selection of non-isolated flows of short and long-term cycles of energy were evident throughout the history of construction and later of architecture. Non-isolated understandings of contemporary energy and matter — insulation or otherwise — should once again guide the specification of materials in the dissipative structure of building and reality.

To help put this into the context of the mineral realm, consider the adobe structures and other earthen structures found throughout the world with regard to De Landa's, Bejan's, and Odum's perspectives. Early and contemporary people shaped mixtures of abundant soil, clays, and perhaps straw. The shaped matter was then baked and dried in the abundant exergy of the sun. Then, since a pile of adobe bricks is not a house, humans assembled the bricks in such a way that the wall thickness became suitable for gravity loads as well as fluxing thermal loads. Over time, this thickness was optimized or, to use Bejan's vocabulary, the thickness "provides easier access to the imposed (global) currents that flow through it." This material practice, in turn, led to the abundance of the humans involved in the system, more directly as shelter.

The emergy-to-exergy ratios of the earthen construction described above are compelling to consider from Bejan's and Odum's perspectives. The exergy functions of thermal modulation as well as housing and other metabolic functions — grain storage, cooking, defense, etc. — were well served by this approach to construction: a series of linked thermodynamic practice that were self-organized so as to increase the escalating power of the human builders.

The emergy flow of this earthen material system, however, is quite different. The approach used abundant soil, with plenty remaining to serve other ecological functions. The soil was then extracted, mixed, and shaped, all with the metabolic inputs of people in archaic contexts. Solar energy, a transformity of one solar emjoule, was used to dry and bake the brick. The heat gradient that could otherwise kill someone in these regions was repurposed to do the thermal work of brick baking that in turn permitted other work to be done through building use. Once installed, the thickness, density, and diffusivity of the earth brick wall all engaged the flow of solar energy to modulate the thermal milieu of the interior, which ideally was cool during the day and warmer at night. The mass served to capture and channel solar energy, and later, re-emit it when more useful, i.e., when it had greater exergy. How the flow of solar thermal energy, the flow of soil (think from geologic formation to erosion to the flow into the wall and eventual erosion again), and the metabolic flows of food and information in the earthen building are powerfully shaped by the selection of the earthen brick all were inseparable from its thermal behavior. It is a powerful, convergent energy system for a human habitat, built on feedbacks and thermodynamic depth. Any architectural agenda for thermal flux would be similarly non-isolated.

Vegetal Insulation

In the course of human development, many early vegetal insulating materials were simply remnants of other agricultural and early industrial efforts: a superb example of pre-modern exergy-matching designs. For what agricultural processes and systems are thermal modulating materials merely a by-product? Straw, sawdust, cork, hemp, flax, coco, and corncobs are examples of vegetal insulation materials that persisted through the twentieth century. Vegetal insulative materials were often combined with mineral matter for both thermal and structural purposes, as is the case with cob construction techniques.

It is difficult to fully determine which of the bio-degradable vegetal materials and by-products might have been used in ancient constructions, but the archaic tendency to throw just about anything into a Roman wall, for instance, is indicative of a habit that may have yielded thermal benefits. While not often directly considered an insulation material, the various reed canopy shade structures used throughout low-latitude cultures certainly influenced the thermal milieu of buildings and adjacent spaces. Reeds were also woven into mat formats, often pressed into a plaster layer for installation. With no thermal processing required, the reed mats have very low transformity, yet exert decent exergy functions for compelling emergy-to-exergy ratios.

Celotex was an early popular form of insulation produced from an agricultural by-product, bagasse, the fibrous material remnant of sugarcane processing. The fibers were then felted into 8-foot-wide panels of varying thickness, from 11 to 125 millimeters thick.[12] Other fiber-based panel insulation products that use various materials persist today: hemp, flax, cotton, coconut, and Giant Chinese silver grass, for example.

Cork, another vegetal material, has been used throughout recorded history. An early account from Pliny the Elder, who was rather taken by the cork oak tree in his *Natural History*, documented the use of cork as roof insulation among a number of other uses.[13] Archaeological remains in the Mediterranean indicate the use of cork as amphorae seals and women's winter footwear.[14] The Portuguese,

A Material History of Insulation in Modernity 143

Cork Trees *(Quercus suber)* **Stripped Bare**

producers of about one-third of the world's cork, benefited from agrarian laws from 1209 that protected its millions of acres of cork oak forests — not unlike ancient cork oaks that were protected as religious symbols.

As its name suggests, cellulose insulation is based on the cellular wall material of plants. The most pervasive organic material on Earth, cellulose insulation is readily sourced and is one of the many ways to extract exergy from the low-transformity of vegetal matter. It has been in industrial use since the early twentieth century in North America and Sweden.[15] Contemporary cellulose insulation — in either loose or board formats — is produced from milling scrap paper mixed with boric acid to prevent ignition and fire. Cellulose insulation materials are hydrophilic and therefore provisions must be made to ensure the cellulose stays dry. While the kraft process used in the conversion of wood into wood pulp requires considerable input of thermal energy, cellulose insulation — often a by-product of other industrial paper production processes — is one of the few insulation materials without intense thermal processes required for its production as an insulation material. Other cellulose-based materials include sawdust and wood chips, again industrial by-products turned into insulation material as a form of sound exergy-matching design.

Dried cereal stalks from wheat, rice, and rye — straw — have been used throughout human history as a rapidly growing, cellular solid insulation material. Whether in mats or bales, the cellulose of these stalks provides entrained air cavities along with minor structural properties. In *The Idea of a Universal Architecture*, Vicenzo Scamozzi describes the use of straw and reeds as part of cold-storage design in seventeenth century Italy. Of the deep, circular vessels for snow and ice on the north slopes of hills, Scamozzi observes that "the walls should be built of bricks and lime or clay with a roof covered with a layer of straw or thick layer of reeds for insulation."[16] Scamozzi was well attuned to ground-coupled strategies for cooling. Besides his interest in cold storage techniques in the Veneto, he also incorporated deep well cooling strategies, as in the central vertical axis of the Villa Pisani: the telluric, penumbral counterpoint to the bright exhaust of the oculus above as separated by the stone grate that ornaments this flow of cooling air.

As part of the European expansion into the western United States, mechanical hay balers were developed, and their usage spread in the second half of the nineteenth century. A new construction technique was perhaps inevitable when taking into account the land rush catalyzed by the 1904 Kinkaid Act. The sandy ground of these western Great Plains produced sod that was inadequate for the sod construction common in other parts of the Great Plains, those with more suitable soil conditions. With few available trees, pioneers in western Nebraska stacked the mechanically baled blocks not unlike load-bearing masonry. Straw is of course rapidly renewable and has relative low thermal conductivity rates. The interior and exterior surfaces of the straw bales were typically plastered. Whereas the original straw bale constructions were load-bearing, most contemporary straw buildings use the bales as infill for a lumber or timber structure. When humans are suddenly placed in unfamiliar, if not dire, milieux, by design they quickly seek powerful emergy-to-exergy ratios in order to survive.

Benefiting from centuries of sediment and root growth, sod construction can be found throughout the world, especially in agricultural contexts. Sod construction consists of sod "bricks" excavated from rich soils. The roots bind the soil composition. The result is a thick, massive, and often rather moist construction technique typically employed when other construction materials are scarce. Peat is another common vegetal insulation product that was used in both brick form and in loose formats in other parts of the world with similar extraction and construction techniques.

Closely related, several cultures use living turf as a form of insulation. Icelandic turf buildings, Irish agricultural buildings, and contemporary vegetated roofs and walls all share the same operative logics. In all of the above vegetal cases, the extraction of insulative capacity from a living system provides unique ecological properties and possibilities. All these vegetal buildings appear as folds or other very direct transformations in a terrain. In the Icelandic examples, an effort was made to create cave-like thermal behaviors in otherwise flat ground.

Zostera marina

One perhaps less familiar vegetal material that deserves slightly more extended attention is seagrass or eelgrass (*zostera marina*). Early European occupants of New England used the seaweed to trap air into various construction cavities, a practice referred to as "banking up."[17] In the coastal regions of New England, the eelgrass was air-dried before being stuffed into wall cavities. Eelgrass — otherwise a dissipated clump of waste biological matter and captured energy — was re-appropriated and additional exergy was extracted where initially none was apparent. This maximizes the production of entropy for any quantity of eelgrass that in turn eventually helps maximize the entropy production for any unit of heating in a building. One early known example, the Old Pierce House in Dorchester, Massachusetts — less than a half-mile from Boston Harbor — was built in 1683. 1893 renovations of that house revealed dried eelgrass between the timber framing of the structure. The house stands today.

In the same year as the Old Pierce House renovation, an MIT and ETH Zurich-trained chemist — Samuel Cabot — developed one of the first insulation "products." Observing the abundance of the eelgrass on a Boston beach while on a walk, he was motivated to industrialize the piles of "waste" eelgrass, subsuming an economic proposition in an ecological function. Was he somehow exposed to the Old Pierce House and its eelgrass stuffing? He ultimately matted dried eelgrass between heavy kraft paper faces and stitched the assembly together. Known as Cabot's Quilt, his product was produced for some fifty years, well into the 1940s. The exergy to emergy ratio for this product is quite impressive, especially when considered as a non-isolated thermodynamic system.

Cabot's promotion of the eelgrass product reveals certain propensities of insulation corporations at the time. A 1913 advertisement boasts: "Cabot's Quilt: The Scientific Sound-deadener and Heat insulator."[18] A 1928 company pamphlet, "Build Warm Houses," also extols the virtues of the thermal and acoustic properties of the material.[19] This pamphlet introduced drawings of typical assemblies and details for architects. A few case studies, proudly including structures in the Arctic and Antarctic, add to the enthusiasm of the catalog.

Cabot's Quilt came in 84-foot long rolls along with nailing caps to fix the matt in place. The rolls were available in a single-ply format (1/3 inch) for thermal insulation, double-ply for acoustic insulation, and triple-ply intended for cold storage applications. A 1929 *Popular Science* article reports on the use of a similar assembly in London.[20] Cabot catalogs offered the quilts as distributed in "the United States, throughout the entire British Empire, including Canada, Australia, New Zealand; in Sweden, Finland, Holland and Argentine Republic."

The *zostera marina* for Cabot's Quilt came from the shores of New England in its first decades of production. By 1907, though, the company began to import bales of the dried material from Nova Scotia and other northern countries with similar coastal climates, such as Germany.[21] Given the harvesting of the material in Nova Scotia, it was perhaps inevitable that another company, Guilfords Limited in Dartmouth, Nova Scotia, began to produce a similar seagrass product, Seafelt.[22]

The harvesting practices and schedule is worthy to consider in a book on the topic of dissipation and the energetics of architecture. The harvesting of *zostera marina* occurred at the end of the local land farm harvesting and fishing season, so labor and equipment were available beginning in July of each year to focus on seagrass gathering. In Yarmouth Nova Scotia, a 75 meter rock wall was built in adjacent waters next to Little Thrum Cap that was designed to capture the seagrass before they reached the beach.[23] The wet harvest was then dried on the recently cut agricultural fields. Once dried, the *zostera marina* was baled and stored in barns before shipment to Boston throughout the fall season.

In terms of dissipations of energy and the power of a system, the *zostera marina* is a deeply compelling non-isolated example to consider. The low-emergy *zostera marina*, having accumulated its solar exergy intake, floated to shore. To maximize the exergy extraction and transformation of the inherent energy in the aquatic plant, humans "banked" the material by packing it around houses, barns, and store houses for thermal buffering in both winter and summer. This maximized the power of the system by protecting accumulated resources and extending work cycles. Further, the harvesting of *zostera marina* occurred in an otherwise fallow period in the production cycle. As such, the region collectively extracted further work from not just extant labor and equipment but from the region's bio-geophysical resources, further maximizing the power of their system. Finally, the exported material did important thermal "banking-up" work in commercial and residential buildings in distant cities such as Boston and New York. The capacity of this system/

The Earliest Instruments of the Insulation Apparatus
Early Cabot's Quilts installed in Williamsburg, VA

148 Insulating Modernism

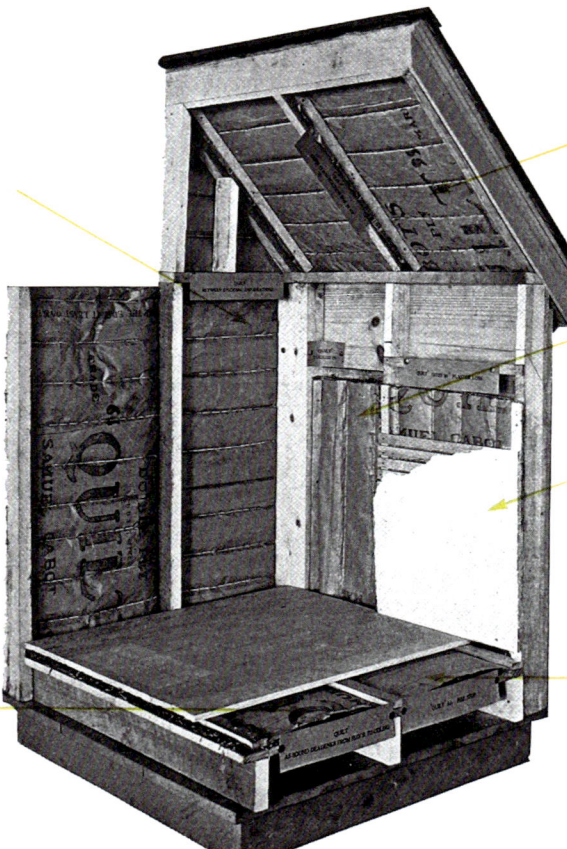

Cabot's Quilt built into wall of house between studs and boarding, making walls heat-, cold- and sound-proof.

Cabot's Quilt installed between floors as sound-deadener.

Cabot's Quilt fastened with battens between rafters on underside of roof for protection against the hot sun in summer and to make upper stories warmer in winter. Cabot's Quilt seals up all air leaks and completely stops drafts.

Cabot's Quilt installed on inner side of the wall studding beneath the lath. This gives a wider air space than some other methods of application and is more sound-proof.

Plaster applied on lath of Cabot's Quilt. The Quilt prevents the plaster from slopping through too far between the lath, thus saving 56c in plastering cost (by actual test), for every $1.00 spent for Quilt.

Cabot's Quilt installed under floor to make sun-room, glassed-in porch or other room without cellar beneath it, as warm as though it had heated space below.

Thickness for thickness, Cabot's Quilt is shown to be the best insulator in a SPECIAL BULLETIN on the figuring of steam and hot water radiation, compiled and published by the Illinois Master Plumbers' Association in 1928. A limited number of copies of this bulletin are available at our office and will gladly be forwarded on your request. Make your application early so as to be sure to secure one.

The Earliest Instruments of the Insulation Apparatus
Cabot installation maquette

A Material History of Insulation in Modernity 149

Cabot's Quilt Grass

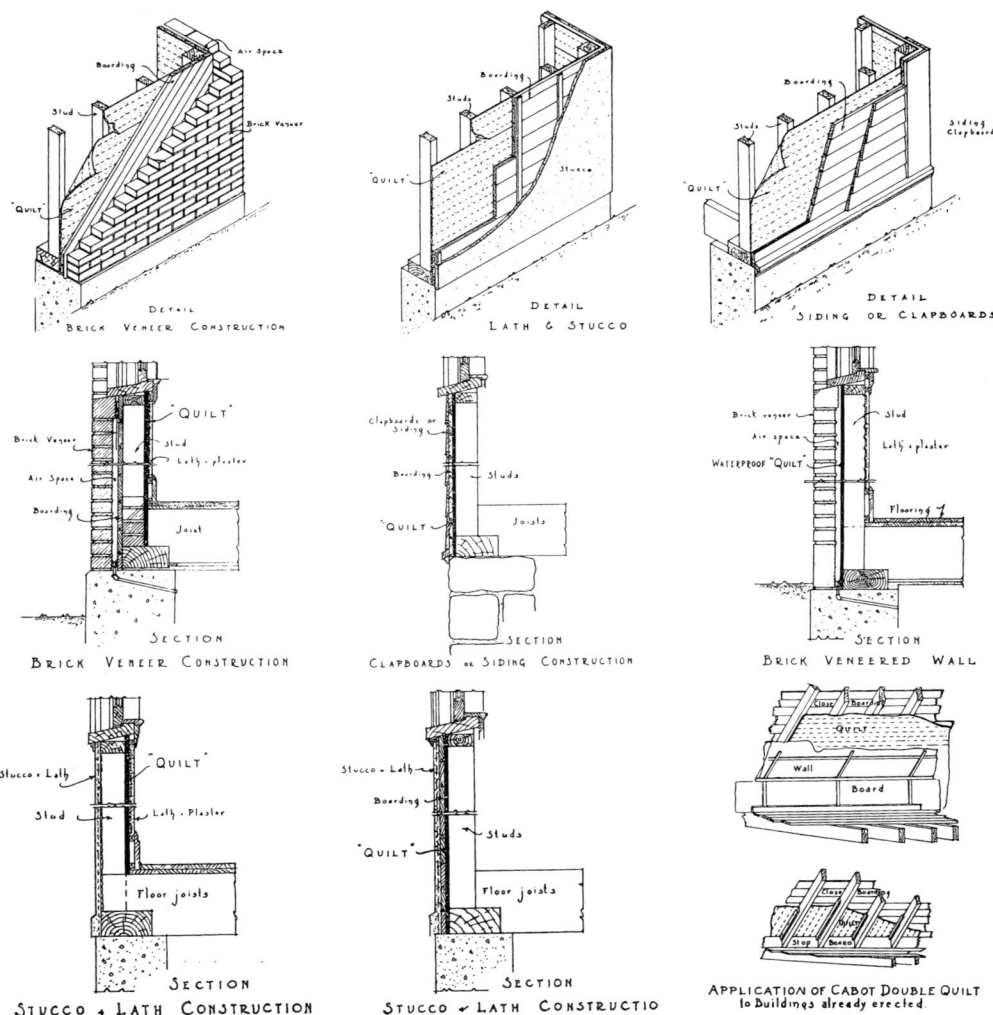

Cabot Library of Early Batt Insulation Details
Even in its earliest manifestations, insulation was made to fit extant buildings but rarely prompted a wholesale reconsideration of construction based on thermodynamic principles. It was primarily from manufacturer brochures such as this that architects became incorporated into the insulation apparatus and insulation became incorporated into buildings.

A Material History of Insulation in Modernity   151

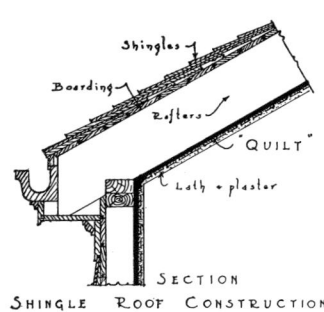

SECTION
SHINGLE ROOF CONSTRUCTION

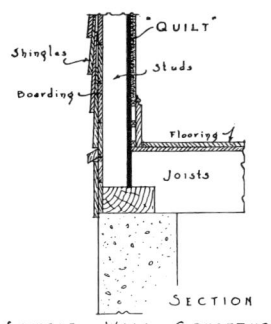

SECTION
SHINGLE WALL CONSTRUCTION

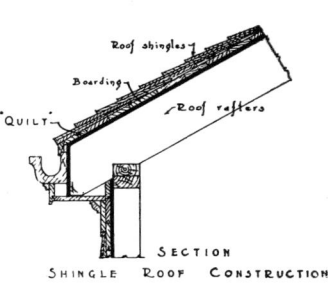

SECTION
SHINGLE ROOF CONSTRUCTION

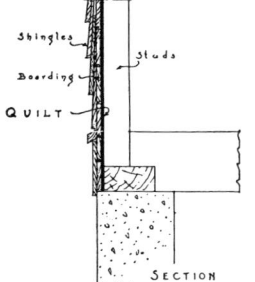

SECTION
SHINGLE WALL CONSTRUCTION

Cabot's Quilt Averages Highest

Assuming that all other insulating materials are just as good as Cabot's Quilt, the following table shows why Cabot's Quilt is the best insulator made.

	Quilt	Cork	Hair	Pulpboards	Flax
Flexibility	100%		100%		100%
Permanence	100%	75%	50%	40%	60%
Fire Resistance	90%	10%	10%		
Decay Proof	100%	75%		20%	
Vermin Repellance	100%	100%			
Average	98%	60%	43%	27%	28%

These figures in all cases give the benefit of the doubt to the materials listed, yet Cabot's Quilt averages 98%, the nearest competing material being only 60%.

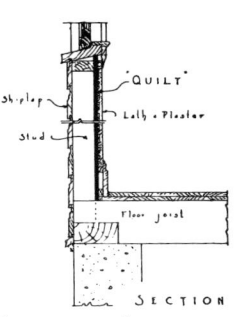

SECTION
SHIPLAP CONSTRUCTION

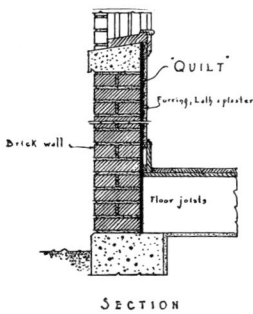

SECTION
BRICK WALL

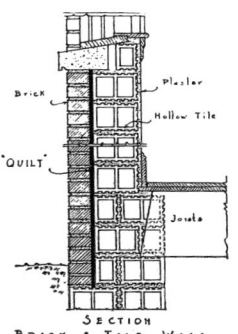

SECTION
BRICK & TILE WALL

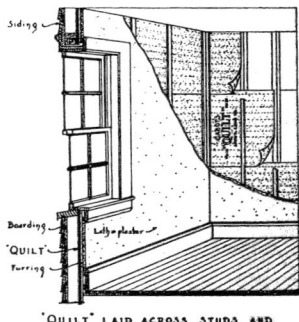

"QUILT" LAID ACROSS STUDS AND
FINISHED WITH PLASTER OR WALL-BOARD

region to capture maximal exergy from this modest-emergy plant is a strong example of a maximal power system with implications for design at multiple spatial and temporal scales.

It was only another biological agent — known as wasting disease — that determined the fate of this product and clipped the power of the system. In the 1930s, a pathogen outbreak in the seaweed species along the Atlantic coasts of North America and Europe killed upwards of 90% of the *zostera marina*.[24] The instigating pathogen discolored the eelgrass and thus reduced its photosynthetic capacity.[25] The resulting carbon imbalance killed the plant and, by extension, the use of seaweed as an insulating material. The wasted seaweed wasted Cabot's Quilt.

Animal

As with the mineral and the vegetal realms of insulation materials, it is useful to consider for which husbandry processes and systems are thermal modulation materials merely a by-product. Animal pelts, as mentioned earlier, were perhaps the first durable thermal modulation for humans. Various animal hairs, too, have been in continuous use through human history as part of various felts and textile products. Whether yurts on the steppe, the tapestries in a Scottish castle, or contemporary insulation products, hybrids of vegetal and animal products are recurrent insulation materials.

Sheep wool, to name one archaic material, is still in use today. The otherwise distinct worlds of clothes and buildings converge in wool. For the farmer especially, the exergy and emergy ratio of this material is compelling to consider. An excellent example of self-organized exergy-matching design, this insulation product, otherwise a "waste" product of the sheep, was used for human and building insulation. Capturing it and channeling it into wall cavities, however, yields an entirely different, and decidedly non-isolated, approach to the thermal modulation of cavities in the construction of a building. The immediacy of the sheep-farm-farmer-farmhouse metabolism is closely linked.

The smooth entanglement of animal hair into felted, supple, solid mats pervades both sedentary and nomadic cultural and material practices throughout the world. The lining of yurts, tapestries in castles, and felt hats, for example, point to the extensive role felt materials play in the human modulation of flow. For Gilles Deleuze and Félix Guattari, the matted technics of felt were a powerful analog for a set of intellectual habits, especially when compared to more linearly striated processes (such as fiberglass production, discussed later in this chapter, or the problematically linear habits of mind that yielded R-value characterizations of fiberglass insulation batts, as discussed in the first chapter).[26]

There was a significant, inevitable shift in the material culture of thermal modulation as materials and processes became increasingly industrialized and transitioned into products. While the effects of this shift are perhaps obvious in the evolution of thermal modulation materials, its implications for the material culture and practices of architecture are less immediately obvious. I will expand on these implications in the next section of this material history, as they are as consequential as the proliferation of insulating products in the modern period of architecture.

Industrialization & Commercialization of Insulation

Just as the preindustrial history of insulating materials began with mineral sources, so does the industrial history of insulating materials. The first industrial insulating products were the product of great concentrations of minerals, and production began, in physical and conceptual ways, to hybridize material categories in this paradigm. The earliest industrial insulating product — mineral wool — is a prime example of these industrial tendencies and the associated shift in relative exergy and emergy quantities in modern, industrialized approaches to thermal modulation.

For what industrial processes and systems are thermal modulation materials merely a by-product? Slag wool, also known as mineral wool, was first developed in Wales in 1840 by Edward Perry.[27] Industrial production and commercialization of the product occurred first in Osnabrück, Germany in the early 1870s. By the early 1890s insulating batts of mineral wool began to appear in North American building industry journals. It remains a fundamental insulating product format, one that many synthetic products would adopt during the twentieth century.

Mineral wool is a product of molten rock, heated in masonry furnaces. Jet nozzles blow a fierce air stream through the molten rock, producing thin, biopersistent fibers that appear somewhat like wool. This product combines the visual, mechanical, and thermal properties of minerals with that of animal hair. Hence the names mineral wool, stone wool, and slag wool, the latter referring more to the high-temperature processing of its material engenderment.

Note the concentration and flow of minerals in the case of slag wool. Slag wool presupposes a concentration of masonry that is necessary to fabricate a furnace capable of the 1600 °C temperatures required to liquefy stone. The concentration of great amounts of fuel for the furnace is also necessary. The jet-stream apparatus presumes another investment of matter and energy. In each of these preparatory

Early Standardization of Insulation Practices, ca. 1932

The first inclusion of insulation in the first edition of Ramsey and Sleeper's *Architectural Graphic Standards for Architects, Engineers, Decorators, Builders, and Draftsmen* was in the above detail for "Sheet Metal Facing."

steps, flows of short and long-term cycles of energy are channeled in such a way that the resulting matt of fibers productively modulates thermal fluctuations. According to Odum and Bejan, the short-term thermal modulations must ultimately be at least as powerful as the preparatory energies in order to not only emerge but to persist over more than a century of building construction. But given the multiple, contradictory demands that drive the production and use of products, we cannot assume that the most ecologically powerful product will necessarily prevail in the short term. This complication of otherwise self-organized processes is clearly also expressed in how, and in what format, insulation entered the culture of building in early modernity.

Early Insulation Products, Formats, & Procedures

In this material history of insulation, the format of the new insulating products, not just the products themselves, is important to consider. The emerging late nineteenth and early twentieth century insulation products were developed in a context of well-established construction practices. In the North American context, balloon and platform wood framing was used in most small-scale buildings. Load-bearing masonry characterized nearly all medium and large-scale buildings. Some alternate modes of concrete or steel structures were burgeoning practices, but generally, all early industrial insulation products were developed for the formats of these predominant construction types.

For this reason, it must be understood that the early insulation products had an ameliorating function: how to improve the thermal behavior of these extant systems. The Cabot's Quilt drawings are emblematic examples. These adaptations of insulation to extant construction were not, of course, totalizing systems that took as their mission discerning the fundamental principles of what might be an optimal approach to the thermodynamics of architecture in short and long-term energy cycles. Rather, these early adopters naturally sought incremental improvements to the technological momentum of other extant building practices. The aim then was, as it is today, to improve a thermodynamically flawed system from a thermal flux point of view — to address the hemorrhaging of heat — rather than the deduction of astute thermodynamic architectures. Within the modern, highly additive and layered, approach to construction, it was easier conceptually and technically to just replace one product for another. This reflects a replacement mentality consistent throughout modernity rather than a reconstitution mentality. This habit of mind and practice unfortunately persists in architecture and the building industry today. How, why, did we get to this state of mind and practice?

One of the more consistent chronicles of normative building practices in modernity, *Architectural Graphic Standards*, is useful to consider in this context.

Architectural Graphic Standards

The first edition of *Architectural Graphic Standards* in 1932 included brief references to insulation.[28] It is no surprise to see that the first reference to insulation in *Architectural Graphic Standards* is an illustration of an insulated metal panel. No architectural product more directly relates to the thermal logic of a refrigerator nor more aims to reduce the appearance and behavior of a building to the isolating fantasies of refrigeration engineers than the insulated metal panel. The early appearance of insulated metal panels, the first in *Architectural Graphic Standards*, is but one manifestation, one formal representation of the refrigeration basis of the insulation apparatus in North America.

This 1932 drawing is an exemplar of how insulation entered the discipline and profession of architecture. As a book that reflects both an effort to standardize and catalog certain architectural practices, it, as such, also often reflects reductive and acquiescent habits of mind. In terms of energy, this reductive paradigm unfortunately occurred precisely at the time when buildings began an enormous transformation in terms of their operational and captured formations of energy.

This insulated metal panel drawing illustrates one half of a refrigeration logic that would quickly come to co-determine not only the design and operation of buildings, but the operations of architects and the building industry as well. This seemingly innocuous drawing associates the building with the refrigerator and presages a paradigm of design and constructions that is increasingly isolated in both intentional and unintentional ways. It is drawings such as this — and their implications — that begin to reposition architectural assumptions into a very particular and peculiar pathway. This drawing — as well as many related advertisements and product brochures — begins to form and deform architectural habits of mind about thermal flux and energy exchange more broadly.

For architects, the metal panel illustration — and others like it — came first, followed by a series of assumptions about heat in walls and ultimately about how energy is characterized in architecture more broadly. More specifically in terms of energy, this illustration reflects a broad diffusion of an insulation logic wherein thermal transfer was reduced to conductivity concerns. As presented, the issue of insulation was clearly one of how to best integrate burgeoning products, rather than an explication of thermodynamic principles. By its 1936 edition, *Architectural Graphic Standards* represented more than just a few details.[29] It also represented

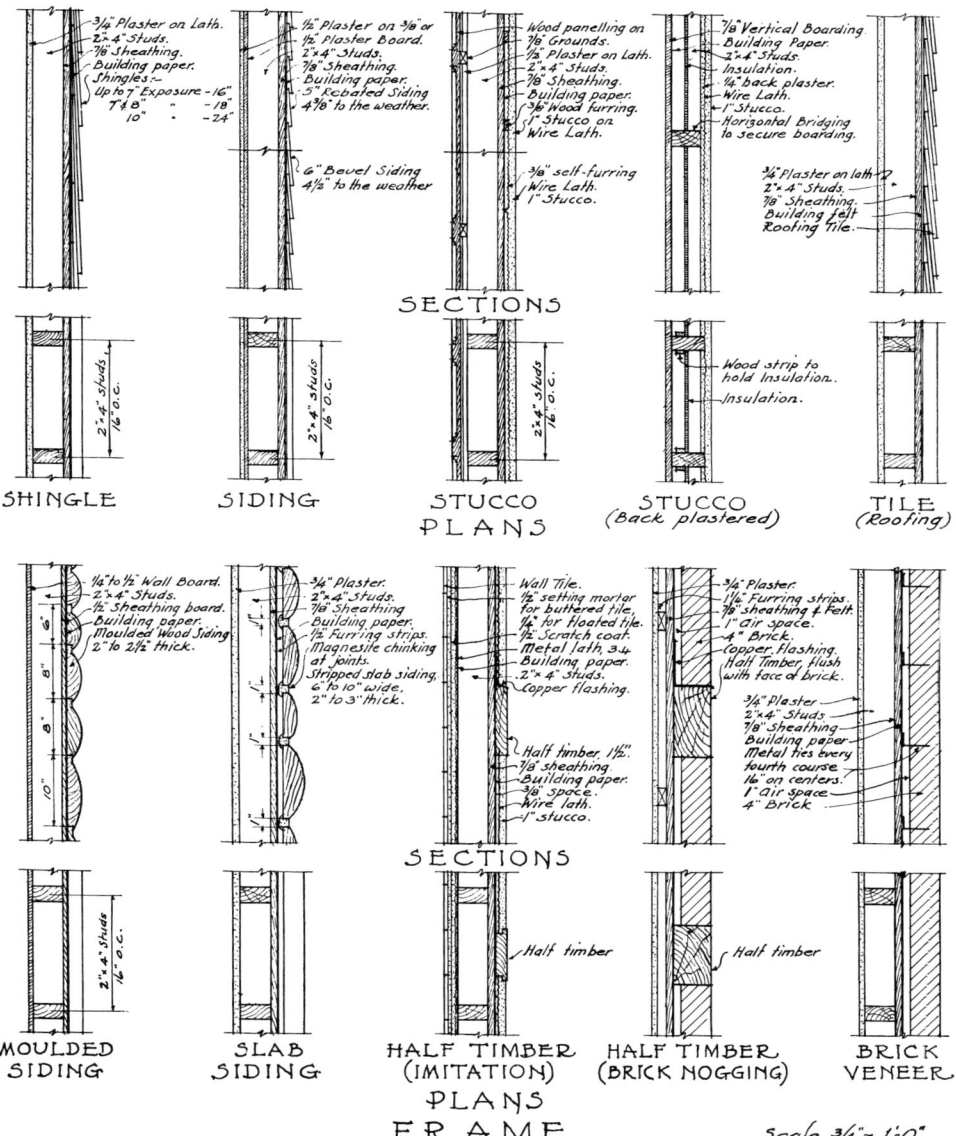

Early Standardization of Insulation Practices, ca. 1932
The only other inclusion of insulation in the first edition of Ramsey and Sleeper's *Architectural Graphic Standards for Architects, Engineers, Decorators, Builders, and Draftsmen* was in the above detail for the ostensibly reflective insulation in the "Stucco (Back Plastered)" detail.

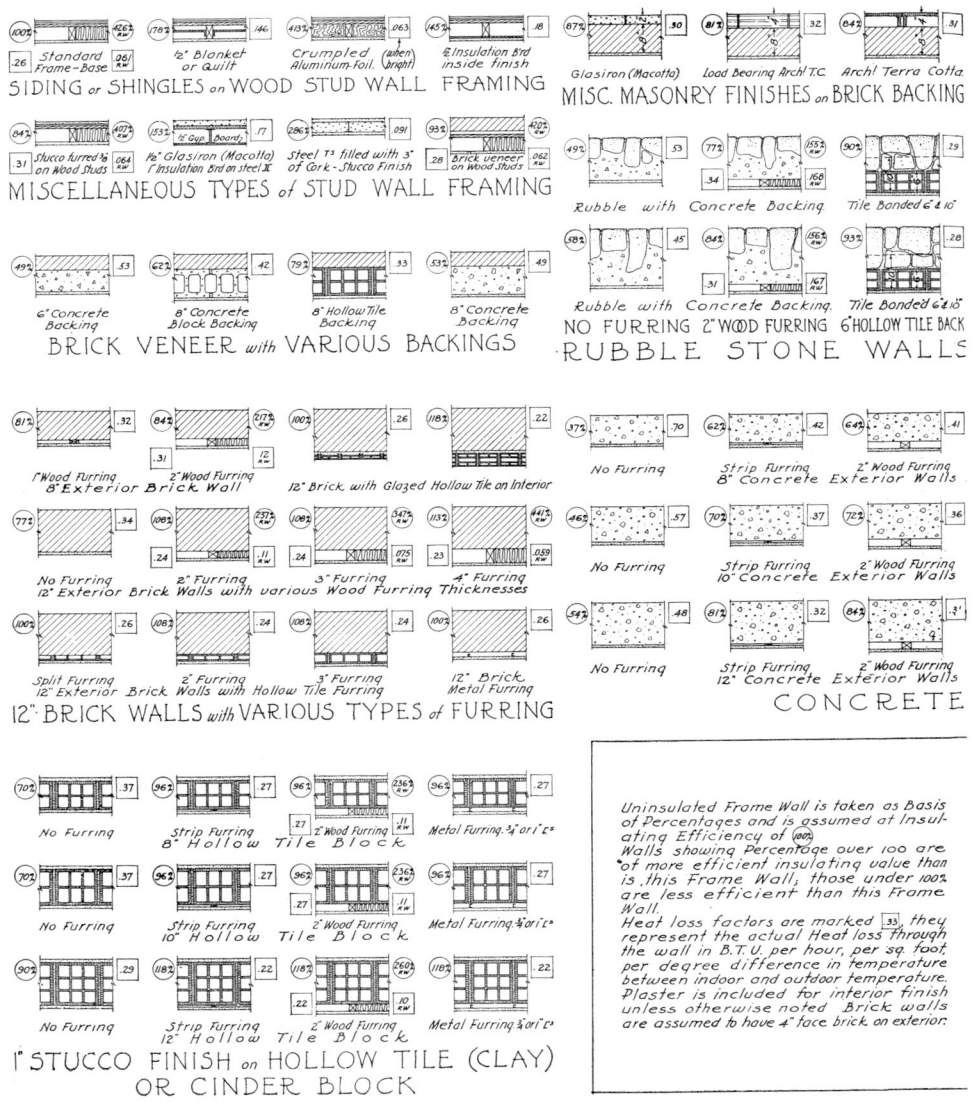

Insulation Types Abstracted and Quantified, ca. 1936
The above spread from the second edition of Ramsey and Sleeper's *Architectural Graphic Standards for Architects, Engineers, Decorators, Builders, and Draftsmen* provides the "Insulating Efficiency" of various material assemblies for buildings.

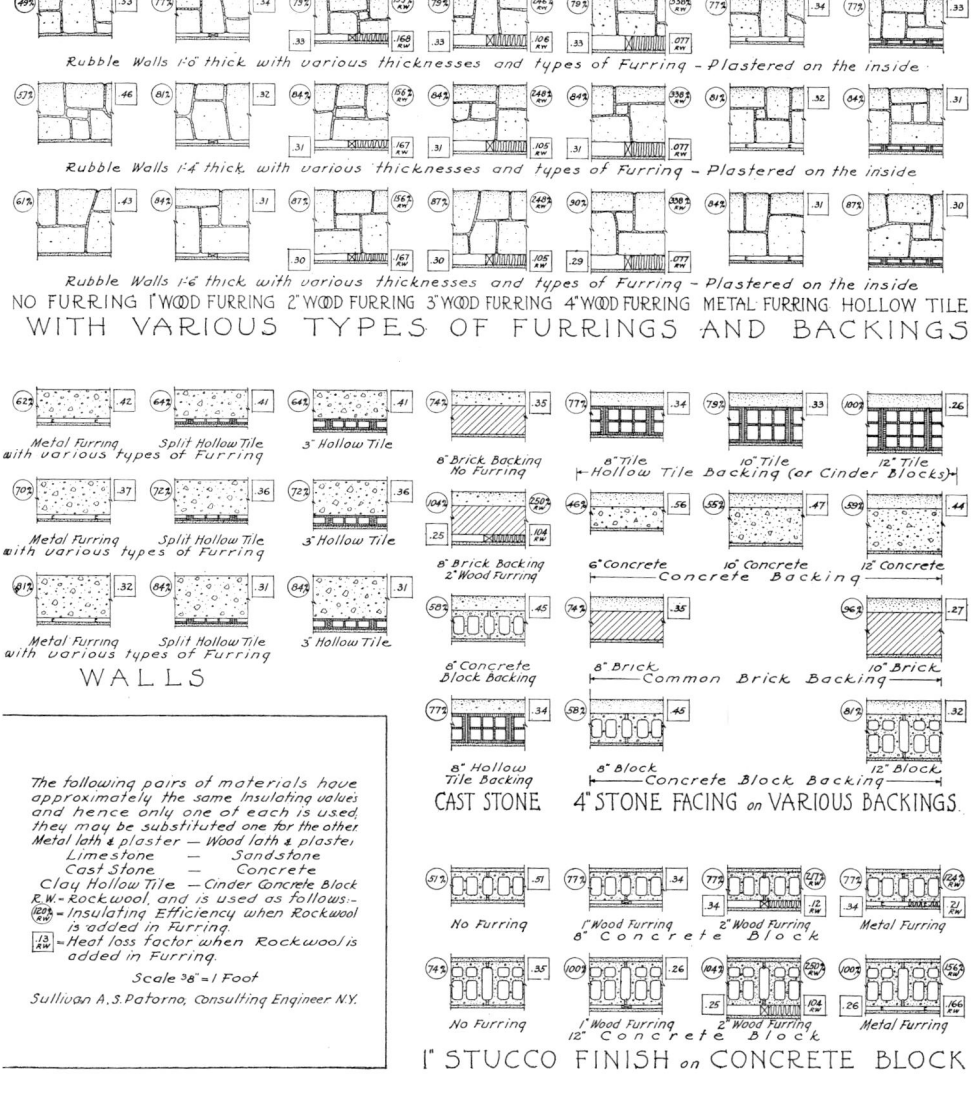

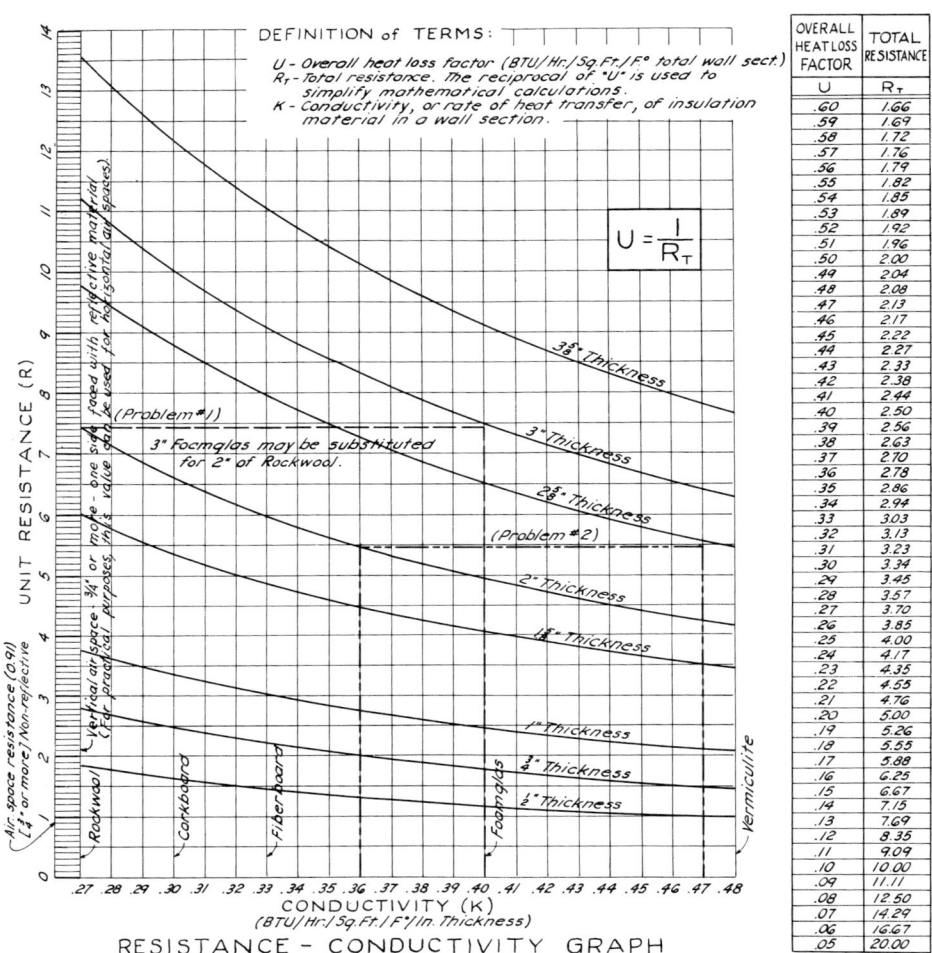

Evolution of the Insulation Apparatus, ca. 1956

The "Resistance-Conductivity Graph" from the 1956 fifth edition of Ramsey and Sleeper's *Architectural Graphic Standards for Architects, Engineers, Decorators, Builders, and Draftsmen* reflects both a deepened engagement with heat transfer and the narrowness of its thermodynamic consideration.

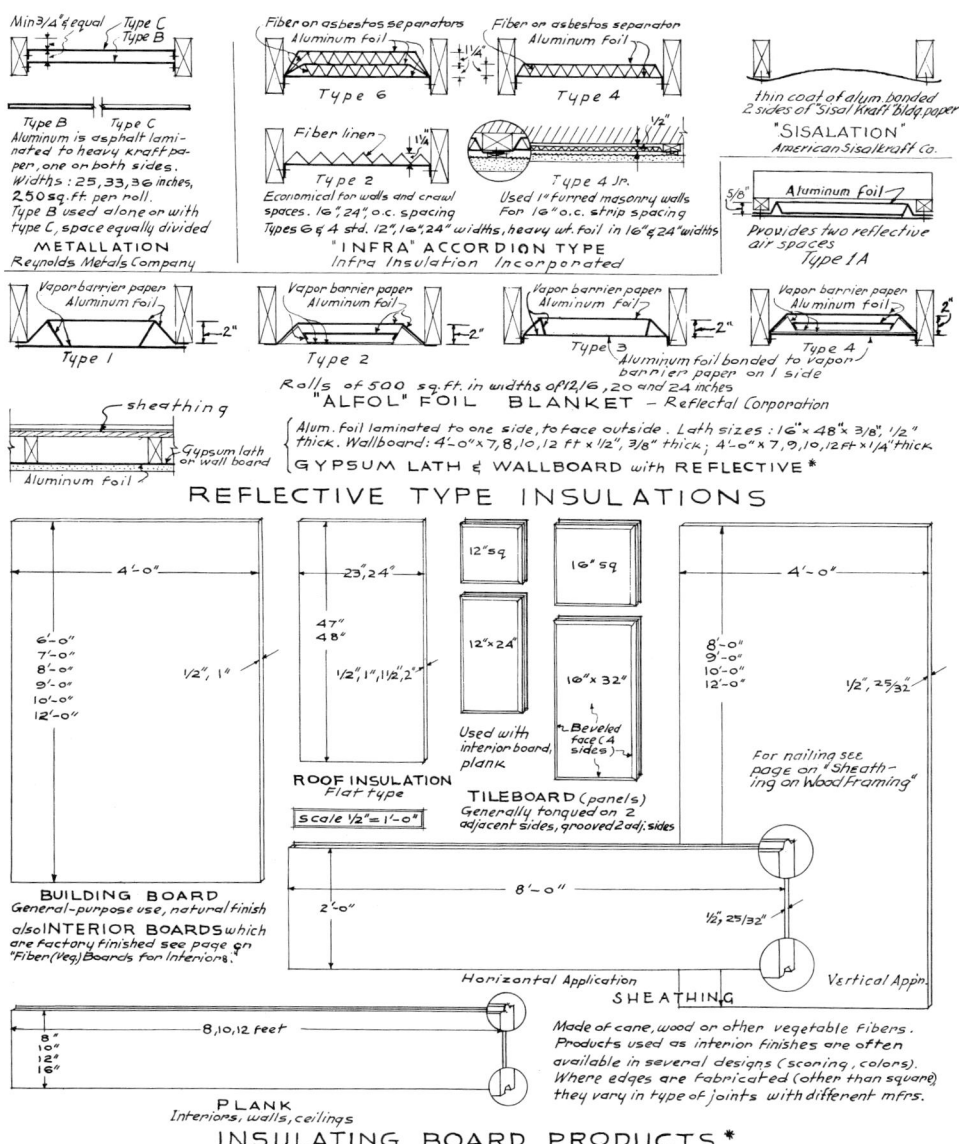

Reflective Insulating, ca. 1956

The 1956 fifth edition of Ramsey and Sleeper's *Architectural Graphic Standards for Architects, Engineers, Decorators, Builders, and Draftsmen* also included a section on reflective insulation formats and details, a response to the post-war interest in radiant thermal systems for buildings.

prescribed habits of mind, pathways of knowledge, and habits for architects to integrate along with details.

As Jane Murphy has noted, the subsequent editions of *Architectural Graphic Standards* included increasing amounts of information regarding insulation in building assemblies, an increase which reveals a burgeoning interest in insulation in the practice of architecture.[30] Yet at the same time, within that interest, there is little consideration for how heat transfer itself might more fundamentally transform an architect's design or specification of the material organization of a building with respect to its thermal behavior or potential. The focus was instead the technocratic, isolated task of where to place insulation and how much insulation to place. This represents a recurrent habit of mind, pathway of knowledge, and prescribed practice about insulation and heat transfer.

By 1956, a more complex characterization was evident in *Architectural Graphic Standards*. A chart in the 1956 edition cross-referencing relative capacities of different materials also helped disseminate the R-value concept. This edition also included a section on reflective, or radiant, insulation, reflecting again a more nuanced characterization of heat transfer but one that was nonetheless invested in presenting how new thinking about heat transfer is stuffed into extant modalities of construction.

As *Architectural Graphic Standards* chronicles, the formats of early insulation products — batts and boards, for instance — were designed to fit typical sizes of construction cavities or to add a minimal new layer to the construction assembly of a building. Early details for insulation installations thus entered the profession through manufacturers. It was largely the role of manufacturers to educate architects about the function and installation of the burgeoning products. As such, architects, too, began to fit prescribed slots and functions within the building industry and the insulation apparatus.

Boards, Batts, & Prosumers

The following would become a recurrent pattern in the twentieth century: manufacturers developed "new" products, marketed them to architects, and architects learned to incorporate the new products and practices. Here, the term "incorporate" is a double entendre. Not only are products incorporated into the design and construction of buildings, but architects become incorporated into the interests and motivations of the manufacturers, and their associated corporations. Architects are central agents, critical nodes, in the burgeoning supply and value chains associated with these building products. While schools of architecture began to emerge at this time, manufacturers were primary purveyors of certain forms of knowledge that were so central to modern architecture and its construction practices.

As agents in this mode of exchange, architects are specific types of professional consumers. They serve both to produce need and professionalize need. Through marketing brochures and advertisements, architects learn that their clients and buildings need a product, like insulation. Over time, these needs become standard practices and pedagogies. While the need in many cases might be necessary or genuine, this particular approach to the production and professionalization of need is unnecessary. Architects, for instance, might reverse the flow of information by discerning more totalizing responses to the problem of energetic dissipation in building and either work with the industry to yield more sane solutions or take the role on directly. Until that happens, architects will often be but a prosumer, negotiating conflicting needs and, exhausted by the process, eventually confusing the ecologically and architecturally optimal with the less bad.

This pattern of exchange placed architects in an acquiescent, capitulating position. It is a pattern — a habit of mind and practice — in which architects must routinely implement corporate interests, forever negotiating between the will of the client and the demands and limited capacities of manufacturers, developers, and regulatory bodies. This mode of acquiescence — evident in everything from the tale of insulation practices in this book, to refrigeration and refrigerators,[31] to the tired euphoria of other "new" materials and technologies that are often engaged for short-term market differentiation,[32] to Daniel Abramson's account of architects' role in the planned obsolescence of not only buildings but entire swaths of cites,[33] to include but a few examples — does not typically serve the built/un-built environment or the practice of architecture well. It is ultimately a limited and limiting mode of action.

With respect to the architectural modulation of heat velocities, a less limited — and less isolated — role for architects suggests a very different understanding of the fiduciary responsibilities of the architect. A more totalizing and rigorous perspective on fiduciary duty would suggest that simply doing one's best in an acquiescent and capitulating context of received conditions is simply not the highest standard of care for neither buildings and clients, nor built and un-built environments. The isolated role of architects, as partially represented by how heat transfer knowledge and products were introduced to the profession, will always yield limited efficacy in the building industry. A less isolated view of material practices is necessary to fulfill not only basic fiduciary duty but, more importantly, the great architectural potential inherent in the thermal flux that remains latent in the patterns of architecture and the building industry that emerged in the twentieth century. Rather than just accepting the terms of available products, practices, and pedagogies, an architectural agenda for energy would seek alternative means and ends for energy that are far more ecologically and architecturally powerful. To this end, it is useful to look more closely at the material culture of modern insulation products.

Mid-Century Professionalization of Insulation

The taxonomy at the beginning of this chapter presents a timeline of the number of insulating products and manufacturers. It illustrates the significant jump in insulating practices that occurred post-World War II. A powerful mixture of several factors and motivations charged this boost in insulation practice. The post-war chemical industry sought new markets as the co-determinants of air-conditioners and insulation reinforced one another in the concept of artificial weather, while environmental "control" courses were introduced in architectural pedagogies, and the concept of insulation resonated well with the isolation of suburban migration and Cold War fears: a mangle of practices that partially explain an interest in insulation during a period that consumed more energy than any prior in human history.

The professionalization of insulation theory and practices was acute in this period. As established in the first chapter, the first task of any profession is to establish need. This was undoubtedly the case with the professionalization of insulation products. Need was established through trade publication and advertisements in the first half of the twentieth century. A quote offered in the first chapter bears repeating here:

> Insulation can be thought about in two ways. You can either follow your own theories or the theory of some architect, who cannot possibly know about insulation unless he is making a specialty of the refrigerating trade; or you can buy manufactured insulation from a reputable manufacturer who will guarantee that it will permit only a certain rate of heat transmission..."[34]

Architects did not in fact know much about insulation, and the insulation industry was happy to identify and fill this gap in condescending and isolating ways, as the above quote suggests. The primary purpose of a quotation such as the one above is to establish professional need, and this specific quotation establishes the need for a new professional ostensibly equipped to provide some now necessary service.

In this dynamic, architects became servants to the terms and agendas of adjacent professionals and manufacturers. This acquiescent pattern emerges again and again with each new product and system. No longer insulating materials, insulating products must be managed by industry, not by architects. In this state, the representation and specification of architecture is managed by a new industry that tells you what you need, why, and how much it will cost you. The professionalization of multiple aspects of architecture rarely, of course, serves its practice or outcomes. It is, rather, most often a capitulation of architecture's multifarious opportunities and obligations. More discriminating and discerning architects could engage their own research program about what might constitute an architectural agenda for thermal flux, or to study any of the other products and practices imposed during modernity.

The narrowly trained refrigeration engineer or insulation salesman may know rates of heat transmission, but proves, by making myopic claims about insulation expertise, he knows little of what is transmitted in a broader sense. This narrow view, especially in the early days of insulation, had some disastrous consequences. In certain circumstances, the narrow professionalization and consideration of insulation became iatrogenic.

Iatrogenic, Carcinogenic

Asbestos is perhaps the best evidence of the isolating and counterproductive habit of mind that took hold in the industrial phase of insulation. When considered in isolation, the fibrous mineral material seems to have an astounding set of attributes, such as its suppleness, resistance to fire, low conductivity, and resistance to animal incursions. These properties have been the source of astonishment and continued use for thousands of years. However, we now know that the size and biopersistence of its mineral fibers prove to be carcinogenic for humans. Asbestos was one early and famous example of broader iatrogenic practices that emerged in modernity from the rush towards technological innovation, market differentiation, and the naiveté of progress ideologies in early modernity. Why isolate carcinogens from heat flux? Why isolate long-term well-being from short-term thermal comfort? Only in an isolated habit of modernist thought could such chronic separations be permissible. (Should you think the iatrogenics of asbestos are an isolated instance, just consider the modernist externalities that engendered Superfund sites and climate change, to name just two.)

Humans have used asbestos, whose Greek root refers to *inextinguishable*, in its mineral forms for thousands of years. Early inhabitants of contemporary Finland used asbestos in various ceramic implements for cooking and eating whereas in Cyprus it was used for clothing, cremation, and lamp wicks.[35] Herodotus documented the Greek use of asbestos for oil lamp wicks.[36] The Roman sacred fire maintained by vestigial virgins likely had an asbestos wick.[37]

An emblematic 1909 article in *Popular Mechanics* extolled the thermal-proofing, fireproofing, and vermin-proofing wonders of whole houses constructed of asbestos in Australia.[38] Unlike ants, vermin, and even fire, the capital-driven motivations of burgeoning industries could not stay away from this material. In buildings, asbestos was used in everything from insulation to furniture to wallboard to gaskets.

As early as 1897 an Austrian doctor made the first diagnosis of an ailment caused by asbestos.[39] By 1906, the first death attributed to asbestos was confirmed.[40] The epidemiology of asbestos was thoroughly established by the 1920s, but widespread production continued well into the 1980s and 1990s. In the late 1980s, the United States Environmental Protection Agency proposed an Asbestos Ban and Phase Out Rule that was later undermined in court, leaving small amounts of asbestos in a range of products. In other countries, such as Australia, the material has been banned altogether. In modernity, the disproportionate enthusiasm for the capabilities of its technics never disclosed the problems associated with its externalized culpabilities. Again and again, in multiple ways, modernity constantly made the future a colony of the present.

In medicine, iatrogenic practices refer to complications that inadvertently emerge from the treatment of some other ailment. Buildings, like bodies, are complex entities and the outcome of new approaches can be difficult to discern. In the case of asbestos insulation, the effort to keep people comfortable also induced various cancers, mesothelioma and/or asbestosis in those people. Why isolate short-term energy cycles from the dissipations of cancer? This is a counterproductive medical outcome that was unwittingly designed into the fabric of many modern buildings.

Sick Building Syndrome

Like asbestos, insulation-driven indoor air quality problems are a recurring topic in iatrogenic attempts at achieving human comfort within the thermodynamic platitudes of energy efficiency and energy conservation. A narrow view of insulation — the more the better! — coupled with infiltration concerns — the tighter the better! — can trigger many problems even as its aims to solve the problem of heat flux through a building envelope.

A Material History of Insulation in Modernity   167

The control of heat gain or losses is but one topic of concern, as Michelle Addington has noted:

> Further compounding the energy usage issues were the inherent problems with indoor air quality resulting from the zealous implementation of energy conscious designs. Buildings were tightened and sealed without any corresponding reassessment of their thermal behavior. As a result, poorly conceived insulation strategies coupled with reduced ventilation contributed to a marked increase in indoor air pollution even though outdoor pollution had been steadily declining. Sick Building Syndrome, Legionnaire's disease, asthma and hypersensitivity pneumonitis are but a few of the illnesses that are directly attributable to poor indoor air quality.[41]

Addington has documented the ventilation-driven factors that engendered aspects of sick building syndrome and, in doing so, helped articulate the co-determinant dynamics of insulation and air-conditioning master modernization, if not iatrogenic, actants in architecture.[42] With respect to iatrogenics, the history of air-conditioning is summarized by a history of air exchange expectations. Shifting air exchange standards from the American Society of Heating, Refrigeration, and Air Conditioning Engineers reveal certain relationships between insulation, air exchanges, and side building syndrome. Over the last two hundred years, various standards have cited between 4 cubic feet per minute (cfm) and 30 cfm.[43] The shifting ventilation rates are an index of both energy consciousness and ventilation load concerns. But they also reflect the increased management of loads related to off-gassing materials, volatile organic compounds, office equipment exhaust, and mold propagation that are all amplified by the impulse for super-insulated, super air-tight buildings that are logical outcomes of a narrow view of insulation practices.

Sick Ecology Syndrome

The iatrogenics of architecture, however, are not limited to specific material practices such as asbestos, or to more systemic outcomes such as sick building syndrome. The larger counterproductive logic of negated efficiencies of many contemporary building/energy systems, not to mention the rift between short and long-term energy cycles, all pose difficult challenges for human well-being over longer periods of time. A narrow system boundary — whether around an insulating product, a building envelope, or a building as a whole — will reveal little about the energy dynamics of a thermal assembly or a building.

Isolating the thermal behavior of building materials from longer-term energy flows can prove problematic, if not outright counterproductive. For instance, a constricted code and pedagogy-driven focus on the insulation of heat flux in short-term energy cycles will inevitably suggest that more insulation is better. Cost might temper the extension of this logic, but claims about R-80 or R-120 insulation in roofs in northern climates are not uncommon today in contemporary practice. While logical perhaps on its own terms, the relationship of short energy cycles to long energy cycles must be considered in a claim about the energy behavior of buildings. In the narrow view of insulation, the tendency is to keep larger environmental and ecological obligations externalized: short-term energy cycles dominate longer-term energy cycles, resulting in a sick ecology syndrome. This severely limits an architect's capacity to design maximal power objects *and* systems.

Conclusion: Structural Solutions or Power-operated Solutions

The material culture of insulation shifted significantly in modernity. Relatively benign thermal modulation materials and practices in the preindustrial world became more powerful and complicated products in the nineteenth and twentieth century with mixed ecological outcomes. Their thermal modulation power increased, but so did their capacity to sway the trajectory of buildings. No longer the benign by-products of agricultural or oceanic flux, the insulating materials became anything but neutral, thermally or otherwise.

If you have any doubt, consider the response to the 1973 petroleum embargo in architecture. As a response to petroleum-triggered energy questions, architects were taught to incorporate ever-increasing amounts of insulation. Ironically, the most popular insulation products used in this period — fiberglass insulation, polyisocyanurate, polystyrene, and various spray foams — *were all petroleum-based products*. Oh, the folly of externalization! This is a paradigm of throwing petroleum on the fire of architecture's modern metabolism and, of course, an excellent example of the mindlessly isolated praxis that this book aims to amend with non-modern concepts and practices for the thermodynamic purpose of architecture: the dissipation of available energy gradients in the most powerful, if not magnificent way possible. 1973, again, was hardly a pivot point in the energy praxis in architecture. Again, it only served to perpetuate and intensify problematic habits of mind and practices as they pertain to architecture.

While the development of thermal insulation — and petroleum — products have made buildings more powerful, how they did so and, more importantly, how they could do so deserves increased attention from architects. Narrow concerns and the associated narrow system boundaries of insulation products and practices cannot be all an architect takes into account when they consider the flux of heat or dissipation more generally.

To Build or Burn

To couple the petroleum-based products of the oil embargo with the role of fire discussed at the beginning of this chapter, it is useful to consider again (and again and again and again) Reyner Banham's famous parable about a savage tribe's encounter with a pile of fallen timbers and their quandary to build or burn the timbers.[44] For poignant rhetorical purposes, Banham emphasized two choices for the tribe: to build wind and rain shelters, a structural solution, or to burn the wood, a power-operated solution. A noble tribe in Banham's view would ration an astute portion of timber for both construction and fire. A more savage group, the precursor to our own contemporary tribes, would tend to either only burn or only build with the resource. Banham's personal enthusiasms, at his point in the energetic pulsing cycle of modernity, favored fire.

What is at stake in Banham's parable is how to extract maximal exergy from the emergy accumulated in the pile of timbers. Any reflexive response to the parable is inevitably a hybrid of the structural and power-operated modes that Banham offers. The particular mixture of this hybrid is highly contingent on a range of systems variables and requires extensive thought about the relative roles that emergy and exergy perform in architecture. The discernment of relative emergy and exergy yields should be a central concern for contemporary architects engaged in an energy agenda for buildings.

A Non-modern, Non-isolated Example

It is useful to now consider a non-modern building that diverged from the general assumptions of the insulation apparatus. The building might seem familiar: Frank Lloyd Wright's 1936 Jacobs House. In the context of this book, the tectonic and energetic aberrations observed about this building in extant literature become productively unfamiliar.

The solid wood cases studied by the Norwegians researchers discussed in the last chapter are interesting to consider in the North American context. Non-vernacular, if not canonic, examples of similar construction typologies are rare in the history of early twentieth North American architecture; consider, however, the example of Frank Lloyd Wright's Jacobs House in Madison, Wisconsin (1936).

As the first of Wright's Usonian houses, this small house tested a number of novel approaches to the construction and thermal behavior of American housing. In the 1930s, at age sixty-nine, Wright remained bemused by the status of house construction in the United States, as well as by the ubiquitous pattern of urbanization linked to the single-family house. Thus, the Usonian houses and Broadacre City reflected overt challenges to the paradigm of housing and settlement in the United States.

At the scale of the house, Wright's provocation was posed in terms of space, cost, structure, and thermal behavior. This necessitated several experiments in the design and construction of the Jacobs house. Starting from the foundation, Wright selected a "dry wall footing" that aimed to eliminate more conventional foundation stem wall and spread footing. This system is essentially a slab on a grade with a

Herbert Jacobs House

Jacobs House
Entry

thickened perimeter that acts like a grade beam: a single-pour solution. With a gravel bed under the slab, Wright assumed limited heaving of the grade, even though it is above the frost line. Notably, this slab assembly also contained a primary heating system, Wright's approach to thermally active structures: his "gravity heat" system. The steam pipes below the on-grade 4 inch slab heated not only the house above, but without insulation, for better or worse, they heated the ground below the house as well, reducing surface freeze-thaw cycles and heave issues as one consequence.

Cross-laminated Wood Walls

Another unique approach to building construction and thermal behavior explored in the Jacobs House was the wall assembly. The wall is a sandwich of vertical and horizontal softwood boards. The first layer, the middle of three, is a vertical board of lower-grade 7/8 inch pine and faced with building paper on each side. Each face is then clad with a 7/8 inch pine board held in place with a 5/8 inch redwood batten and screws, with all screw slots turned horizontal at 12 inches on center. The bottom edge of each horizontal board is either shiplapped or grooved to receive the tongue of the board below. The top edge of each horizontal pine board is sloped for drainage. This continuous load-bearing assembly bears (as indicated in the construction drawings) on the slab edge, slotted into an inverted zinc "V" section and some expansion material that separates the wood base from the concrete slab. The original owner, Herbert Jacobs, noted a 3 inch slot the in width of the center boards that fit atop a 3 inch fin embedded in the slab elsewhere shown as a galvanized metal spline.[45] The thickened base also consists of a five-layer sandwich, perhaps anticipating the inevitable weathering. The top of the assembly consists of a continuous milled nailing plate upon which the roof assembly bore and into which the ceiling board and batten system engaged.

The unusual approach to the Jacobs wall construction has justifiably drawn attention from observers of modern architecture. The wall strikes Michael Cadwell, for instance, as "odd" and "bizarre;" an approach with "questionable insulating and structural properties."[46] While he acknowledges that Wright's approach was surely better than the typically un-insulated cavity wall counterparts, he nonetheless characterizes the Jacobs wall as deficient from an insulation point of view: "it does not meet today's insulating standards."[47] But if anything, this observation only evinces a partiality towards the insulation apparatus, not a concern for the actual thermal behavior of the building and much less for the total thermodynamics of the system. In the terms of the insulation apparatus, the Jacobs wall is certainly deficient, but that does not mean Wright's general experiment is not thermally viable, and should not be so easily dismissed. The wall is deficient only if one equates thermal comfort with insulation. This false isomorphism of comfort, energy, and insulation has crippled aspects of modern architecture. Contemporary architects need not be so acquiescent.

Fortunately, Cadwell's amiable curiosity about Wright's approach allows him to speculate further about how the solid wood walls might transfer heat from the radiant slab.[48] Indeed, Cadwell's intuitions serve him well and point to criteria beyond the insulation apparatus as the means to experience and evaluate this house. There are a number of more nuanced thermal behaviors evident in the Jacobs House that Wright intuited and specified that collectively exceed what any insulation calculation ever could. When combined with the radiant strategy and a few observations about human physiology (chapter 3), the Jacobs House sandwich walls appear as a refreshing and remarkable aberration in the insulation apparatus in modern architecture.

In similar terms, Cadwell begins to see the linked thermally active capacities of the masonry masses and concrete slab.[49] The great mass — even excessive mass given the diminutive size of the house — of the masonry central fireplace behaves at once as a giant masonry radiator in the winter and an absorber of heat in the summer. The thermal diffusivity of the wood, concrete, and masonry are strategic, complimentary agents in all seasons, whether mechanically activated or if other dynamics drive the system. Although Wright aimed to banish radiators from the Usonian house, both the meandering concrete slab and the convolutions of the masonry mass are nothing but radiators. In an effort to eliminate conventional radiators for cost and aesthetics, Wright turned the bulk of this house into a giant, architectural radiator. He knowingly made the house a sink and source of great thermal flux. This strategy for accumulation and distribution for thermal flux is as architecturally essential as it is for its accumulation and distribution of structural loads. This convergence of matter and energy is profoundly architectural and exhibits multiple aspects of an architectural agenda for energy rarely, if ever, present in the insulation apparatus.

Likewise, Cadwell's misgivings about the structural capacity of the solid wood assembly soften as he describes the structural strategy in more integral terms by including the following: the role of the massive masonry components, the orthogonal convolutions of the wood wall, and the depth of the bookcases that stiffen long walls in the living room and bedrooms. In fact, Wright indeed had a non-isolated view of the wood and masonry walls from a structural and thermal perspective. Rather than a liability, the Jacobs House is one the few modern examples of an ar-

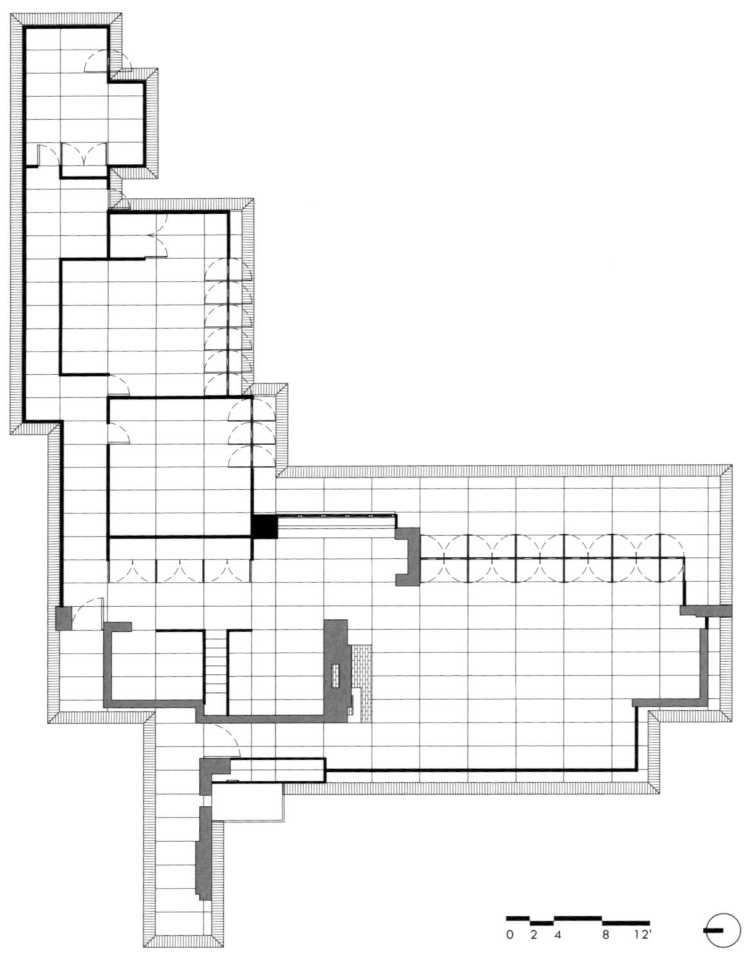

Jacobs House
Plan

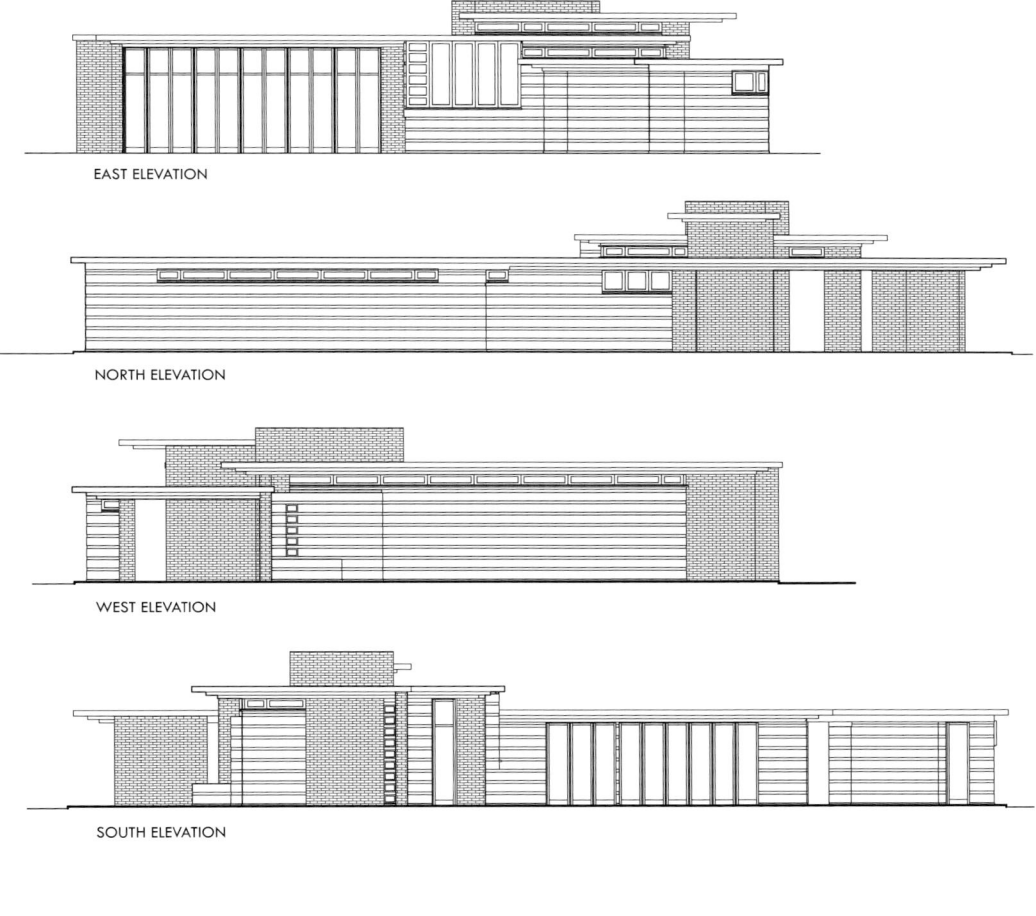

Jacobs House
Elevations

chitect thinking in non-isolated ways about the architectural capacities of even the most rudimentary and seemingly known materials.

Another astute chronicler of modern architecture, Ed Ford, is even more suspicious, if not outright dismissive, of Wright's approach to the Jacobs House construction and thermal behavior. "By the current standards of performance," Ford writes, "the Usonian system has some obvious shortcomings."[50] Ford first cites the lack of insulation in the wall and slab as one major shortcoming.[51] Likewise, the radiant heat system and the slab on grade are both precarious liabilities for Ford. In the case of the Jacobs House, the lack of slab insulation is more of a problem than the wall, in Ford's estimation, especially when considered in tandem with the radiant system. Structurally, the roof was a confounding assembly of 2 x 4s stacked edge to edge with no mechanical connection between them. These are all valid observations in terms of structural and normative insulation practices.

But where these astute observers of modern architecture see the Jacobs thermal design as deficient, I think Wright was clearly attending to a vastly different set of principles than those embedded in the insulation apparatus alone. If submitted to the terms of the insulation apparatus, the Jacobs House will fail. Yet this bias is a function of how one looks at a building and not a function or property of the design itself. The perceived deficiencies of the house are more of a consequence of deficient terms and criteria used to discuss them, which neglect how the house might actually perform thermally. While I will not claim that Wright fulfilled the intuitions I see as central in his conception of this early thermal experiment, I likewise cannot claim that the literature about the thermal design of the Jacobs House has observed the house on its own terms, or its actual thermodynamics. Other perspectives are necessary to grasp the flux of heat in this small, experimental house.

Functionally & Symbolically Indispensible

More focused on heat than the conventions of construction, Luis Fernández-Galiano sees more of the actual thermodynamics present in the Jacobs House. He describes Wright's attempt with the "gravity heat" thermally active surface slab as an effort to make the thermal milieu of the house less present visually through the elimination of radiators, and more apparent physiologically as part of a modern agenda for heat in architecture. The original radiant system involved steam heat in a few 2 inch pipes looping each wing of the house.[52] The system was later switched to a hydronic system as the fuel source changed to coal from oil.

A Material History of Insulation in Modernity 177

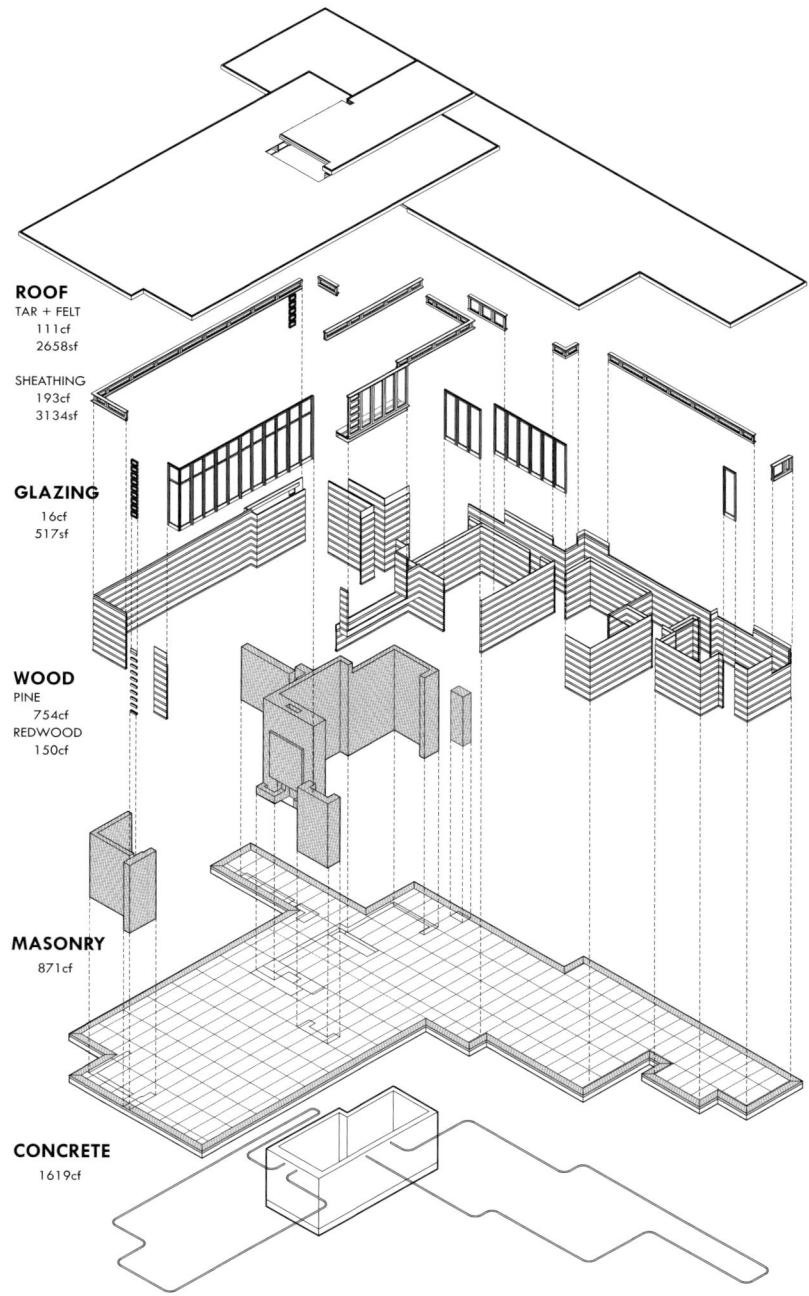

ROOF
TAR + FELT
111cf
2658sf

SHEATHING
193cf
3134sf

GLAZING
16cf
517sf

WOOD
PINE
754cf
REDWOOD
150cf

MASONRY
871cf

CONCRETE
1619cf

Jacobs House
Elevations

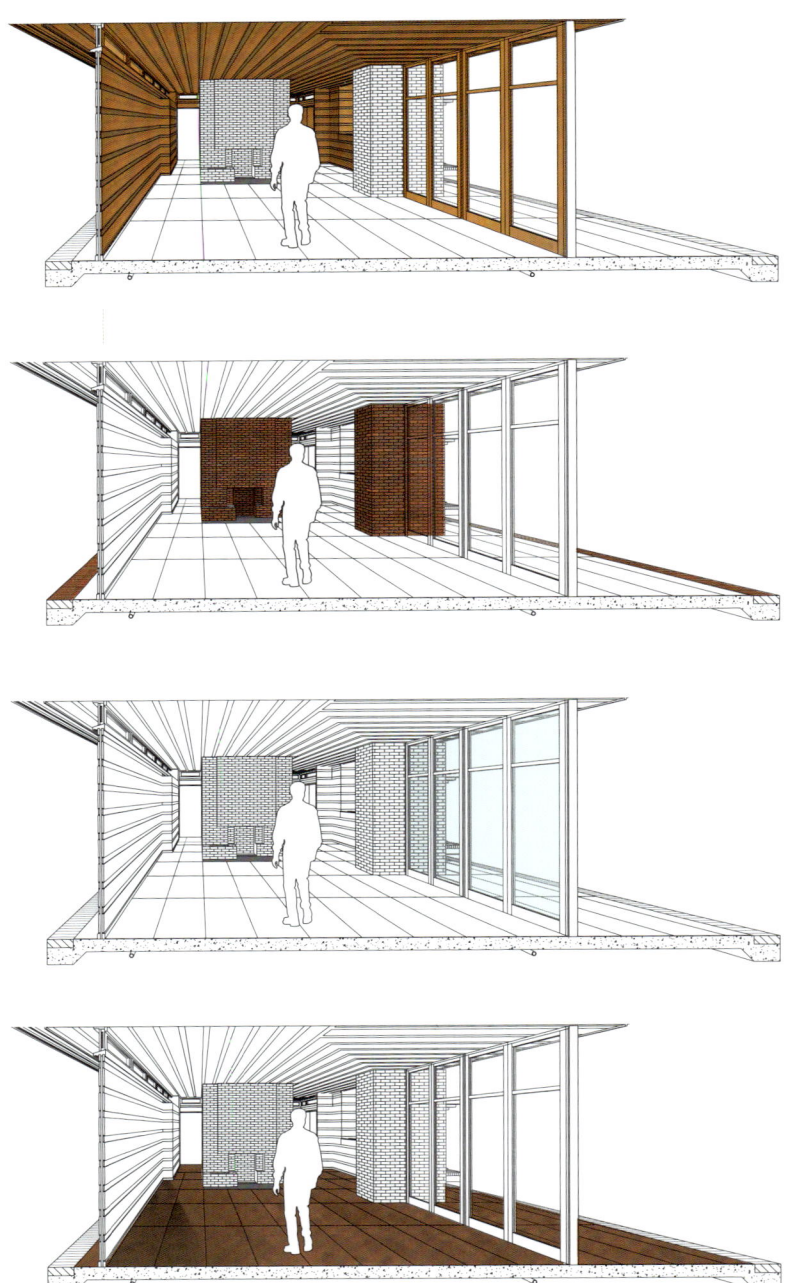

Jacobs House
Exploded material systems and thermodynamic figuration

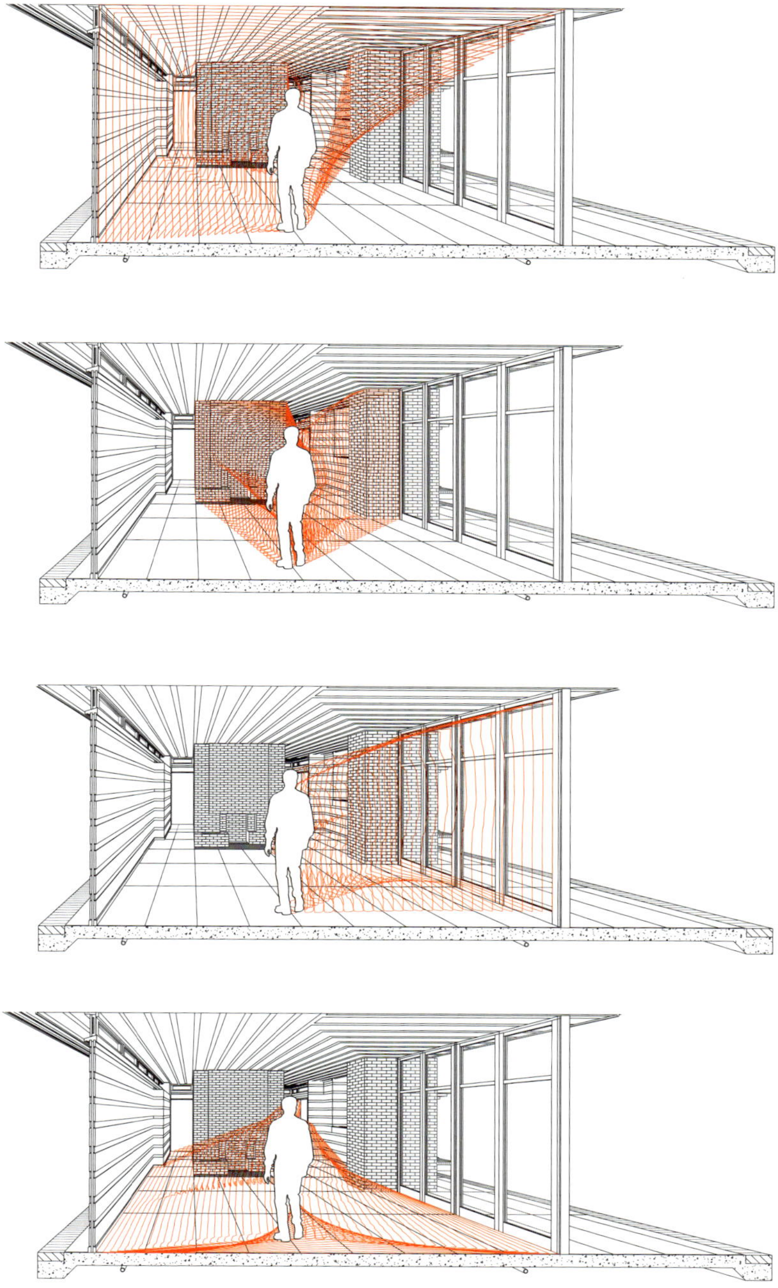

A Material History of Insulation in Modernity

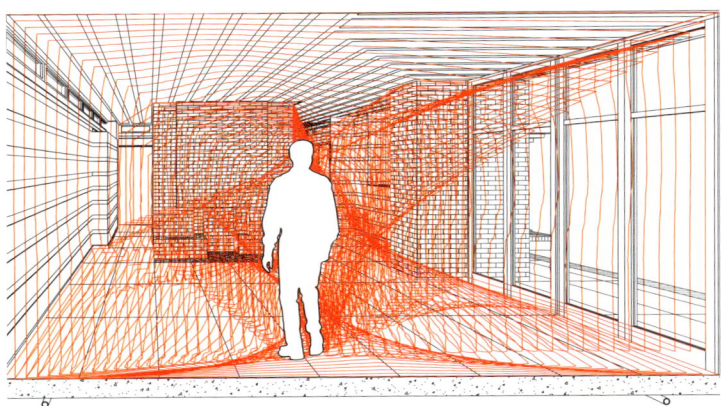
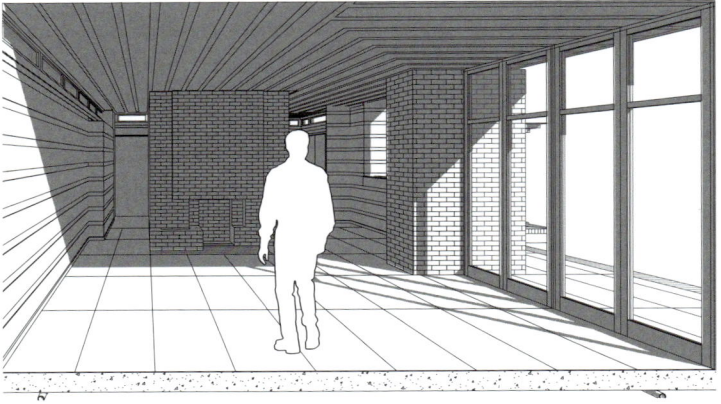

Jacobs House
Thermodynamic figuration for Herbert Jacobs

Fernández-Galiano saw the telluric fireplace, however, as "as functionally redundant as it was symbolically indispensible."[53] For Fernández-Galiano, the fireplace is an architectural compensation, decidedly not an amplification of the building's thermal agenda. This view underestimates the role of the mass, which appears almost as backwards as the interior photograph in Fernández-Galiano's account of the building.[54] For any human body in the living room, the mass of the fireplace is thermally active in both summer and winter, in both its power-operated and non-power-operated modes. Far from redundant, the mass of the floor, fireplace, and wall, when coupled with sun, fire, and human bodies, shape a novel perspective on the thermal milieu of a house: an architectural agenda for heat. This comports with Wright's exposure to the Ondol variants he experienced in Japan.

When the solid wall, the radiant slab, and the fireplace are combined as a convergent strategy, it is clear Wright had very specific and worthwhile ideas about heat transfer and, perhaps, intuitions about specific heat in mind. Wright likely had little more than obscure thermal experiences and intuition — coupled with characteristic hubris — to guide his design and specification. Thermal experiences in Japan and the Desert Southwest certainly expanded his thermodynamic imagination for this seemingly most familiar of architectural materials.

While minor and under-developed in the North American context, the Jacobs House represents an architect thinking much more deeply about heat flux than the insulation apparatus would typically engender or what a house a little more than eight feet tall might demand structurally. Wright's intuitions about how radiant heat and the modest but non-trivial heat capacity of the solid wood walls might engage with a human occupant are worth considering. It was, in fact, not the wood wall that caused discomfort for the owner, but downdraft from the clerestory windows.[55] But even this draft left Herbert Jacobs only mildly discontent, as he noted: "I should point out that sweaters and a roaring fire in very cold weather were a tradeoff that we gladly paid in exchange for the year-round delight to the eye of the glass walls of [the] living and bedroom and for the lower cost of a small heating system."[56] Indeed, Jacobs had cogent insights on the thermal milieu of his residence. For example, in response to expertise advice, Jacobs — a journalist — had this to say about the fireplace:

> I am aware that engineers gravely assert that fireplaces, unless provided with special heat exchange devices, are inefficient to the point of drawing in so much cold air that they neutralize the effect of the heat they give out. I do not believe, however, that they pay sufficient attention to the radiation effect of a fireplace. People sitting in front of one

are cheered by the sight of the blaze, and directly warmed by the heat waves radiated from the fire. If their backs happen to be nestled in a large chair, they are quite comfortable.[57]

Wright's ambition to merge such intuitions with the fundamental structural strategy of the floor and walls demands studied consideration as well, not as an artifact to repeat, but more as a habit of mind to emulate. Wright's non-isolated conception of the fundamental structural and thermal demands of a modest house was as rare in modern architecture as the systems he designed. Jacobs himself made observations about the efficacy of the approach under a range of conditions:

The massive fireplace, the brick piers, and the cement floor combined also to create a 'thermal mass' which helped to equalize temperatures, especially in the hot, humid days of a Midwest summer. They retained enough coolness from the night aided by air currents from the many windows, to keep the house from being stifling hot.[58]

Wright or Wrong?

While Wright did not have the means to execute or quantify his intuitions, his experience with radiant transfer from the hydronic system and solar gains, as well as the diffusivity and effusivity of wood, certainly seems to have motivated such an unusual departure from the norms of construction. Given Wright's investment in alternate modes of heat transfer, it is difficult, if not reductive and depreciative, to see his solid wood walls as ignorant deviations from the history of twentieth century wood cavity construction. Given the configuration and specification of materials, the orientation, location, and diurnal specification of program in respect of solar cycles and the thermally active role of slab and fireplace, the walls play more of a role than isolated observations about a lack of insulation suggest.

Today, with the burgeoning role of cross-laminated timber structures, for instance, and a greater understanding of what even a simple piece of wood can do thermally, Wright's approach does not seem as bizarre or deficient as it might otherwise be characterized. The deficiencies of the Jacobs House appear only in regard to the insulation apparatus, not in terms of an architectural agenda for heat transfer or thermal comfort. Indeed, Wright's thermal imagination, even if not technically perfected, points to an enlarged scope and capacity for the architect and the building in the context of heat transfer. It reflects architectural resistance to reductive platitudes about thermal resistance. Discussing this building's merits only relative to R-values is more insufficient than the building's thermal performance. Rather than pandering to professionalized need in the Jacobs House, Wright refreshingly considers architectural alternatives to the prescriptions of industries through both architectural and thermodynamic speculation.

With the Jacobs House, Wright certainly did not perfect certain intuitions about heat transfer in the context of housing. However, the architectural experimentation — as evident in subsequent instantiations of the solid wood wall and radiant system — points to an architectural line of thought and practice of heat flux deftly converged with construction through design. Wright's approach is as architecturally specific as it is compelling in its thermodynamic intuition. In this regard, in no way is the Jacobs House deficient in the history of architects' experimenting with alternate modes of heat modulation through the configuration of matter. Quite the contrary: he is one of the few modern architects stubborn — or recklessly fearless — enough to determine an architectural agenda for heat flux not shaped by the professionalized need and products imposed by adjacent industries. So it is not that Wright was wrong, it is just that he was not wrong in the way the insulation apparatus would portray it. Likewise, Wright did not necessarily get it right, but the Jacobs House does evidence certain non-modern habits of mind that are the right start to an architectural agenda for energy in all its dissipations.

Integrated Exergy & Emergy Analysis of the Jacobs House

Once considered beyond a reductionist perspective about insulation quantities, the exergy and emergy analysis of the Jacobs House has other lessons to teach. For example, the actual construction ecology of the project is interesting to consider. The etymological root of both ecology and economy — *Oikos*, Greek for house — is very poignant in the case of the Jacobs House. For reasons that have

Jacobs House Materials	volume (ft³)	volume (m³)	density (kg/m³)	total weight (kg)	transformity (sej/unit)	emergy	reference
Masonry	1,221.17	34.58	1,920.00	66,392.99	4.8×10^{12}	3.18686E+17	Brown and Buranakarn, 2003
Concrete	1,303.11	36.90	2,370.00	87,452.92	1.81×10^{12}	1.5829E+17	Simoncini, 2006
Pine (includes sheathing)	947.10	26.82	400.00	10,727.55	2.4×10^{12}	2.57461E+16	Odum, 1996
Redwood	149.90	4.24	450.00	1,910.11	2.4×10^{12}	4.58427E+15	Odum, 1996
Glass	15.65	0.44	2,580.00	1,143.35	4.74×10^{12}	5.41948E+15	Haukoos, 1995
					total	5.12726E+17	

equal thermodynamic and fiscal outcomes, Wright specified a cunning set of materials and, by extension, material geographies for the Jacobs House.

By orders of decreasing volume, the pine wood in the project would have come from northern Wisconsin, ca. 200 miles from Madison. The masonry units for the hearth were culled from piles of discarded units that did not meet the specification for Wright's SC Johnson and Son Administration Building, then under construction in Racine, Wisconsin, 100 miles from Madison.[59] This cycling of the Racine project "waste" to Madison "use" was obviously an issue of economy but has meritorious ecological efficacy as well. Likewise, the plate glass for the house was re-purposed from department store storefronts in Madison, cut to fit new frames.[60] The more perdurable redwood attachment battens would have come from northern California, some 2,200 miles from Madison. The extended durability, and thus work, of the redwood attachment batten and fascias — especially given its relative quantity in the house — presents a perhaps worthwhile exergy density to emergy density ratio. The redwood is a good example that an emergy perspective does not always mandate local materials but rather more powerful exergy and emergy ratios.

The combined, interrelated behavior of the slab, masonry, and wood walls must also be understood together to grasp the efficacy of the strategy as conceived. For most of the year, the solid wood structure is an adequate strategy for the building. On the coldest days, other systems such as the fireplace or Herbert Jacobs' sweater would be necessary.

When considered in composite, as a convergence of tectonic and thermodynamic facts, the Jacobs House appears in new ways. It is foremost an architectural experiment that deeply questioned fundamental assumptions of contemporary house design and construction. If the experiment were expanded in scope or otherwise perfected through further inquiry, then architecture would undoubtedly have benefited from the architectural agenda for energy that is evident in this seemingly aberrant house. The fact remains, however, that dissipation plays a non-trivial role in the behaviors of this house and to reduce its vitality and potential to the platitudes of more normative insulation expectations robs the discipline of vitality and potential.

Foams & Forms

In all cases, the material history of insulation is a history of cellular solids: the selection of materials with variable of mass-to-entrained air ratios. The class of materials known as cellular solids is indeed extremely interesting to plumb in the context of energetic dissipation. However, as Peter Sloterdijk has articulated at length, the co-isolated but communicating cells of a cellular solid material (wood, aerated concrete, or soapy bubbles) provide a profound model by which to characterize social relationships as well. As such, the dissipation of thermal flux in matter immediately brings up questions about the co-isolated but communicating cultural, social, and professional agents in the larger foamed apparatus of insulation. As Sloterdijk observed, "What is currently being confusedly proclaimed in all the media as the globalization of the world is, in morphological terms, the universalized war of foams."[61] The next chapter will consider another aspect of the foamed apparatus of insulation: the various roles pedagogy and physiology have played in the insulation of modernism.

Notes

1. Alfred J. Lotka, "Contribution to the Energetics of Evolution," *Proceedings of the National Academy of Sciences*, vol. **8 (6)**, June **1922**. p. **148**.
2. Ulrich Beck, *World at Risk*, London: Polity Press, **2009**.
3. Richard Wrangham, *Catching Fire: How Cooking Made us Human*, New York: Basic Books, **2009**. p. **57**
4. Leslie C. Aiello and Peter Wheeler, "The Expensive-Tissue Hypothesis: The Brain and the Digestive System in Human and Primate Evolution," *Current Anthropology*, vol. **36 (2)**, Apr. **1995**. pp. **199–221**
5. Marcus Vitruvius Pollio, *Vitruvius: The Ten Books on Architecture*, trans. Morris Hicky Morgan, New York: Dover Publications, **1960**. p. **39**
6. Adolf Loos, "The Principle of Cladding," *Neue Freie Presse*, **4** Sep. **1898**.
7. Ibid.
8. Manuel De Landa, *A Thousand Years of Nonlinear History*, New York: Zone Books, **1997**. pp. **26–27**.
9. Ibid. p. **27**.
10. Bruno Latour, *We Have Never Been Modern*, Cambridge, MA: Harvard University Press, **1993**.
11. Adrian Bejan, "Constructal-theory network of conducting paths for cooling a heat generating volume," *International Journal of Heat Mass Transfer*, vol. **40 (4)**, **1997**. pp. **799–816**.
12. Torben Dahl and Ola Wedebrunn, "Thermal Strategies: Towards a Modern Insulation," in Jos Tomlow, ed., *Climate and Building Physics in the Modern Movement: Proceedings of the 9th International DOCOMOMO Technology Seminar*, Paris: Docomomo International, **2006**. p. **72**.
13. Pliny the Elder, *Natural History*, Book XVI.
14. Dávid Bozsaky, "The historical development of thermal insulation materials," *Architecture: Periodica Polytechnica*, vol. **41 (2)**, **2010**. p. **50**
15. Margit Pfundstein, *Insulating Materials: Principles, Materials, Applications*, Basel: Birkhäuser Edition Detail, **2008**. p. **47**.

16 Vicenzo Scamozzi, *The Idea of a Universal Architecture III: Villas and Country Estates*, trans. Henk Scheepmaker, Amsterdam: Architectura & Natura Press, 2003. p. 262.
17 Sandy Wyllie-Echeverria and Paul Alan Cox, "The Seagrass (*Zostera Marina* [Zosteraceae]) Industry of Nova Scotia (1907–1960)," *Economic Botany*, 53 (4), 1999. p. 420.
18 "Cabot's Quilt," Advertisement, *The American Architect*, 12 March 1913. p. 7.
19 "Build Warm Houses," Samuel Cabot Inc., marketing brochure, 1928.
20 "Walls Insulated with Lining of Seaweed," *Popular Science Monthly*, September 1929. p. 67.
21 Wyllie-Echeverria and Cox, p. 422.
22 Ibid. p. 420.
23 Ibid. p. 423.
24 C. E. Addy and David A. Aylward, "Status of Eelgrass in Massachusetts during 1943," *The Journal of Wildlife Management*, vol. 8 (4), Oct. 1944. pp. 269–275
25 Charles E. Renn, "The Wasting Disease of Zostera marina. I. A Phytological Investigation of the Diseased Plant," *Biological Bulletin*, vol. 70 (1), Feb. 1936. pp. 148–158
26 Gilles Deleuze and Félix Guattari, *A Thousand Plateaus: Capitalism and Schizophrenia*, Minneapolis: University of Minnesota Press, 1987. pp. 474–500.
27 Bozsaky. p. 51
28 Charles George Ramsey, Harold Reeve Sleeper, *Architectural Graphic Standards for Architects, Engineers, Decorators, Builders, and Draftsmen*, New York: John Wiley & Sons, Inc.; London: Chapman & Hall, Ltd., 1932.
29 Charles George Ramsey, Harold Reeve Sleeper, *Architectural Graphic Standards for Architects, Engineers, Decorators, Builders, and Draftsmen*, second edition, New York: John Wiley & Sons, Inc.; London: Chapman & Hall, Ltd., 1936.
30 Jane Murphy, "Insulation and Modernism in America: 1900–1955," in *Proceedings of the 2007 95th ACSA Annual Meeting in Philadelphia, PA*, Washington, DC: ACSA Publications, 2007.
31 Ruth Schwartz Cowen, "How the Refrigerator got its Hum," in Donald MacKenzie and Judy Wacjman, eds., *The Social Shaping of Technology*, Philadelphia: Open University Press, 1985. pp. 202–218.
32 Kiel Moe, "Automation Takes Command: The Nonstandard, Un-Automatic History of Standardization and Automation of Fabrication in Architecture," in Rob Corser, ed., *Fabricating Architecture: Selected Readings in Digital Design and Manufacturing*, New York: Princeton Architectural Press, 2010.
33 Daniel M. Abramson, "Obsolescence: Notes Towards a History," *Praxis: Journal of Writing + Building* 5, 2003. pp. 106–112.
34 J. H. Bracken, "The Storage of Ice," *Ice & Refrigeration*, vol. 38 (4), April 1910. p. 211.
35 Malcom Ross and Robert P. Nolan, "History of asbestos discovery and use and asbestos-related disease in context with the occurrence of asbestos within ophiolite complexes," *Geological Society of America*, Special paper 373, 2003. pp. 447–470.

36 Ibid. p. **449**.
37 Ibid. p. **449**.
38 "Australian Asbestos Houses," *Popular Mechanics*, Oct.**1909**. p. **445**.
39 Bozsaky. p. **51**
40 Ibid. p. **51**.
41 D. Michelle Addington, "Good-Bye Willis Carrier," in *Proceedings of the 85th ACSA Annual Meeting and Technology Conference*, Washington, DC: ACSA Publications, **1997**. p. **88**.
42 D. Michelle Addington, "The History and Future of Ventilation," in Jack D. Spengler, Jonathan M. Samet, and John F. McCarthy, eds., *Indoor Air Quality Handbook*, New York: McGraw-Hill, **2001**. pp. **2.1–2.16**.
43 Addington, "The History and Future of Ventilation." pp. **2.7–2.9**.
44 Reyner Banham, *The Architecture of the Well-tempered Environment*, London: The Architectural Press, **1969**. p. **19**.
45 Herbert Jacobs, *Building with Frank Lloyd Wright: an Illustrated Memoir*, San Francisco: Chronicle Books, **1979**. pp. **33–35**.
46 Michael Caldwell, *Strange Details*, Cambridge, MA: The MIT Press, **2007**. pp. **60–62**.
47 Ibid. p. **62**.
48 Ibid. p. **62**.
49 Ibid. p. **56**.
50 Edward R. Ford, *Details of Modern Architecture*, Cambridge, MA: The MIT Press, **1990**. p. **337**.
51 Ibid. pp. **337–338**.
52 Jacobs. pp. **59–60**.
53 Luis Fernández-Galiano, *Fire and Memory: On Architecture and Energy*, trans. Gina Cariño, Cambridge, MA: The MIT Press, **2000**. p. **251**.
54 Ibid. p. **252**. The photograph of the Jacobs hearth is mirrored as published.
55 Jacobs. p. **54**.
56 Ibid. p. **54**.
57 Ibid. p. **54**.
58 Ibid. p. **54**.
59 Ibid. p. **54**.
60 Herbert Austin Jacobs, *Franks Lloyd Wright: America's Greatest Architect*, New York: Harcourt, Brace & World, Inc., **1965**. p. **133**.
61 Peter Sloterdijk, *Bubbles: Spheres Volume 1: Microspherology*, Los Angeles: Semiotext(e), **2011**. p. **71**.

"*Every concept, every institution, every practice that interferes with the continuous deployment of collectives and their experimentation with hybrids will be deemed dangerous, harmful, and — we may as well say it — immoral.*"[1] **Bruno Latour**

Physiology, Insulation, Climate, and Pedagogy

Architects are ostensibly concerned with heat transfer because they are concerned with human comfort as much as resource considerations (and not merely with the satisfaction of technocratic tautologies and energy certification checklists.) What, then, about the human body in the dynamics of heat and human comfort? The specificity and dynamics of bodies and other larger collectives appear as curiously, but not surprisingly, isolated from considerations of heat flux in the development of insulation theories and practices in modernity. Despite pretensions about human occupancy and comfort, architecture generally suffers little direct engagement with the specificity of human physiology. The exceptions to this observation, while few in number, are rich examples. Relatedly, while the incorporation of insulation yielded slightly more sensitivity to issues of climate, the role of insulation as an operative agent in the body, building, and climate collective of energy in architecture was not always as evident either, especially in its pedagogical manifestations.

The curiously sporadic appearance of human physiology and climate in the isolated abstractions of heat in the development of insulation is the focus of this chapter. The apparatus of insulation — and its co-determinant air-conditioning apparatus — developed more universally applicable heat transfer theories first and later asked questions about human comfort and physiological response. The early twentieth century transfer of insulation practices from steam pipes and cold storage rooms to buildings for human occupation did not adequately account for some of the more nuanced physiological responses and potential forms of feedback possible when the body is more centrally positioned in the composition of thermal architecture.

The human body can be seen either as introducing complexity to analytical approaches to heat flux (yet another set of intricate, transient parameters!), or as a starting point of specific pathways between heat flux and human comfort. In the steady-state conductivity approach, the body does introduce complexity to the matter of heat transfer. However, this complexity is in fact a source of great architectural and ecological power. If the human body is an incredibly nuanced sensor, and often controller, of the interior milieu, particular forms of thermal reception and emission could be much more fundamental to the heat flux dynamics as considered by architects.

The physiological apparatus of the human body might better serve as the starting point of an architectural agenda for energy and heat flux than the historical contingencies of conductivity and refrigeration research, as the first chapter discussed. Such an approach would more fundamentally account for factors such as the human body's bias towards radiant transfer or the role of thermal effusivity, for instance.[2] Conductivity would remain an important concern, but would be more productive in consideration of thermal architectures that invite greater physiological interaction with buildings: the architectures, cultural traditions, and sources of delight that emerge from bare feet, radiant symmetries, or the thermal aedicule of the Japanese Kotatsu are a few examples. Not only would thermal diffusion and thermal effusion be important, but the diffusion and effusion of humidity would be significant as well. If these concerns are taken as the center of an agenda for energy in architecture, there would be substantive implications for the design of heat flux in buildings. What is at stake is a more hybrid, composite characterization of the thermodynamic milieu of bodies and how buildings might be designed accordingly, rather than conditioning bodies to fit the machines and designs of modern building science.

This chapter considers some of these emblematic, overlooked physiological premises of human thermal comfort — not as a response to imposed systems, but as the start of an agenda for thermal architectures. I will take two approaches to this topic in this chapter. The first considers the role — or lack thereof — of physiology in twentieth century architectural practices and pedagogies regarding thermal energy. How content related to energy, systems, and ostensibly human comfort was introduced to architecture students shaped systemic aspects of contemporary architecture and the way we think about contemporary notions of performance. This narrative largely bears on the degree to which reductive practices dominated pedagogy in this area and the degree to which more curious designers and instructors discerned alternate paradigms for energy in architecture. Related, this more architectural approach focused on the agency of climate as part of a more cogent exergy concern for buildings.

This first approach helps identify a series of "unsaid" aspects of the body-building-climate thermal milieu that is the focus of the second approach. This latter, smaller portion of the chapter explicates some key physiological and material properties that can augment our understanding of insulation — or, more specifically, its alternatives.

Archaic & Ancient Pedagogies

Hippocrates documented the earliest known lessons on the relationship between physiology and climate, ca. 400 B.C.E., in *Upon Air, Water, and Situation*.[3] For Hippocrates, the character and proclivities of people were largely tied to their climate. From his perspective, physical conditions of health and disease were linked to place as much as the personal behaviors of individuals. Thus, as considered in the first third of his treatise, the orientation of a city with respect to various winds determined much about the character of its people.

Our earliest lessons on the relationship between physiology and the thermal milieu of buildings come from Vitruvius. In his description of "The Education of the Architect," Vitruvius remarks that among many necessary skills, the architect should "have some knowledge of medicine."[4] He soon expands on this assertion, stating that the architect must account for "the questions of climates, air, the healthiness and unhealthiness of sites, and the use of different waters."[5] Far more than just a basis for geometric mimesis, the human body, for Vitruvius, helped organize everything from material selection to (like Hippocrates) city planning and the well-being of the state.

Modern Physiology

While there are numerous medieval sources on surgery and anatomy, equivalent sources on the relationship between bodies and buildings or bodies and climate are rare in the later ancient and medieval period. In contrast to ancient and medieval contexts, physiologists Charles Edward Amory Winslow and Lovic Pierce Herrington cite 1628 as the origin of modern, scientific physiology.[6] In that year William Harvey published a Latin text on the circulation of blood.[7] While a contemporaneous architect, such as Vicenzo Scamozzi, incorporated the latest scientific knowledge in both his treatise and his buildings, this applied knowledge remained focused on the building and its performances. Aspects of his scientific interests — optics, cold storage — were of benefit to bodies but were documented as building-related phenomena. It would not be until the nineteenth century that the subject became more directly the focus of architectural consideration and publication.[8]

Nineteenth Century Buildings & Comfort

The earliest modern account (1825) related to comfort, physiology, and buildings is Thomas Tredgold's inclusively titled *Principles of warming and ventilat-*

ing public buildings, dwelling houses, manufactories, hospitals, hot-houses, conservatories, &c.; and of constructing fire-places, boilers, steam apparatus, grates, and drying rooms.[9] Tredgold's subject is admittedly broad, but his physiological and architectural observations were often both specific and accurate, given their subject's nascent status. For instance, Tredgold made the observation that radiant heat can keep people warm even if the air temperature is low.[10] Related in scope, Charles Tomlinson's *A rudimentary treatise on warming and ventilation; being a concise exposition of the general principles of the art of warming and ventilating domestic and public buildings, mines, lighthouses, ships, etc.* documents a similar set of observations. Notably, this treatise begins with a range of physiological observations before elaborating on various methods of warming the various typologies mentioned in the title.

Physiological responses to air movement were a central research topic for several scientists and physicians in the first half of the nineteenth century: John Arbuthnot, David Boswell Reid, and Nicholas Gauger are three examples. Arbuthnot articulated the cooling effects of air movement over the surface of perspiring skin.[11] Reid was concerned with temperature but he emphasized air movement as well.[12] As Thomas Bedford observed, Gauger was convinced that it was not room temperature that determined discomfort but rather an "inequality of temperature and want of ventilation that caused numerous maladies."[13] If nothing else, these early studies helped identify how intricate the relationship of physiology, temperature, and comfort could be. This physiologically-driven concern eventually led to the development of a range of measurement devices that began to account simultaneously for temperature, air movement, and radiant transfer.

Measurement & Sensation

With respect to design, physiological comfort requires the difficult step of somehow quantifying the vagaries of human experience into a quantified system of knowledge, with all the attendant problems of abstraction involved. Despite the development of thermal measurement devices — such as those described in the first chapter — physiology research quickly revealed numerous discrepancies between the sensitivity of the body and the measurement of the instrument. For example, in 1826 William Heberden found the measurement of temperature of heat an imperfect measure of what he described as "*sensible cold*...the degree of cold perceptible to the human body in its ordinary exposure to the atmosphere."[14] As a response, the balance of his research was focused on a thermometer that better characterized thermal and air movements as a human body might experience its milieu.

Building on these nineteenth century observations, some novel instruments that helped refine heat transfer knowledge emerged in the early twentieth century. The kata thermometer was devised in 1914 to measure the effects of air movement on human comfort.[15] The kata thermometer was a large bulb thermometer with alcohol inside. The fluid was heated above 100°F and the time required for the internal temperature to drop to 95°F was measured as an index of air cooling effects. The dry version provided both radiant and convective cooling readings while the wet version added evaporative cooling effects.[16]

In the 1930s, globe thermometers appeared for the measurement of mean radiant temperatures. Building on these instruments, A. F. Dufton developed the eupatheoscope at the British Research Station in 1932.[17] This instrument was initially called the eupatheostat.[18] It provided for the concept of an "equivalent temperature."[19] This measure merged air temperature, air movement, and radiant transfer as combined factors, a better indicator of human comfort in a space. Other instruments were devised for measuring the emissivity of various materials.[20]

This progression of physiologically-related thermal measurement devices was coupled with a progression of charts and various scales of warmth. From the measure of air temperature, scales of warmth began to acknowledge the role of air movement, humidity, and radiant transfer as a function of surface temperature. Concepts of effective temperature, equivalent warmth, and corrected effective temperature reflect a progressively refined measure of environmental conditions, focused on physiological sensitivity.

Building scientist Thomas Bedford was acutely aware of the role of physiology in issues of heat transfer in buildings. His 1948 book, *Basic Principles of Ventilation and Heating*, oscillates between the exosomatic functions and equipment related to thermal comfort and the endosomatic processes of human comfort. For example, the structure of his chapters bounces from measurement to perceived scales of warmth. Before introducing any mechanical equipment, Bedford emphasizes the role of the following topics: "Warmth and Comfort," "Sensations of Warmth and Cold," "Stimulating and Pleasant Environments," and the "Quality of Air."[21] Published at the beginning of the post-war period, this was the most robust account of physiology related to buildings at the time and established a robust basis for an architectural agenda for architecture.

Physiology & Early Modernist Architecture

These basic observations from physiologists and certain building scientists rarely impacted architecture, however. For some architects, like Le Corbusier, the role of physiology was more thematic and rhetorical. For others, such as Frank Lloyd Wright, physiological responses to building material and energy systems directly shaped their work but most often in a characteristically intuitive, if not rogue, manner.

I include in the following paragraphs a couple of other notable examples. To be clear, these are the exceptions to an otherwise physiologically and climatically impoverished set of normative practices. The disjunction between these novel practices and the contemporaneous architectural pedagogies is significant. As a first example, physiological processes, design, and construction were inextricably bound in the work of Hannes Meyer.

Meyer's Biology

In his 1928 thesis on "Building," Hannes Meyer made several claims related to physiology, heat transfer, and design. In unambiguous terms, he observed, "building is a biological process. Building is not an aesthetic process. In its design the new dwelling becomes not only a 'machine for living', but also a biological apparatus serving the needs of body and mind."[22] As such, everything from one's sex life to heating to sun exposure to cooking served as motivations for design. Amongst the metabolic flux of his biological perspective on design he stated that architects must "calculate the heat loss of the floor," "the shadow cast by the house on the garden and the amount of sun admitted by the window to the bedroom," relate "the heat conductivity of the outside walls with the humidity of the air outside the house," and finally, "consider the body of the house to be an accumulator of the sun's warmth."[23]

In design terms, his interest in the quantitative explication of physiological and environmental phenomena is easy to characterize as reductive and deterministic.[24] However, the prevailing modernist enthusiasm for managerial quantification does not fully explain his persistent interest between biological processes and formation, not as a metaphorical or analogical paradigm, but in all the literal implications his writing articulates. While his own design methods were deterministic, his motivations and writing do not fully seat into a deterministic discourse but rather into something more vital, showing that physiology co-motivates his work in a way that few of his contemporaries evince.

As with the RFG experiments mentioned in the first chapter, Hannes Meyer was similarly invested in a program of research, analysis, and experimentation that included physiology as constitutive of design. His short text on "Building" suggests that physiology and its relationship to a built milieu was of fundamental importance to Meyer. "Building," Meyer claimed, "is the deliberate organization of the processes of life."[25] The political and war-related disruptions of the 1930s and 1940s in Germany and Europe at large prematurely concluded similar lines of thinking and pedagogy. Similar motivations persisted, though, in the more politically and climatically stable context of Southern California.

Neutra & Physiology

The psychologist Wilhelm Maximilian Wundt's observations on physiology motivated multiple aspects of Richard Neutra's practices and formed a sustained preoccupation highly legible throughout his career.[26] Neutra was preoccupied with how physiology might inform and co-determine aspects of design.[27] As Neutra claims very early in *Survival through Design*, "Physiology must direct and check the technical advance in constructed environment."[28] The relationship of body, mind, and environment was the focus of his concern, and physiology would be the guiding agent. Towards the end of his career, he observed "human biology, human brain dynamics, and glandulor fluctuations of biochemistry have for many years been my study."[29]

As is well documented by Neutra and his observers, Neutra's architecture was shaped around a range of physiological concerns that focused on the energies that act on the nervous system of the body. Given Wundt's influence, Neutra's work has been reduced to various ocular preoccupations or psychoanalysis, but a broad interest in a range of physiological factors is legible in both his writing and built work.

Neutra's position about physiology and progress, if not evident through the title of his book, *Survival through Design*, is made more clear in the prologue to his fourth chapter: "from a baby carriage to a metropolis, our man-made surroundings, top-heavy with technological trickery, have become our model of destiny — and a source of never-ending nervous strain."[30] *Survival through Design* is by far the most thorough and impassioned call to action about the role physiology ought to have in architecture and its pedagogies. As Neutra observes,

> **Future instruction in environmental design and architectural training will instill detailed awareness of the basic physiological actuality that the human nervous apparatus is continuously stimulated through a large number of sense receptors.**[31]

Despite his passions, this type of instruction has not yet occurred in architectural education.

A later text by Neutra offers insight about physiology, climate, and design that deserves attention. Neutra's position on physiology was that it was far from a stable entity, and very closely aligned with evolution, as is clear when he notes, "Mutations mean evolutionary potential, the change for advance which nature herself for eons has capitalized on."[32] He continues: "without deviation from a 'standard' which is anyway never an organic concept, without 'mutation', evolution will freeze and end."[33] For Neutra, design was the central agent of this productive mutation. As pressing then as it is now, the mutation of the energy-physiology discourse, pedagogy, and practice in architecture desperately needs exactly such a mutation in this century.

From Practice to Pedagogy

While examples of physiological motivations for design can be found in idiosyncratic cases such as Meyer and Neutra, the examples are again the exception to the rule that modern architectural training included little work on physiology, to say nothing of issues of energy in general. It is thus illuminating to look at what role physiology, as well as energy and climate, had in architectural pedagogies in the twentieth century, at the time that insulation became a more central agent in practice.

German institutions have the oldest pedagogies related to building physics. In the early twentieth century, Hermann Rietschel at the Königliche Technische Hochschule in Berlin and Friedrich Wilhelm Hermann Fischer at the Technische Hochschule in Hanover both taught topics related to building heating and ventilation.[34] However, even in this context of early pedagogies, in "the 1920s and 1930s few German books on building science written for architects existed," Jos Tomlow notes, "and none of them gave a complete overview of the discipline."[35] While Hannes Meyer's interests in physiology-related modes of heat transfer were part of the Bauhaus pedagogy in Dessau in the late 1920s, courses directly related to building physics would not emerge until much later.

In too many cases, the thermal milieu of architecture remained in construction engineering courses, with little physiological agency evident. As the number and complicatedness of systems expanded in the post-war years, there was increased pressure to incorporate these new demands into architectural pedagogy.

The demand was apparent.[36] How to best achieve this incorporation, however, was far less apparent. This quandary produced highly divergent approaches. The most prevalent pedagogical approach simply deferred to the means and methods of mechanical engineering. The less common but more compelling pedagogical approach evidenced a struggle to discern and define an architectural agenda for energy. To this end, there are many famous and lesser-known pedagogical exceptions in the form of ambitious architectural research laboratories and research stations.

One early method that privileged the architectural opportunities inherent to the energy systems of short and long energy cycles was that of Professor Walter Burkhardt of Alabama Polytechnic Institute (now Auburn University). Burkhardt was the local District Officer for the Historic American Building Survey. In this historic survey work, he was particularly interested in the climatic adaptations that emerged over the decades in central Alabama buildings. It is decidedly important that an architectural agenda for energy emerged out of observation of vernacular responses rather than from the technology transfer from refrigeration. As architect Joseph King has noted,

> Burkhardt's work documented such devices as adjustable shutter and awning systems that had been developed over many decades and in many different site-specific iterations to catch breezes, provide shade from the sun and allow for micro-adjustments of climate in interior spaces. Plan and spatial elements such as dogtrots and porches were also being documented and were used, in addition to building forms and construction materials, to mediate climate.[37]

Both short and long-term energy dynamics are fundamental to this type of pedagogy: the documentation of longer-term self-organized climatic adaptations to short-term climate fluxes. The impact of Burkhardt's pedagogy is most famously evident in the early work of Paul Rudolph, one of Burkhardt's students. In both Burkhardt's more vernacular interests and Rudolph's work in Sarasota, there is a clear bias towards architectural responses to climate rather than the impulse elsewhere to seal and isolate a building so that it can be pumped full of machinery, ducts, electrical illumination, and other professionalized needs.

The Fitch Pitch

This same connection between climate, vernacular response, and preservation as part of short and long-term energy agendas was fundamental to James Marston Fitch's work. Fitch worked as a meteorologist for the United States Army during World War II. After the war, Fitch published several articles on meteorology and climatology for architecture as well as the well-known 1948 book on *American Building: The Forces That Shape It*.[38] The initial publication beautifully conflates the historical process of evolving American building typologies alongside their energetic bases. How could such evolutionarily and thermodynamically imbricated topics be otherwise separated?

The ultimate reception of Fitch's work, though, was as divided as his 1948 book became once divided into separate volumes that counterproductively parsed environmental concerns from historical concerns.[39] Why isolate the long term from the short, the physiological from the cultural/historical? The hybrid collective of typology, climate, and energy evident in the first edition is lost in subsequent editions and led to the dismally divergent fields of building science and historic preservation. This is a persistent habit of mind of modernism: to isolate and segregate what is not segregated in reality, the conceit of an artificial purity and autonomy.

Fitch's most profound observation — even if so simple and direct as to hazard oversight — about energy in *American Building* resonates loud in the present book: that "the function of a building is to control the *rate* at which heat is lost."[40] As an echo of Boltzmann, architecture, like any structure, is a struggle for entropy. Like so many others who immerse themselves in the thermodynamic depth *and* history of buildings, Fitch learned that the velocity of energy is essential to the power efficacy of a system.

Like Burkhardt and Fitch, another keen observer of self-organized building responses to respective climates was Sibyl Moholy-Nagy. Her 1955 *Perspecta* article on "Environment and Anonymous Architecture" was the preparatory work for her subsequent book publication, *Native Genius in Anonymous Architecture*.[41] While by no means a technical assessment of her subject, her focus on climatic, formal, and material responses to varying spatial and temporal environments is closer to an architectural habit of mind — a pedagogy — for energy, heat, and human comfort than parallel technocratic agendas for the same subject. The mission of her observations is unequivocal in her book's opening sentence: "buildings are transmitters of life."[42] A few pages later she observes: "The architect of today has a hard time holding on to this mission."[43] In her view this is because "the academies are closed" and "the great unifying ideas of homogenous societies no longer supply a natural common denominator."[44] These observations that academies and academics are closed systems, locked only from inside the academies, are all the more evident when one looks at the more typical pedagogy of energy systems at the time.

"How Much Must an Architect Learn…"

The sensitive and very architectural accounts of climate, energy, and physiology above were not pervasive in this period, however, and were not fundamental aspects of architectural pedagogies. A deeply problematic, capitulating but typical architectural agenda is clearly manifest in a 1961 *Journal of Architectural Education* article on the topic: Louis Axelbank's article, "How Much Must an Architect Learn about Mechanical and Electrical Services, and Where?"[45] The article summarizes a series of pedagogical positions — from that of Max Abramowitz to Louis Kahn — about how to best incorporate the rapidly burgeoning demands of "mechanical and electrical services" in buildings. The focus of the summary is one of integration: finding space in both the architectural curriculum as well as in the building for the range of services required of post-war buildings. Given constrained time in the curriculum, Axelbank's acquiescence is palpable in his ultimate recommendation: "What is important, then, is to arrive at a consensus on how much an architect needs to know in order to work effectively with the mechanical and electrical engineers (without them, when necessary) and how much of that knowledge must be attained in school."[46]

The preposterous, dismal mid-century modern construal of the energy systems in architecture as a series of proprietary mechanical, electrical, and plumbing systems — MEP systems — had a short but numbing history in architecture; especially when compared to the fecund transformations and uses of energy in built ecologies during the thousands of years that preceded the capitulating turn to MEP systems. Coddled in progress ideologies and fantasies of environmental control, the aberrant rationality of these modernist systems transformed multiple aspects of architecture yet rarely reflected a thermodynamic advancement of built environments: MEP systems routinely did not make buildings as powerful as possible in either architectural or ecological terms.

The separation of structure from energy concerns was unfortunately already well established in the habits of mind of architectural pedagogies by this point, as evident in this article. One part of this article considers the relative roles and time spent on mechanical and structural engineering. Regarding this debate, Axelbank quotes a bright observation from Hugh Stubbins, Jr.: "The architect is not going to be a structural engineer. He doesn't have to know the details required of an engineer and he should have a more imaginative approach than most engineers have."[47] Nowhere in Axelbank's summary of mechanical and electrical services is there any evidence of corollary imagination about how architects might adopt anything but a capitulating role to the means and methods of the more developed disciplines when it comes to energy. To ensure that no architectural agenda shall emerge, Axelbank's final recommendation — followed by generations of faculty — is unequivocal about the architect's relationship to energy, for they should be "taught by one or more competent practicing engineers with years of experience in the mechanical and electrical services of buildings."[48] Max Abramovitz seconded this last observation: "It's best to have the engineers teaching engineering to architects."[49] *As if the topic of thermodynamic dissipation was one limited to engineering!*

The absolute paucity of intellectual engagement and imagination in this report on the state of energy, physiology, and climate pedagogy in the discipline is not surprising. If not just to confirm his complete capitulation, under the subheading of *"Less Important Areas,"* Axelbank lists "solar-heating, high-temperature hot-water distribution, radiant panels for both heating and cooling and luminous ceilings."[50] These topics — perhaps the topics most relevant to an architectural agenda for energy — are included only so as to be deemed less important.

While Axelbank's observations seem resonant with the other pedagogues cited in his article, one anonymous critic responded to the article. Both the critic's observations and Axelbank's rebuttal were included in the *Journal of Architectural Education* article. The anonymous critic questioned the acquiescent, mechanical/equipment orientation of Axelbank's observations and recommendations, as well as his assessment and recommendation:

> In his summary of what the course in engineering should contain, he exemplifies all the shortcomings of many engineers in their view of this field and their inability to understand what the architect needs to know. This view is entirely equipment-oriented, and does not ever mention the following: a. Reaction of human beings to energy levels in the environment — physiological needs. b. Interaction of buildings with climatic factors — outside temperature, wind, precipitation, humidity, cycling, illumination, insolation. c. Mechanical services as they influence architectural design in layout, details, types of construction.[51]

This strong statement constitutes the beginning of an architectural agenda for energy. To start, the anonymous critic employs the word energy, whereas Axelbank never mentions it. Axelbank's rebuttal is as dismissive as it is simple, claiming "there just isn't sufficient time in the curriculum" to address the critic's more comprehensive and architectural agenda for energy.

Research & Experimentation

Axelbank and his anonymous critic made their observations in 1961. Their respective views reflect a polarity of approaches: one more architecturally ambitious than the other. Even a decade before, more architecturally ambitious approaches could be found, not across North America, but in compelling pockets, such as the architectural research labs that emerged in 1950s.

One of the earliest was Gordon McCutchan and William W. Caudill's work at the Texas Engineering Research Station. This research station was connected to the Texas Agriculture and Mining University. The stated mission in one of their publications is worth including in full:

> The Texas Engineering Research Station, in June 1949, initiated a program in environmental engineering — correlated research in which the essential environmental factors are considered simultaneously, through an investigation of the effects which various architectural shapes have upon man's physical environment. This program is directed towards obtaining data from which a scientific approach to total environmental control of buildings can be developed. Logical architecture can result only through careful consideration of environmental factors such as light, air, and sound. Most information available on the subject, however, stems from isolated research, i.e., research in natural lighting without consideration of natural ventilation or sound conditioning. Accordingly, in today's architecture good lighting is often obtained at the expense of ventilation, adequate ventilation is often achieved by sacrificing sound conditioning and so on. In the Texas Engineering Experiment Station's program, the approach is experimental, and calls for the establishment of relationships between full-scale buildings and models for the purpose of predetermining natural lighting performance, natural ventilation performance, and sound conditionings.[52]

McCutchan and Caudill focused on models not just as objects of representation but to demonstrate luminous and air flow behavior as well. This focus on models as design aids and as the primary vehicle for helping establish an agenda for energy and its evaluation in architecture was developed further by Dean Hawkes in his 1970 publication, "A History of Models of the Environment in Buildings."[53] Beginning with environmental models and design aids as early as 1865, Hawkes established a long history of models in his sustained consideration of the environmental traditions in architecture.[54]

One aspect of McCutchan and Caudill's mission statement in particular resonated with many pedagogical assumptions in the second half of the twentieth century. The term "environmental control" is explicit and expressive about certain ambitions regarding the environment and the energy contained therein. Their emphasis on control is curious given the otherwise non-isolated premise of their work.

A command and control mentality was pervasive in post-war technological habits of mind, including architectural pedagogies. Take, for example, a common course name for MEP pedagogies in this period, "Environmental Control Systems." The name was initially borrowed from NASA's research on the hermetically sealed capsules required for space flight.[55] By the late 1960s, courses in most architecture schools taught the scientific fiction of environmental control. Yet whereas NASA was focused on the closed system life support of a capsule in space, hence the 1967 NASA document, "Technical History of the Environmental Control System for Project Mercury," architects were grounded in the non-isolated life support of terrestrial systems.[56]

The technological source and escalation of this borrowed name resonated deeply with the managerial posture of MEP systems. This name also revealed the narrowness of its focus: simplified system boundaries that could be readily quantified and "controlled." By positioning themselves in a discourse on control technique, MEP systems were thus positioned as management implements and architects as mere managers. The control of elusive environmental parameters and systems was no doubt a broad ambition in this period.

Despite their claim of environmental control, McCutchan and Caudill sought methods and models to deal with convergent forms of energy in the teaching/research program. This reflects an idiosyncratic strand of architectural thought in this period that extended earlier observations about physiology, buildings, and climate. Their focus on models as scalar indicators of behavior and performance, rather purely abstracted quantification, further evinces the architectural basis of their agenda for energy and form.

A New Climate for Climate

Before moving to a more famous architectural laboratory that was inspired by the Texas Engineering Research Station — the Olgyay lab at Princeton — it is important to establish, equally, some other adjacent developments related to climate and physiology that also motivated the Olgyay brothers.

The post-war decade was marked by a renewed emphasis on the relationship between climate and building design. Some authors, such as architect and city planner Ernst Egli, emphasized the role of climate as a driver of settlement patterns.[57] Egli's geo-historical approach picked up where Vitruvius left off and included some beautiful observations about self-organized settlement patterns in the archaic world.[58] This strain of macroclimatology in architecture is related to Sidney F. Markham's influential 1944 *Climate and the Energy of Nations*.[59] Other authors focused more directly on the architectural adaptations of buildings to climate through design. Jeffrey Ellis Aronin's 1953 *Climate & Architecture*, for instance, focused on many microclimate impacts, including the impact of Rudolf Geiger's magnificent 1950 edition of his microclimate text, *The Climate near the Ground*.[60]

The aforementioned Thomas Bedford had a particularly compelling characterization in 1948 of Ellsworth Huntington's 1915 book, *Civilization and Climate*. Bedford observed that concentrations of population "during the course of history can be accounted for by a theory of climatic pulsation."[61] Other researchers, such as climatologist Helmut Landsberg, published articles on topics such as "Microclimatology" in architectural journals.[62] Landsberg, incidentally, would later write some early articles on "Man-Made Climate Changes."[63] From popular magazines to academic research, climate was a non-trivial concern for design in this period.[64] Taken together, these sources reflect a greatly deepened awareness and interest in the climatic basis for architecture, one of the foci of the Olgyay brothers.

The Architectural Laboratory at Princeton

Victor Olgyay and his twin brother Aladar Olgyay developed the Architectural Laboratory at Princeton University in the late 1950s. A 1954 report on the *Application of Climatic Data to House Design* consists largely of work the Olgyay brothers completed at the Massachusetts Institute of Technology. It established many of the methods and ambitions that were later developed at Princeton.[65] The scope and activity of the Princeton laboratory brought increasing levels of science and method into the space of design, as documented, for instance, in *Solar Control and Sun Shading Devices*, 1957.[66] These two architectural research laboratories largely constitute the exception to more engineering-focused agendas for energy and climate in mid-century architecture. Later, though, the physiological basis of architecture would become important in this lab.

Post-war Physiology

An overt focus of the Princeton Architectural Laboratory — as well as the resulting publication *Design with Climate: Bioclimatic Approach to Architectural Regionalism* (1963) — was the role physiology plays in human comfort: a major constituent in the *bio* of the *bioclimatic*.[67] Victor Olgyay's perspective on physiology reflected the new, post-war characterizations of physiology. This influence included Fitch's important texts but also European and Australian sources cited above, each contributing to a broad interest in the physiological context of architecture.

For instance, the Olgyays' understanding of the heat loss of the body in different thermal conditions relied on sources such as Thomas Bedford's "Environmental Warmth and Human Comfort" and Winslow and Herrington's *Temperature and Human Life*.[68] The latter reports on the results of an extended research project involving the minds of physiologists, physicists, and environmental hygienists.[69] The chapters cycle through the human metabolism, physiological adaptation to thermal change, the role of clothing, the influence of climate, and, revealing the interests of the sponsoring agency, air-conditioning and the human body.

Some of Olgyay's observations run counter to what we know about energy systems today — such as "man strives for the point at which minimum expenditure of energy is needed to adjust himself to his environment."[70] Today we know that non-isolated energy systems that prevail maximize power intake and use through elaborate feedback mechanisms. This is more of an issue of quality, use, and feedback than mere quantity. But Olgyay worked to assemble multiple factors from multiple sources to help articulate more specifically how a body fits into both interior and climatic milieus.

Despite the post-war interest in the role of physiology and the incorporation of new research in this domain, the impact was rather minimal. Reflecting on the developments in architecture since the first edition publication (1948) of *American Building: The Forces That Shape It,* Fitch's preface to the second volume — *American Building: The Environmental Forces That Shape It* — had the following unambiguous critique:

> That earlier version [of the book] sought to establish a holistic concept of man/environment relationships — a necessary frame of reference within building could be fruitfully analyzed and viable goals for the future established. Unfortunately, it cannot be claimed that either the building field, or the professions which guide it, have found it worthwhile to pursue this theoretical line. Indeed, in many respects, it must be admitted that American architecture pays less attention to ecological, microclimatic, and psychosomatic considerations than it did a quarter of a century ago…Ironically, this failure to achieve experientially acceptable levels of performance takes place at the center of the world's most developed technology. It might almost be said, in fact, that the failure is made possible precisely by the overwhelming presence of that technology.[71]

Fitch could make a similar observation today. In general practice, any substantive evolution in this regard in architecture has been minimal in the sixty years since the post-war interest in physiology.

Post-war Pedagogies

Paralleling Fitch's second volume, in one of few histories of architectural science, Australian professor and historian of architectural science Henry J. Cowan revealingly titled his fifth chapter: "Environmental Design Replaces Structure as the Principal Problem of Building Science."[72] His central concern was expressed in his first subheading: "The Conflict between Structural Lightness and Environmental Efficiency."[73] Of particular concern to him regarding the decreasing weight of building structures through engineering was that "thick masonry walls provided sound and heat insulation."[74] From a thermal point of view, Cowan bemoaned of early modern architects: "few of them (Frank Lloyd Wright is one of the exceptions) had initially a sufficient appreciation of the environmental consequences of substituting thin walls, and particularly glass walls, for the traditional thick masonry walls."[75] The thermodynamic depth of this observation is non-trivial and pushes well beyond Luddite nostalgia towards real progress, given the dissipative structure of architecture, as discussed in the next chapter.

In his summary of interior environmental concerns, Cowan reveals several problematic assumptions about energy and buildings at this later point in the twentieth century. He also identifies conflicting tendencies in building science and design at that time. Despite his observation that "the interior environment of many of the surviving medieval and Renaissance buildings is surprisingly satisfying," he restates his modern concern: "conflicts between the ideals of environmental efficiency and structural economy have not yet been fully resolved."[76] They persist today, although in more ways than Cowan had in mind.

While that conflict no doubt exists, it is the conflict between the principles that he uses to evaluate respective building systems and the thermodynamics of ecosystems that should be of primary concern. The thermodynamically inexplicable focus on efficiency does not necessarily support the actual ecological concerns that can motivate an agenda for energy and material consumption in architecture. This disjunction arises largely from how energy systems are discussed in pedagogy and practice, in ways that occlude the actual non-linear behaviors of the non-equilibrium thermodynamics of small and large-scale architectural systems.

Pedagogies of Waste

Take, for instance, the perennial guilt that is designed to be instilled by persistent claims about just how consumptive the building industry is in North America: statistics about buildings consuming about half the energy in the United States.[77] The thermodynamically operative function of the energy question is NOT (only) the amount of consumption but the amount of feedback inherent to a design. Assuming that energy consumption should be minimized misconstrues the non-isolated thermodynamics of a building and already minimizes the ecological potential of a building's feedback. Thus a preoccupation with energy consumption alone, with efficiency as the countermeasure, is an ecologically perverse practice. As ecosystem scientists tell us, energy systems that prevail maximize their energy intake, use, and feedback. The focus on energy efficiency and conservation reflects more about HOW energy is discussed rather than HOW energy systems tend to behave in short and long-term energy cycles.

The predilection for efficiency and a trimming of waste has far more to do with what Max Weber characterized in the *Protestant Ethic and the Spirit of Capitalism* than it does with anything related to thermodynamics.[78] Applying Calvinist morality to energy systems will not engender sound, coherent energy pedagogies and practices for architects. Not surprisingly, it tends to engender little more than a litany of moralizing rhetoric and claims about energy.

If a Calvinist/Protestant ethic shaped the cultural aspect of architecture's relationship with energy, capitulations to imposed disciplines and systems shaped other important aspects of architecture's pedagogy and practices regarding energy. It is hard to argue that the trajectory adopted by many architectural pedagogies — represented so bluntly by Louis Axelbank in 1961 — set a very architecturally powerful approach to issues of energy, physiology, climate, and design. With courses related to energy and environment in schools of architecture often relegated to engineers, as Axelbank recommended, the possibility of an explicitly architectural agenda for energy and environment could not evolve. The means and procedures of convergent design thinking were thus rarely a concern for these non-trivial topics in architecture. They were instead treated as an administrative and curricular obligation to be met. Despite the compelling, representative exceptions cited in this chapter, the concept of human comfort was not well incorporated into the topic of human comfort, save the quantitative concerns of an organization such as ASHRAE.

In this regard, it is apparent that by 1973 and thereafter, architectural pedagogies and their resultant practices remained locked, as Pynchon would say, only from the inside by isolated concerns and characterizations of energy. While energy pedagogies at that time, and today, reflect a considerable amount of effort, they remain rather closed and very incomplete. Too ready to accept the assumptions and boundaries of simulation software, professors and students routinely perpetuate key problems inherent in the insulation apparatus of modernity.

When human comfort is coupled with the other primary purpose of an energy agenda in architecture — the ecological potential of buildings — physiology becomes even more important. To this end, the balance of this chapter will focus on but one parameter that helps establish the physiological and ecological concern as one basis for an architectural agenda for energy. In other words, human-environment interactions occur at multiple spatial and temporal scales. A thermodynamically cogent agenda for energy in architecture must recursively mind all these scales and cycles of energy. As such, it is useful at the end of this chapter to reflect on a few lesser-known thermal parameters of bodies and buildings that help draw greater attention to this recursive habit of mind regarding energy.

Thermal Effusivity

Thermal effusivity is very relevant to the thermal regulation of buildings and the dissipative structure of architecture in reality, and so bears expansion here. Within the haptic milieu of the human sensory apparatus, the qualitative perception of warmth can be more explicitly understood through the material property of thermal effusivity than the more common focus on temperature might otherwise indicate. I include thermal effusivity as an example of aspects of material properties and physiological response that are manifestly qualitative and, as such, have resisted inclusion in the more reductive quantifications of thermal milieux in architecture and building science. Thermal effusivity is a measure of a material's ready ability to exchange heat with its milieu. It is mathematically expressed as:

$$e = \sqrt{k\rho c}$$

where:

k = thermal conductivity

ρ = density

c = specific heat capacity

What is most important about the thermal effusivity of a material in this history is a.) how it relates to the thermal effusivity of human skin, and b.) how that relationship could further modify assumptions about heat transfer in the context of architecture. When taken together, these two observations can alter assumptions about insulation in a number of cases. When considered correctly, at stake for design are levels of human comfort achieved through modulation of perception, rather than modulation of temperature alone.

The thermal effusivity of various materials quickly reveals how the perception of warmth is related to this quantitative value. As architectural engineer Lisa Wastiels notes, "materials with a higher effusivity, like metals, will be perceived as colder than materials with a lower thermal effusivity, like woods."[79] Owing to the difference in thermal effusivity, a piece of wood and a piece of steel — at the same temperature — will be perceived as warmer and colder respectively. While this is occasionally attributed to the conductivity of the materials, the non-steady-state instance of a hand touching material means that "heat flow is not proportional to the thermal conductivity of the material, as under steady-state conditions, but to its thermal effusivity."[80] The role of specific heat capacity and thermal effusivity is therefore as important as conductivity in this matter.

To connect the heat flux of a building envelope to the heat flux of the human envelope — skin — some knowledge of the thermo-physical properties of biological matter is necessary.[81] When quantitatively expressed, it is apparent that materials with thermal effusivity lower than human skin are favorable to perceptions of warmth; again, even if all material temperatures are the same. In their discussion of the thermo-physical properties of warmth, Yoshiro Obata and his colleagues provide some specific and important thermal effusivity values.[82] The thermal effusivity relationship between the palm of the hand ($1.263 \text{ kJ/m}^2\text{s}^{1/2}\text{K}$) and larch wood ($.403$) compared with the hand and steel (12.53) explicitly articulates what is intuitively known about the relative "warmth" of materials. Likewise, the thermal engagement of the sole of the foot (1.012) with various flooring materials — granite (3.54), maple ($.57$) — similarly confirms what the body already knows. Even just these physiological and material facts alone can completely refigure the role of material specification when thinking about energy in architecture.

As physicist Edgar Marín observes, "while static and stationary phenomena are governed by parameters such as specific heat and thermal conductivity respectively, under non-stationary conditions thermal effusivity and diffusivity are the more important ones."[83] So while material temperature — the aim of conductivity-driven insulation calculations — matters, the thermal effusivity and diffusivity of building's materials also matters. This becomes essential in transient, non-steady-state conditions, such as that of human occupation. If an architect begins to connect short and long-term cycles of energy — and thus begins to incorporate more physiology and more mass into an agenda for energy — then the thermal effusivity and thermal diffusivity of the materials become even more important.

Mid-Nineteenth Century Modernism

In the context of this chapter, it is worth noting that some of the mid-nineteenth century insights into the first law of thermodynamics emerged largely from observations about physiology. The great physiologists Hermann von Helmholtz, Justus von Liebig, and Siegmund Mayer all contributed key insights into the first law and the notion of conservation.

Helmholtz in particular is a fascinating individual to consider at the conclusion this chapter. His early insights dealt with the conservation of energy as it related to muscle metabolism. Not by any coincidence, later work with a range of collaborators and students reached from ophthalmology to "heat death" to electromagnetism. Helmholtz, it seems, could make little distinction between physiology and physics: in his work they appear as but one extended field of energetic interactions. He argued that heat, light, electricity, magnetism, and mechanics could all be related. This, amongst all his contributions, is his greatest contribution to the habits of mind for contemporary designers. If only architects could possess a similar praxis.

In that spirit, the following example of the ancient Roman baths, as well as examples in the next chapter, articulate a less isolated agenda for energy in architecture, including the role of entropic dissipation that emerged so powerfully in the middle of the nineteenth century and disrupted so many assumptions about energy. The full implications of thermodynamics have not yet entered architecture and the matter demands a much broader and more totalizing agenda than what building science and design have thus far considered. This example serves as a concluding example for this chapter.

A Non-modern Consideration of the Roman Baths

> "In the countless quarrels between Ancients and Moderns, the former come out as winners as often as the latter now, and nothing allows us to say whether revolutions finish off old regimes or bring them to fruition."[84] – Bruno Latour, *We have Never Been Modern*

The ancient Roman bathing complexes represent a compelling example of an architecture that is thoroughly thermodynamic in its conception, design, and use but one that resists modern — if not to say reductive — social, formal, and technical characterizations. The Roman baths, for instance, foreground physiological responses to buildings in ways that few modern buildings do and in ways that the modern praxis of heat transfer struggles to adequately describe. At the same time, many of the bathing complexes are ancient antecedents to modern climate adaptation design strategies. Finally, the sheer scale, mass, and material geography of Roman Imperial baths can be only described as geological, a fact that resists the metabolic rifts of modern buildings and the larger thermodynamics of urbanization. Thus the baths, owing to this mix, are better understood through a non-modern perspective: through a conjugate understanding of physiology, matter, space, energy, form, and time that cannot be found in the partitioned forms of knowledge in modernity. In this way, the baths in turn help us better understand a range of contemporary non-modern projects and principles.

The Ancient Roman Bath

The ancient Roman bath was a central architectural presence in the daily life of a Roman. A typical workday ended by noon, followed by a light lunch. After the midday nap, around one o'clock (the eighth Roman hour) a bather would head to any of the 952 baths in Rome, as documented in the 354 AD Regionary Catalogue of buildings in the capital.[85] In this number were the massive, famous Imperial examples of Caracalla, Trajan, Agrippa, Constantine, and Diocletian. There the bather would typically remain for several hours, engaged not only in bathing but also any of the libraries, lecture halls, gardens, markets, concerts, galleries, or other facilities contained in the bathing complex.[86]

The bathing sequence itself, while varied greatly according to the preference of individual bathers, generally proceeded from warm to hot to cold.[87] The sequence began in the *apodyterium,* the heated changing room with a wood equivalent of the contemporary locker room. Here the bather would change from dirty city clothes to a bathing tunic and sandals. Varying degrees of exercise in the *palaestra,* exterior exercise yards, would prime the body for sweat. Sweat was the primary mechanism for cleansing the body, so the bather would generally oil themselves in the *aleipterion*, a warm room for massage with oil, before exercise or before the hotter rooms.

The sequence then shifted to an increasingly warm, but not necessarily linear, series of rooms upon the sounding of the *tintinnabulum*, a bell announcing the opening of the hot baths.[88] The hotter rooms would include the *heliocaminus*, a room for bathing not in water but the sun, and the *tepidarium*, a warm room that might be either humid or dry and provided a transition from other spaces. The hottest rooms were the *caldarium*, a hot, humid room with hot pools, and the *laconium* or *sudatorium*, a hot, dry room much like a contemporary sauna. The hot rooms were heated by the *praefurnium*, a furnace room that fed the famous hypocaust heating system, thus warming both floor and wall. Some *praefurnia* would also heat metal boilers, the primary source for hot bathing water, others might power the *testudines alveoloria* that maintained the temperature of the baths, and yet other *praefurnia* would only operate for room heating. At some point, the sweat, oil, and dirt would be scraped from the skin surface of the bather by a servant, followed by a plunge in the cold pools of the *frigidarium*, a relatively cold room, and the *natatio*, a swimming pool open to the sky. It is important to note that, in especially the large Imperial complexes, there was no fixed sequence to this series of rooms. As the bath plans reveal, the architecture afforded a series of uses, sensations, and experiences but did not overly prescribe them as many accounts suggest.

These variable circuits of sensation were directly afforded by the material, spatial, and energetic disposition of these buildings. Most emphatically, the thermal fluctuations evident in the various circuits through a bath would have been felt primarily as abrupt radiant gains or losses, even if the air temperature might fluctuate less. The thermally active surface floors, walls, and ceilings in all spaces —

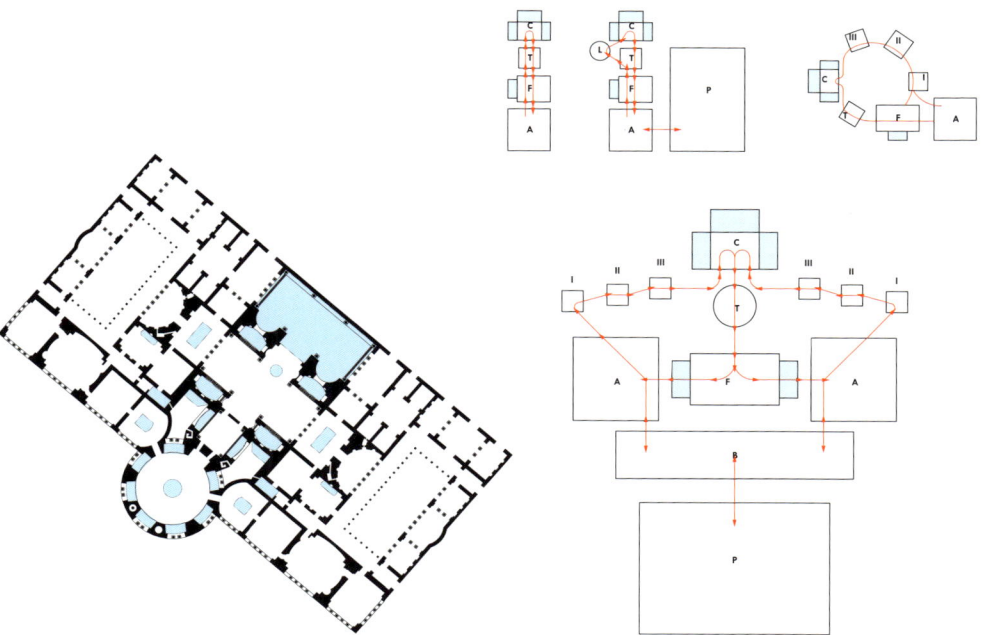

Ancient Roman Baths
Left: Plan, Baths of Caracalla
Right: Krencker's bath typologies

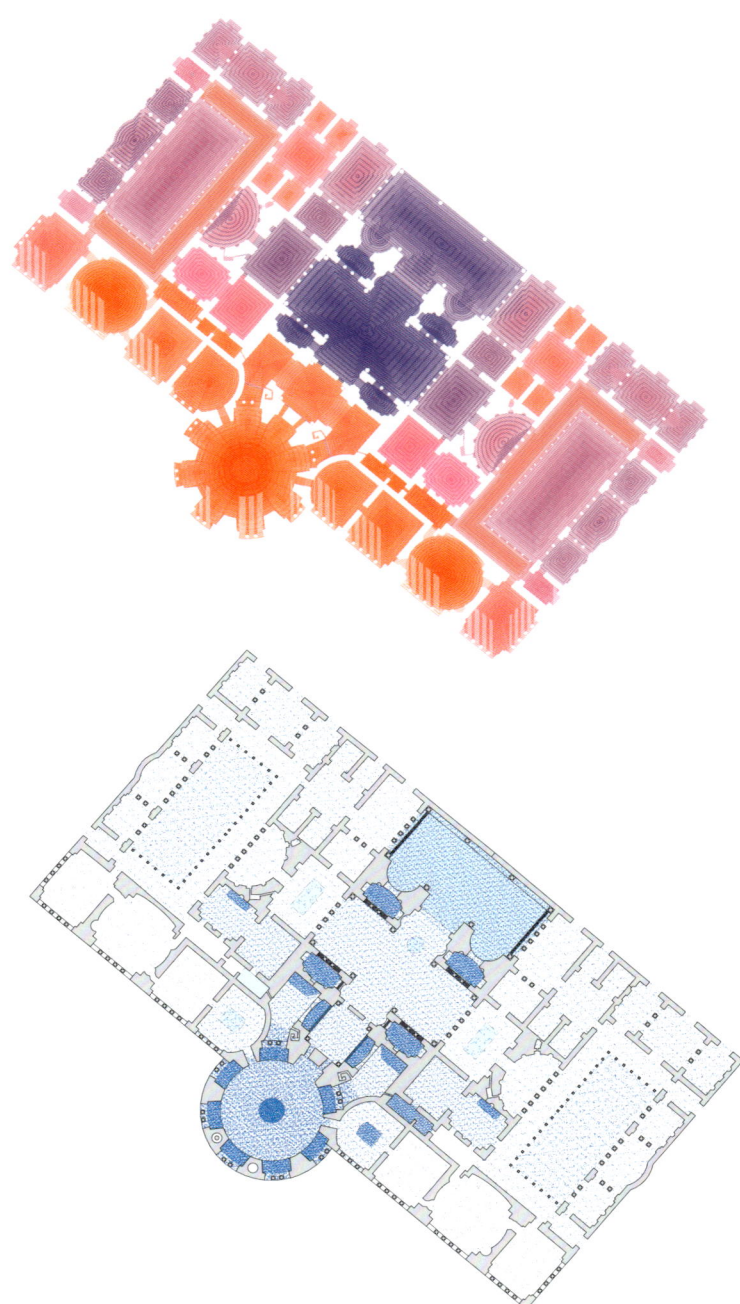

Ancient Roman Baths
Above: Thermal Plan, Baths of Caracalla
Bottom: Humidity Plan, Baths of Caracalla

regardless if intended as heat sources or heat sinks — determined most of these thermal variations.

The Baths of Caracalla

The mass of Caracalla, for instance, is fully implicated here. In the southwest-facing rooms, solar gains and the hypocaust heated the surfaces of each room, charging the mass for long pulsing cycles of radiant losses to bodies. This same mass in colder spaces — when shaded and unheated by the hypocaust — can only be understood as a massive heat sink, as a visit to Caracalla today reveals. The instantaneous radiant transfer of heat to and from the body was a desired form of thermal delight in the bathing sequence, and the architecture hosted multiple thermodynamic itineraries and sensational possibilities.

Modern Eyes on Ancient Baths

The architecture of the bathing complex has received much interest from modern architectural and archaeological observers. On the one hand, the legibility of the Imperial bath plan motivated multiple compositional interpretations from the Beaux-Arts through modernity. On the other, the ingenuity of the Roman bath heating apparatus, when coupled with the plan, engenders one of the more compelling thermodynamic architectures in the canon of architecture. But the depth of this thermodynamic architecture defies both modern formal and technical readings.

When modern observers study the ancient baths, they tend to do so with modern eyes. Take, for instance, modern readings of Caracalla. Whether projected as a series of primary, secondary, and tertiary axes of symmetry, as negotiated by poché, in a Beaux-Arts interpretation or a high-modernist functionalist description of the baths, both cases do not adequately align with the spatial and thermal experience inherent to the baths. The archaeologist Daniel Krencker, for instance, produced diagrams that depict straight, linear sequences that align with various axes in plan.[89] Janet DeLaine likewise offered a strictly functionalist reading of the plan and then mechanically deduced a geometric explanation of the composition.[90] Neither of these functionalist descriptions appropriately reflects the possibilities inherent in the plan of the building from an architectural or thermodynamic point of view.

A more thermodynamic reading of the Caracalla plan — one based on a more composite reading of space, matter, energy, and time — however, resists reductive explanations. A polyvalent view of the plan must negotiate more simultaneous terms and this forces another understanding of the plan. One key to this reading is to start with the ambulant human body and imagine the physiological sensations and disjunctions that emerge in the various possible sequences through the matrix of possible circuits in the plan. The baths offer a range of physiological sensations and experiences, linked to a range of hygienic and social praxis, and enabled by cunning thermodynamic organizations and techniques.

Rather than the initially perceived clarity of the plan as a Beaux-Arts plan, the plan can be better understood as a matrix of co-isolated but communicating rooms motivated by the thermodynamics of rapid physiological modulation. A proliferation of minor passages and optional routes embedded in the poché — each undermining the major axes of the plan — suggests that an individual could have achieved their own sequence according to their custom. A bather might thus discern a range of thermodynamic, spatial, luminous, and moisture sequences. In this view of the Caracalla plan, the only relevant datum of the plan is the southwestern solar orientation of the flank of hot spaces snapped to the southwest face of the building. When this flank of spaces is coupled with the parallel subterranean *praefurnium* service galleries and the profuse amount of daylight and solar gains, the plan is understood as more of a matrix of passages towards and away from heat, humidity, shade, and light. Even along the hot flank of spaces, there was a range of thermal and humidity milieux. To march a modern observer through a rigid, deterministic sequence of axially determined spaces at best raises a number of questions about the many minor deviations and aberrations evident in the plan. While the plan has primary spatial and structural axes, these do not necessarily align with the primary thermodynamic circuits available to the bathers.

In short, Caracalla is an architecture that primarily affords physiological coherence with only local spatial and structural coherence. The enabling architecture of this coherence is inherently thermodynamic: the apparatus of heating, in conjunction with solar orientation, determines the composition of the building. As ancient bath scholar Fikret Yegül observes, "The development of the heating systems of Roman baths provided the basis for the order and organization of spaces and had a direct bearing upon their planning."[91] In short, baths were thermodynamically planned, physiologically used, and thus not (only) functionally determined in modern terms.

Thermodynamic Urban Planning

The plan of a bath and the composition of its heating and water apparatus was not the only thermodynamic factor in the baths. Given the magnitude of the Imperial bathing complexes and the heating load thus implied, over time the bath-

Physiology, Insulation, Climate, and Pedagogy 215

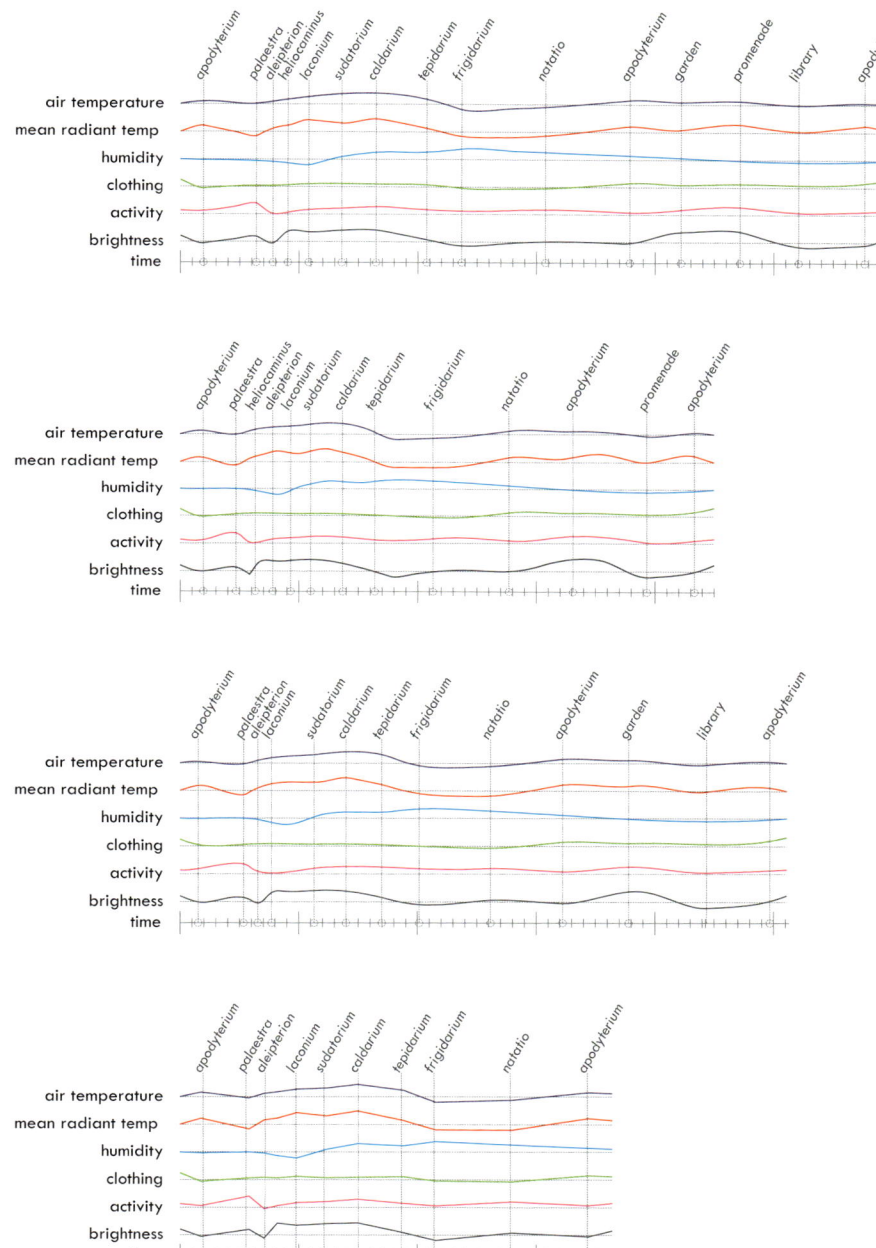

Ancient Roman Baths
Physiology fluctuations of various bather sequences

ing complexes, as Vitruvius notes, were oriented with the flank of hot bathing spaces facing southwest.[92] This filled this flank of hot spaces with light and solar heat gains for the afternoon hours, the period of greatest bathing intensity. This thermodynamic principle of organization determined the orientation of the remaining axially aligned spaces and consequently for the mass of the complex. It thus inserted a thermodynamic figure into the urban morphology of Rome, one of several constantly negotiated urban determinants. The evolution of the Imperial *thermae* in Rome, in fact, reflects this thermodynamic evolution towards optimal solar gains as the type evolved over the centuries.

Caracalla Geology

Given the size and magnitude of the Imperial baths the resources inherent in the baths also demand articulation. While the thermodynamics of large spatial and long temporal scale systems remains the most opaque to architects, the question

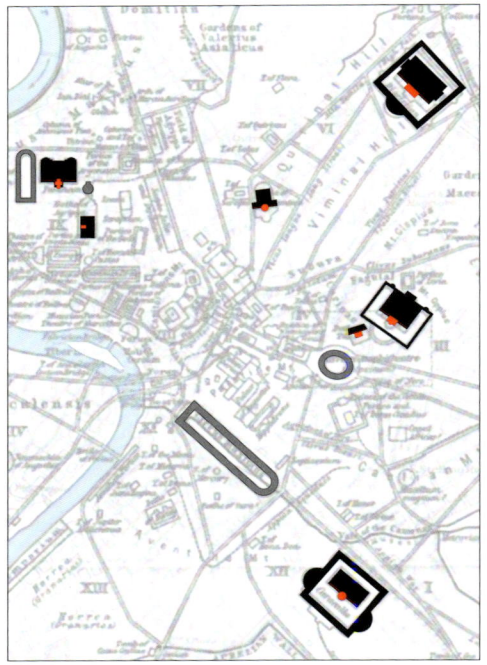
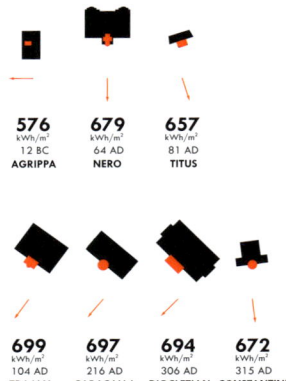

Ancient Roman Baths
Thermodynamic evolution of Imperial Roman baths

Physiology, Insulation, Climate, and Pedagogy 217

Ancient Roman Baths
The Baths of Caracalla are an architectural and geological movement.
Above: View through *caldarium* / Bottom: View through *frigidarium*

of material geography, duration, and use all are fundamentally thermodynamic problems that should not be isolated or externalized from other building-related thermodynamic propositions.

Both the sheer material reality of the baths as well as the water and fuel supply required of the complex has received recent attention. Janet DeLaine provides a rigorous account of the construction of Caracalla.[93] For instance, she calculated the various material quantities, costs, transportation requirements, and sources surrounding Rome. When this type of accounting is coupled with varied attempts to estimate the fuel required for the baths, a less isolated thermodynamic consideration of the baths is in place.[94] In other words, Caracalla is no long understood as an isolated object composition but a inherently regional force in the urbanization of Rome.

Hypocaust: Ancient vs. Modern

Modern considerations of the Roman bath thermal apparatus generally, and unwittingly, re-enact an "Ancients versus Moderns" debate. Apologists for both sides can be found in the literature but only modern perspectives can be found that consider the thermal questions of the baths in multiple and simultaneous ways. In other words, the baths are generally described in technical, historical, or architectural terms but rarely in composite. The folly of modern, partitioned knowledge instigates at best isolated considerations of the bath complexes.

For instance, the basic structure of the hypocaust heating system is by now fairly well established but remains incomplete. The heated spaces of a Roman bath bore a raised floor, the first part of the *praefurnium* exhaust pathway. The second, contiguous part of the pathway was in the wall, a series of vertically and horizontally interconnected hollow tiles known as *tubuli*. Some heated spaces would have just a raised floor with more conventional flues for vertical exhaust. Still other examples connected floor, walls, and vaults through hollow, curved intrados tubes and even hollow voussoirs have been found.[95]

The increasing number of thermally modulated surfaces would have lowered the operating temperature of the *praefurnium* as the work is distributed over a large area and removes thermal asymmetries. The evolution of hypocaust designs is thus best understood as an exergy-matching design. The designs evolved so that low-temperature wood fires would heat water. The exhaust of the fire was captured and channeled into the surface of the floor, walls, and ceilings of the baths for quicker dissipations as well as charged the massive structure of the building for longer-term storage. In each stage, the Romans were extracting heat no longer available for work in the previous step in the cascade, all in service of hygiene and thermal delight. The inputs and outputs of the thermodynamic system are best considered in non-isolated terms.

In the past century, there has been a steady production of research related to the hypocaust thermal apparatus in a range of adjacent disciplines from archaeology, engineering, and architecture. In the last couple of decades, for instance, computational fluid dynamics have been employed to depict the possible thermodynamic behavior of the baths.[96] These digital models extend the query initiated by earlier analog numerical models of ancient baths that began to emerge in the 1950s.[97] When this range of analytical work is coupled with evolving work on the cultural and political context of the baths, the background for a less isolated understanding of the baths is in place but even now remains incomplete.

One of the obstacles to a more complete characterization of the baths is that modern characterizations of the Roman baths tend to treat the baths *and* bathers as if there where modern entities. For instance, the quantitative models adopt a modern comfort model — treating the ancient bather as a modern ASHRAE subject — and treat the hypocaust as if it were a modern HVAC apparatus.[98] Certain problems emerge when these modern assumptions and habits map onto the baths, however.

Foremost, the boundary conditions of the bathing complex are not a simple matter to discern. Too much remains unknown about the architecture of the baths to adequately quantify the typology and too little is known about the exact bathing habits to adequately characterize the typical bather. Likewise, the behavior and number of bathers both affecting the physiological response of the bathers is not simple to discern. To get a glimpse of some of the issues that haunt modern analog and digital accounts of these buildings, consider the following among the known unknowns of the baths:

1. As noted, if these studies include the physiology of the bather at all, they base their analysis on modern "comfort" standards. Yet, the primary purpose of the ancient baths, as now, was to deliberately exceed "comfort" for set periods of time in terms of temperature, sweat, and activity. As with the Turkish bath or the Finnish sauna, bathers specifically sought the sensation and physiological benefits of conditions beyond what would have been an already stretched zone of comfort in the Roman Empire. Rather than avoiding excess heat, the bather seeks it. Rather than avoiding sweat, the Roman bather exerted and sweated as the primary means of cleansing. So it is not a question of finding a numerical model wherein the non-modern bath comports with the modern comfort models of heating and refrigeration standards. In no

way could bathing be construed as aspiring for comfort in modern terms. Looking at the ancient bather with the eyes and limits of modern standards does not at all begin to address the physiological context of bathing, what should be the fundamental unit of thermodynamic analysis in a bath.

2. The number and activity of people in any bathing space would vary significantly hour to hour in the baths. The modern comfort models in the literature assume a single, stationary nude male. However, an ambulant bather would have been moving from space to space that may have been crowded or un-crowded for variable amounts of time, a sequence with its own specific physiological responses and adaptations that have little to do with the construal of stationary nude males. Further, it is known that a light tunic was worn in many parts of the bathing ritual.[99] Clothing, activity, and other bodies would all modulate the thermal and humidity effects of baths on the bather, especially at the high bath temperatures where the influence of radiant, convective, and evaporative heat transfer shifts and fluctuates the most.

3. In a more subtle but not-trivial way, any vaporization and/or condensation of water in the baths for the various cold, warm, and hot pools in the baths would influence convective flows of heat and humidity as well as the mean radiant temperature depending on the location of a bather, much less the effect of a submerged bather. Yet the role of the *testudines alveoloria* that maintained the pool temperatures in conjunction with the *praefurnia* that provided the initial hot water for the baths is generally absent in the various forms of analysis.[100]

4. In the analysis of one article, the role of the baths' massive structure was ignored.[101] In others, solar gains were privileged over convective losses.[102] In yet others, the radiant effects of surface temperature were overlooked in the attempt to determine the air temperature of the baths, despite that the relevant temperature is the operative temperature sensed by the bather.[103] This last study, focused on a small bath of at most a residential scale, cannot adequately account for the more complex and interesting combination of factors in larger spaces with more people, more volume, and more complicated boundary conditions. In short, isolated characterizations persist for the intricate and difficult thermodynamic architecture of the Roman bath.

5. It is very difficult to ascertain which, if any, vaults in a bathing complex would have extended the hypocaust up into the vault itself as a sixth thermally active surface, as in the Stabian Baths at Pompeii. Adding this sixth radiant surface to the space would have certainly affected the already complex heat transfer interactions. In hot rooms, the ceiling, whether vaulted or

domed, activated or not, would no doubt have been warm from stratified air, but the activated type would have been undoubtedly warmer.
6. While the baths have great mass and could be assumed to be treated as quasi-stable steady-state in certain circumstances, given the obvious role of solar orientation, the shifting use of bathers, water temperatures, and convection, the heat transfer mechanisms must be considered severally and in a way that resists steady-state accounts.
7. Perhaps the biggest gap in knowledge deals with exhaust. It is at best debated how the hypocaust was actually exhausted, with what frequency, and if the exhausts were operated with as much attention as the furnaces and the bronze discs that modulated the oculus exhaust of the *laconium*. The reconstruction of a small bath in Turkey foregrounded this important topic that would have a profound effect on the rate of heat gain or loss in the system.[104] Likewise, these unknowns would as well determine the convection pattern of exhaust gases in the hypocaust and would thus indicate which parts of a room might heat up faster and to higher temperatures. The mass of the floor system would have guaranteed rather even temperature distribution and thus surface temperatures, but the furnace end of the surface would inevitably have been warmer than a distant portion away from a *tubuli* exhaust outlet. The studies presume a constant temperature throughout the hypocaust but establish no technical means for achieving this. Unequal air pressures are inevitable in a number of *tubuli* explanations and this would result in uneven temperature distributions. This is one of the more confounding aspects of the hypocaust system: how even temperatures could be created yet still maintain sufficient draft. In short, it is not possible to assume multiple aspects of the hypocaust performance given the dearth of evidence about the roof and exhaust architecture of the baths. In many cases, these studies yield results with a level of exactitude that far exceeds the possible certainty of input.

This moderate, but abbreviated, list of known unknowns is suggestive of why modern models have not, and cannot, satisfactorily account for the roman baths as a thermodynamic architecture. Modern studies use modern eyes to explain fundamentally non-modern techniques and architectures. These known unknowns point towards the difficult whole of the bath: the modern difficulty of accounting for the ancient thermodynamic architecture that nonetheless operated as a coherent thermodynamic monster of space, body, matter, energy, and time.

The difficulty of these incomplete modern accounts has not prevented condescension, bred of assumed certainty and exactitude, among modern observers: "It is hardly safe to assume that the Romans had any idea, in the modern sense," architect Edwin Thatcher noted, "of the physiological principles of human heat loss and heat generation."[105] The Romans certainly did not have a modern sense of the physiological sense of human heat loss and gains and it just might be this lack of isolated knowledge that enabled them to evolve such a complete thermodynamically and physiologically active architecture.

A less isolated, less modern account of the baths based on the physiology of the bather and the resultant thermodynamics is necessary to resolve lingering questions in the textual and material legacy of the baths. From physiological responses to heat, light, and humidity to urban morphology to material geographies, the Roman baths were thoroughly thermodynamic and it proves as difficult as it problematic to isolate these dissipations from each other. Consequently, it is exactly the topic of dissipation that constitutes the content of the next chapter.

Notes

1. Bruno Latour, *We Have Never Been Modern*, Cambridge, MA: Harvard University Press, 1993. p. 139.
2. Kiel Moe, *Thermally Active Surfaces in Architecture*, New York: Princeton Architectural Press, 2010.
3. Hippocrates, *Upon Air, Water, and Situation*, London, 1752.
4. Marcus Vitruvius Pollio, *Vitruvius: The Ten Books on Architecture*, trans. Morris Hicky Morgan, New York: Dover Publications, 1960. p. 5.
5. Ibid. p. 10.
6. Charles Edward Amory Winslow and Lovic Pierce Herrington, *Temperature and Human Life*, Princeton: Princeton University Press, 1949. p. vii.
7. William Harvey, *Exercitatio anatomica de motv cordis et sangvinis in animalibvs*, Frankfurt: Sumptibus F. Fitzeri, 1628.
8. Vicenzo Scamozzi, L'idea *della architettvra universale, di Vincenzo Scamozzi, diuisa in x libri*, Venice: Expensis avctoris, 1615.
9. Thomas Tredgold, *Principles of warming and ventilating public buildings, dwelling houses, manufactories, hospitals, hot-houses, conservatories, etc.; and of constructing fire-places, boilers, steam apparatus, grates, and drying rooms; with illustrations experimental, scientific, and practical...*, London: Printed for J. Taylor, 1824.
10. Ibid. p. 9.
11. John Arbuthnot, *An Essay Concerning the Effects of Air on Human Bodies*, London: Tonson and Draper, 1733.
12. David Boswell Reid, *Illustrations of the Theory and Practice of Ventilation: With Remarks on Warming, Exclusive Lighting, and the Communication of Sound*, London: Longman, Brown, Green, & Longmans, 1844.
13. Thomas Bedford, *Basic Principles of Ventilation and Heating*, London: H. K. Lewis & Co. Ltd., 1948. p. 4.

14 William Heberden, "An Account of the Heat of July, 1825; together with some remarks upon sensible cold," *Philosophical Transactions of the Royal Society of London*, vol. 116 (1/3), 1826. p. 71. (The emphasis is his.)
15 Henry J. Cowan, *Science and Building: Structural and Environmental Design in the Nineteenth and Twentieth Centuries*, New York: John Wiley & Sons, Inc., 1978. p. 222.
16 Ibid. p. 222.
17 Ibid. p. 222.
18 A. F. Dufton, "The Eupatheostat," *Journal of Scientific Instruments*, vol. 6, 1929. p. 249.
19 A. F. Dufton, "The Measurement of Equivalent Temperature," *The Journal of Hygiene*, vol. 33 (4), Nov 1933. pp. 474–475.
20 T. Napier Adlam, *Radiant Heating, Radiant Cooling, and Snow Melting: A Practical Treatise on American and European Practices in the Design and Installation of Systems for Radiant, Panel, or Infra-Red Heating, Snow Melting and Radiant Cooling, Including Step-by-Step Procedure, with Typical Problems Solved by the Application of Simplified Working Data, Charts, and Tables*, New York: The Industrial Press, 1949. pp. 40–45.
21 Thomas Bedford, *Basic Principles of Ventilation and Heating*, London: H. K. Lewis & Co. Ltd., 1948. p. 4.
22 As included in Ulrich Conrads, *Programs and Manifestos on 20th-Century Architecture*, Cambridge, MA: The MIT Press, 1975. p. 117.
23 Ibid. p. 119.
24 K. Michael Hays, "Diagramming the New World, or Hannes Meyer's 'Scientization' of Architecture," in Peter Galison and Emily Thompson, eds., *The Architecture of Science*, Cambridge, MA: The MIT Press, 1999. pp. 233–252.
25 Ibid. p. 120.
26 Sylvia Lavin, "Open the Box: Richard Neutra and the Psychology of the Domestic Environment," *Assemblage*, no. 40, Dec. 1999. pp. 6–25.
27 Michael J. Ostwald and Raeana Henderson, "The Modern Interior and the Excitation Response: Richard Neutra's Ocular-centric Phenomenology," *Architecture Research*, vol. 2 (3), 2012. pp. 27–35.
28 Richard Neutra, *Survival through Design*, New York: Oxford University Press, 1954. p. 4.
29 Richard Neutra, "Design is a Human Issue," *Annual of Architecture, Structure and Town Planning*, Calcutta, vol. 3, 1962. p. B3.
30 Richard Neutra, *Survival through Design*. p. 23.
31 Ibid. p. 144.
32 Richard Neutra, "Design is a Human Issue." p. B2.
33 Ibid. p. B3.
34 Jos Tomlow, "Building Science as Reflected in Modern Movement Literature," in Jos Tomlow, ed., *Climate and Building Physics in the Modern Movement: Proceedings of the 9th International DOCOMOMO Technology Seminar*, Paris: Docomomo International, 2006. pp. 7–8.
35 Ibid. p. 9.

36 Louis Axelbank, "How Much Must an Architect Learn about Mechanical and Electrical Services, and Where?" *Journal of Architectural Education*, vol. 15 (4), Winter 1961. pp. 24–29.
37 Joseph King, "Twitchell and Rudolph," in Christopher Domin and Joseph King, eds., *Paul Rudolph: The Florida Houses*, New York: Princeton Architectural Press, 2002. p. 26.
38 James Marston Fitch, *American Building: The Forces That Shape It*, Boston: Houghton-Mifflin Company, 1948.
39 This book was later split into two volumes: James Marston Fitch, *American Building 1: The Historical Forces That Shape It*, Boston: Houghton Mifflin Company, 1966; and James Marston Fitch, *American Building 2: The Environmental Forces That Shape It*, Boston: Houghton Mifflin Company, 1972.
40 Fitch, *American Building 2: The Environmental Forces That Shape It*. p. 195.
41 Sibyl Moholy-Nagy, "Environment and Anonymous Architecture," *Perspecta*, 1955, vol. 3. pp. 3–77.
42 Sibyl Moholy-Nagy, *Native Genius in Anonymous Architecture*, New York: Horizon Press, 1957. p. 11.
43 Ibid. p. 14.
44 Ibid. p. 14.
45 Axelbank. pp. 24–29.
46 Ibid. p. 25.
47 Ibid. p. 25.
48 Ibid. p. 27.
49 Ibid. p. 28.
50 Ibid. p. 27.
51 Ibid. p. 28.
52 Gordon McCutchan and William W. Caudill, *An Experiment in Architectural Education through Research*, College Station: Texas Engineering Experiment Station, Texas A. & M. College System, 1951. As quoted from the inside cover.
53 Dean Hawkes, "A History of Models of the Environment in Buildings," *Land Use and Built Form Studies*, Working Paper no. 34, 1970.
54 Hawke's subsequent publications include *The Selective Environment*, *The Environmental Tradition*, and *The Environmental Imagination: Technics and Poetics of the Architectural Environment*.
55 The earliest instantiation of the term "environmental control system" I have found is 1955, in: "SAE Golden Anniversary Aeronautic Meeting Summaries of the Papers Presented to the Meeting Held in Los Angeles, October 11–15, 1955," *Aircraft Engineering*, 1956, vol. 28(1). pp. 28–31.
56 Frank H. Samonski, "Technical History of the Environmental Control System for Project Mercury," *National Aeronautics and Space Administration Technical Note NASA TN D-4126*, October, 1967.
57 Ernst Egli, *Climate and Town Districts: Consequences and Demands*, Zurich: Verlag für Architektur – Erlenbach, 1951.
58 For example, Egli's analysis of the Egyptian town of Kahun yields some perhaps obvious observations about the climatic determination of Egyptian ar-

chitecture and urbanization (ibid. pp. **85–88**), but he provides many rich examples.
59 Sidney F. Markham, *Climate and the Energy of Nations*, London: Oxford University Press, **1944**.
60 Jeffrey Ellis Aronin, *Climate and Architecture*, New York: Reinhold Publishing Corporation, **1953**; Rudolf Geiger, *The Climate near the Ground*, Cambridge, MA: Harvard University Press, **1950**.
61 Thomas Bedford, *Basic Principles of Ventilation and Heating*, London: H. K. Lewis & Co. Ltd., **1948**. p. **3**.
62 Helmut Landsberg, "Microclimatology," *Architectural Forum*, vol. **86**: **114**, March **1947**.
63 Helmut Landsberg, "Man-Made Climatic Changes," *Science*, New Series, vol. **170**, no. **3964** (Dec. **18**, **1970**). pp. **1265–1274**.
64 The relationship of climate and architecture during this period is only briefly mentioned here. For much more extensive research on this issue, see Daniel Barber's related work on this topic.
65 Housing and Home Finance Agency, *Application of Climatic Data to House Design*, Washington, Housing and Home Finance Agency, Office of the Administrator, Division of Housing Research, **1954**.
66 Victor Olgyay and Aladar Olgyay, *Solar Control and Shading Devices*, Princeton: Princeton University Press, **1957**.
67 Victor Olgyay, *Design with Climate: Bioclimatic Approach to Architectural Regionalism*, Princeton: Princeton University Press, **1963**.
68 Thomas Bedford, "Environmental Warmth and Human Comfort," *British Journal of Applied Physics, vol.* **1**, **1950**. pp. **33–38**.
69 Charles Edward Amory Winslow and Lovic Pierce Herrington, *Temperature and Human Life*, Princeton: Princeton University Press, **1949**.
70 Olgyay, *Design with Climate*. pp. **14–15**.
71 Fitch, *American Building: The Environmental Forces That Shape It*. pp. **vii–viii**.
72 Henry J. Cowan, *An Historical Outline of Architectural Science*, second edition, New York: Elsevier, **1977**.
73 Ibid. p. **106**.
74 Ibid. p. **106**.
75 Ibid. p. **107**.
76 Henry J. Cowan, *Science and Building: Structural and Environmental Design in the Nineteenth and Twentieth Centuries*, New York: John Wiley & Sons, Inc., **1978**.
77 See, for example, how an advocacy group such as Architecture 2030 restates this litany: http://architecture2030.org/the_problem/buildings_problem_why (consulted October **27**, **2013**).
78 Max Weber, *The Protestant Ethic and the Spirit of Capitalism* (**1904/05**), trans. Stephen Kalberg, New York: Oxford University Press, **2011**.
79 Lisa Wastiels, Hendrik N. J. Schifferstein, Ann Heylighen, and Ine Wouters, "Relating material experience to technical parameters: A case study on visual

and tactile warmth perception of indoor wall materials," *Building and Environment*, vol. **49**, **2012**. p. **360**.
80 Edgar Marín, "The role of thermal properties in periodic time-varying phenomena," *European Journal of Physics*, **2007**, vol. **28** (3). p. **433**.
81 Atsumasa Yoshida, Kakeru Kagata, and Tetsuya Yamada, "Measurement of Thermal Effusivity of Human Skin Using the Photoacoustic Method," *International Journal of Thermophysics*, vol. **31**, **2010**. pp. **2019–2029**.
82 Yoshiro Obata, Kazutoshi Takeuchi, Hiroshi Imanishi, Yuzo Furuta, and Kozo Kanayama, "Engineering Evaluation of Tactile Warmth for Wood," *Int. J. Soc. Mater. Eng. Resour.* vol. **10**, no. **1** (Mar. **2002**). p. **16**.
83 Marín. p. **443**.
84 Bruno Latour, *We Have Never Been Modern*, Cambridge, MA: Harvard University Press. p. **10**.
85 Axel Boëthius and John Bryan Ward-Perkins, *Etruscan and Roman Architecture*, Baltimore: Penguin, **1970**. p. **271**.
86 Fikret Yegül, *Baths and Bathing in Classical Antiquity*, Cambridge, MA: The MIT Press, **1992**. p. **33**.
87 The two best general sources on the Roman baths in general are the above book by Yegül and Inge Nielsen, *Thermae et Balnea*, Aarhus: Aarhus University Press, **1990**.
88 Yegül. p. **38**.
89 Daniel Krencker, *Die Trierer Kaiserthermen,* Augsburg: Dr. Benno Filser Verlag, **1929**.
90 Janet DeLaine, "The Baths of Caracalla: A Study in the Design, Construction, and Economics of Large-scale Building Projects in Imperial Rome," *Journal of Roman Archaeology Supplementary Series*, no. **25**, **1997**.
91 Yegül. p. **356**.
92 Marcus Vitruvius Pollio, *Vitruvius: The Ten Books on Architecture*, trans. Morris Hicky Morgan, Cambridge, MA: Harvard University Press, **1914**. Book V, Chapter X, p. **157**.
93 On material supply and processes, see DeLaine, pp. **85–193**. For water supply, see Leonardo Lombardi and Angelo Corazza, *Le Termi di Caracalla*, Rome: Fratelli Palombi Editori, **1995**.
94 The most recent attempt at quantifying the fuel load of a bath is the dissertation of Ismini Miliaresis, *Heating and Fuel Consumption in the Terme del Foro at Ostia*, PhD thesis, University of Virginia, **2013**. The collective attempts at fuel loading estimations, though, remained mired by the limitations of the varying thermodynamic expressions of the baths and other partitioned explanations.
95 Yegül. p. **365**.
96 For example, see Taylor Oetelaar, Clifton Johnston, David Wood, Lisa Hughes, and John Humphrey, "A computational investigation of a room heated by subcutaneous convection — A case study of a replica Roman bath," *Energy & Buildings*, vol. **63**, **2013**. pp. **59–66**; James W. Ring, "Windows, Baths, and Solar Energy in the Roman Empire," *American Journal of Archaeology*, vol.

100, no. 4, Oct. 1996. pp. 717–724; Tahsin Basaran, "The heating system of the Roman baths," *ASHRAE Transactions*, vol. 113, pt. 1, 2007. pp 199–205.
97 Edwin Daisley Thatcher, "The Open Rooms of the Terme del Foro at Ostia," *Memoirs of the American Academy in Rome*, vol. 24, 1956.
98 Thatcher was the first to integrate these modern standards into the analysis of an ancient architecture, but nearly all subsequent characterizations do so as well.
99 Yegül. p. 34–5.
100 Jane DeRose Evans, "Heating and Water Supply Systems," in Jane DeRose Evans, ed., *A Companion to the Archaeology of the Roman Republic*, Hoboken, NJ: Wiley-Blackwell, 2013.
101 Tahsin Basaran and Zafer Ilken, "Thermal analysis of the heating system of the small bath in ancient Phaselis," *Energy and Buildings*, vol. 27, 1998. pp. 1–11.
102 Thatcher. pp. 218–219.
103 Oetelaar et al. pp. 59–66.
104 Fikret Yegül with Tristian Couch, "Building a Roman Bath for the Cameras," technical report, *Journal of Archaeology* 16, 2003. pp. 169–175.
105 Edwin Daisley Thatcher, "Solar and Radiant Heating — Roman Style: The Open Rooms at the Terme del Foro at Ostia," *The Journal of the American Institute of Architects*, March 1958. p. 121.

"The general struggle for existence of animate beings is therefore not a struggle for raw materials — these, for organisms, are air, water, and soil, all abundantly available — not for energy, which exists in plenty in any body in the form of heat albeit unfortunately not transformable, but a struggle for entropy, which becomes available through the transition of energy from the hot sun to the cold earth."[1] **Ludwig Boltzmann**

The Architecture of Dissipation

The *transfer* of heat through a wall, roof, or foundation is but one form of dissipation that matters in the context of a building's overall energetics: its total energy system. Much larger and longer scale forms of *transition* and *dissipation* are inherent in every building but are infrequently considered in contemporary architecture. This collective range of scales, however, at last pushes the topic of energy in architecture closer to the thermodynamics that drive and support life. "Life," ecologists James J. Kay and Eric D. Schneider note, "is a response to the thermodynamic imperative of dissipating gradients."[2] Architecture, like life itself, is a form that exists to dissipate available energy gradients in the most powerful way possible. This universal thermodynamic principle, in all its manifold implications for design, is the core of any vital architectural agenda for energy.

Typically, architecture and building science treat these larger spatial and longer temporal scale forms of energy transition and dissipation as externalities. But why externalize that which helps you understand the dynamics of architecture and life itself? Given their magnitude, reach, and impact, these multiple and simultaneous forms of energetic transition and dissipation should not be isolated from the concerns of contemporary architects, for they have great architectural potential. Anything less is a rather cynical fiduciary, political, ecological, and architectural closed system that at best mocks the thermodynamics of life. Whether the most archaic or the most contemporary, buildings and cities — as manifestations of self-organized processes — exist to dissipate energy. Whether it is the evolution of a species, the evolution of civilizations, or the evolution of an architectural agenda for energy, designs that dissipate energy gradients by maximizing their power and entropy will prevail given perennial thermodynamic indicators and tendencies. This thermodynamic endgame for non-isolated systems is abundantly clear. What is less clear is how to design and direct architectural and urban outcomes to achieve this end.

In many cases, there is far more at stake architecturally — in terms of both ecology and urbanization — in these larger scales of dissipation than in the thermal transfers of a building envelope alone or other preoccupations associated with closed systems. The ambition here is not to shift energy discourse in architecture solely to these larger formations of energy, but rather to recursively oscillate be-

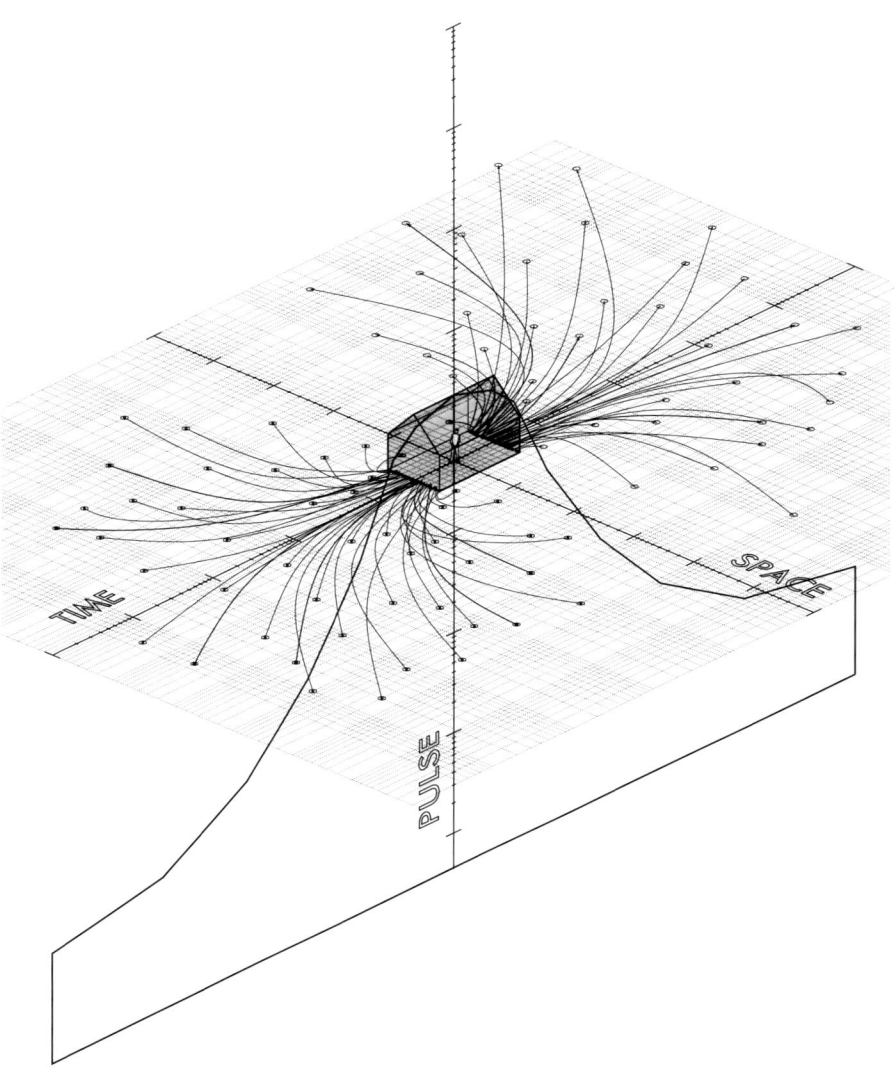

Architecture is a Design of Spatial and Temporal Pulses of Energy
Every building is but the hardened edge of multiple, simultaneous dissipations of energy. The relative magnitude and velocities of this multi-scalar convergence of designed dissipation establish the total thermodynamic milieu of a building. Modern architects were constitutively trained to consider the object, not the system; to resist dissipation in the object rather than conceive of powerful dissipative flow systems.

tween the small and large scales and forms of energy inherent in a building. Modern architecture tended to isolate the large to focus on the small scale of energy in buildings. A more conjugate, non-modern praxis for the total dissipations of a building is thus essential to the greater thermodynamic depth inherent to architecture and urbanization.

Dissipation

This chapter is concerned with the dissipation of energy. Dissipation refers to the flux and transitions (movement) of energy in a system. Recall that energy reflects the capacity to do work. Dissipation therefore refers to the shifting of work capacity through a system. In the process, available energy will become bound energy, that is no longer available for work.

There is nothing shocking about this explanation of energetic dissipation, especially given its roots in the classical thermodynamics of the nineteenth century. What is more evolutionary for the energy discourse in architecture is the role dissipation plays in the non-equilibrium thermodynamics of open systems, as Boltzmann and Lotka began to articulate. How the perception of dissipation shifts from the fate of heat death of thermodynamic equilibrium in closed systems to the very impetus and engine of thermodynamic vitality has great consequence for design today. Dissipation and dissipative structures are the thermodynamics that float the complex behaviors of life and they reveal inordinately consequential lessons about how, and why, to dissipate energy in one way over another through design.

Thus, I contend that the key to a more totalizing architectural agenda for energy in the non-equilibrium thermodynamics of open systems — the thermodynamics that apply to buildings and cities — is to finally ascertain, if not design, the dissipation of energy at multiple, simultaneous scales and forms. The dissipative structure of a building should be the central concern of design. This is the prime purpose and ultimate energetic explanation of any organization of matter, energy, and life, not the least of which includes architecture and urbanization.

Given the manifold dissipations of energy in buildings in cities, it simply makes no sense to mechanically maintain curiously isolated system boundaries focused on one form of energy — such as isolated from other scales of consideration — for this would occlude an understanding of how and where to best intervene and design the energetic dissipations that are constitutive of buildings.

What is ultimately at stake in this book, then, is the radical distinction between *isolated perspectives on energy* and *non-isolated perspectives on the dissipative structure of architecture*. Understanding the dynamics of dissipation is critical to grasping this distinction and the importance of an isolating versus a non-isolating agenda for energy. So whereas the insulation apparatus was historically structured on concepts and practices of isolation, this chapter will consider alternatives to these concepts and practices: non-isolated buildings and the dissipative structures that presuppose them. It is time to de-isolate both epistemologies of energy in architecture and the energetics of buildings.

The enlarged, more totalizing thermodynamic agenda for energy — an un-insulated/non-isolated agenda — of this chapter builds on the insights of scientists and related thinkers, such as Ludwig Boltzmann, Alfred J. Lotka, Georges Bataille, Howard T. Odum, Robert Ulanowicz, Ilya Prigogine, Eric Schneider, James Kay, and Adrian Bejan: all keen observers of the thermodynamics that both trigger and buoy the operations of life. The power of their insights into the far-from-equilibrium thermodynamics of systems and design has yet to fundamentally reshape architecture's understanding of energy and its associated formal and ecological potentials.

Reckoning

Curiously, observations on non-equilibrium thermodynamics developed exactly parallel in time to the apparatus of insulation, yet the latter was routinely isolated from the import of these emergent descriptions of reality. These historically simultaneous but divergent epistemologies about energy in architecture now demand reckoning.

Given the historical arc of insights and oversights of insulation practice presented in the previous chapters, the intent of this chapter is to frame greater ambitions for the transitions and dissipation of energy that pertain to buildings. Astute designs for the total dissipation of energy in buildings stand to radically transform the purpose and activity of design towards far more ecologically and architecturally powerful modalities. Formulations and formations that exceed the unfounded modern managerial ethos of energy efficiency are not only possible, but are urgently needed in the twenty-first century. We need to move swiftly towards vital and fecund non-modern futures not only for buildings and cities, but also for all those unwittingly specified environments that constitute the externalities of modernity. We can no longer insulate architecture from the delirious possibilities of the dissipations inherent in its constitutive necessary excess.

Insulating concepts and practices were developed to minimize the dissipation of heat in short-term energy cycles. But the paucity of concern about the dissipation of energy over long-term energy cycles in insulation practices constitutes another chronic omission in the development of insulation. This omission and its implications are the focus of this chapter. To understand this chapter, one must first understand the multiple forms of diffusion and dissipation that are inherent in buildings. These multiple forms of energy are best understood through their relative durations and velocities: from nearly instantaneous solar radiation transfer, to the slower accumulation of energy in the mass of a building or a city over long periods of time, the rise and fall of civilizations and to the conveyor belts of plate tectonics.

To design with energy in mind is a proposition about the modulation of various energy scales, forms, and velocities: faster here, slower there, almost paused, or stored, but never ceasing. Further, the velocity design of energy dissipation presumes knowing when to accelerate, dampen, or store multiple, simultaneous forms of energy. The amount of time energy spends in an energy hierarchy is one ecological indicator of system maturity.[3] In certain cases, rapid dissipation of energy might be ecologically astute. While in other cases, the aim is how to best dissipate energy so as to increase entropy production in the most powerful ways possible. No doubt, these principles are essential to maximum power and maximum entropy production designs. While easier said than done, it is crucial to grasp the fundamental role of dissipation in the thermodynamics of buildings, designed towards maximal ends.

Fourier, Reconsidered

To begin this endeavor, it is useful to go back to the origins of heat transfer theory: to Jean Baptiste Joseph Fourier. Central to the primary motivations and contentions of this book, and to this chapter in particular, is Ilya Prigogine and Isabelle Stengers' identification of Fourier's derivation of heat propagation in solids in the early nineteenth century as the birth of the "science of complexity."[4] In science, Fourier's observation and derivation yielded much more than just the explication of conduction, although you would not grasp that from the apparatus of insulation. His partial differential equations triggered the cascade of science towards theories of irreversible processes and ultimately those about the dissipative structure of the universe. In architecture, a narrow appropriation of Fourier's observations on conduction and diffusion was one trigger for the cascade of building science towards frequently isolated and narrow concerns about energy. It is time for architects to revisit Fourier's insight, especially in the context of non-equilibrium thermodynamics.

The persistent focus on conventional notions of conductivity in building science is perhaps the best example of the delta between what Fourier's early experiments wrought in architecture and what they could provoke for the future of design. There was more to Fourier's derivation than heat moving from one end of a bar to another and there is more to heat transfer than what building science represented in modernity. His observations about both the diffusion and dissipation of heat are important in the context of buildings at multiple and simultaneous spatial/temporal scales. The Prigogine-Stengers claim about Fourier's inauguration of the science of complexity is also important. If nothing else, it suggests that there is far more at stake in the dissipation of energy in a building envelope than any isolated equation or reductive value might suggest.

Building insulation practices never incorporated, or fully participated in, the profound cultural and intellectual transformations that were triggered by Fourier's observations. Insulation and the science of complexity are seemingly quite discrete topics. However, their shared origin in Fourier's work is more than suggestive of the intricate consequences they imply for design. The transition of heat through a material or material assembly and the more systemic dissipation of energy in the production of that material or material assembly should be directly related because they are related in reality. There is no thermodynamic, ecological, or architectural reason to isolate the former from the latter. Both are relevant but, most importantly for the vitality of architecture and life itself, the latent complexity of adaptive feedback in that system is only possible when they are considered together in a recursive methodology.

To be absolutely clear, the thermal conductivity, thermal heat capacity, and thermal diffusivity are indicators of small-scale heat flux. The dynamics of exergy consumption, the accumulation of emergy, maximum entropy production, residency time of energy in a system, and the resultant adaptive feedback of a system are indicators of larger scale heat flux for dissipation in an architectural agenda for energy. One of these scales or one of these parameter sets means relatively little without the others in the overall energetics of buildings and cities. The latent complexity involved in these simultaneous scales and forms of dissipation should be of extreme interest to architects and their architectural, urban, and ecological ambitions.

Again, the complexity discussed here should not be confused for the mere complications of technique or shape that are so often associated with complexity in architecture. To be more scientifically precise, complexity here is a measure of the capacity for adaptive feedback in a system. This vital complexity could occur in the most simple and in the most intricate and complicated of building formations. Shape matters, but not in the conservative, parochial way it typically does in design discourse today. What is of consequence here for any architectural formation is the complex, adaptive feedback capacity of that formation. The shape spaces, supply chains, material geographies, logistics, and fate of a building are just as relevant as the materials, systems, operations, and shapes of a building as an object. The capacity of feedback amongst material, climate, physiology, and space are of course deeply intertwined but so are the seemingly more temporally and spatially remote forms of energy so often externalized in modern and contemporary architecture. This would be inherent to the formation of novel thermodynamic architectures for this century.

Thermodynamics & Dissipation

Building on Prigogine and Stengers' claim that Fourier signaled the birth of complexity science, Luis Fernández-Galiano and historians of science identify Sadi Carnot's 1824 publication of his *Réflexions* — coupled with the 1850 publications of Joule and Clausius — as the critical transition from a mechanistic epistemology to one in which entropy and the dissipative structure of reality began to take hold in multiple disciplines, from physics to biology to art.[5] This epistemological shift, well documented in multiple sources but summarized briefly below, was essential to the full development of thermodynamics and its full fluorescence into the non-equilibrium thermodynamics that characterize life. A brief review of the development of these laws, and the laws themselves, are important to understanding the thermodynamics of architecture from a non-isolated perspective.

The term thermodynamics, initially *thermo-dynamics*, was coined by William Thompson (also known as Lord Kelvin) in 1848 and defined in 1854. In 1852, he articulated concepts of universal heat death and the dissipation of energy. Thompson observed: "When heat is created by any unreversible process (such as friction), there is a dissipation of mechanical energy, and a full restoration of its primitive condition is impossible."[6] The core of the first and second laws is embedded in this observation: the quantity of energy in a closed system will remain constant whereas the qualities of that quantity of energy will not remain constant.

Not only will the qualities of energy change, but they will move in a particular direction. In 1865 Clausius made two contributions to this evolving understanding of thermodynamics, "*1.) Die Energie der Welt ist constant. 2.) Die Entropie der Welt strebt einem Maximum zu.*"[7] In other words, the energy of the world/universe is constant and the entropy of that system strives towards a maximum. This latter is the core of the second law. As Fernández-Galiano states, "The philosophical and scientific importance of the second principle can hardly be overestimated."[8] In architecture, this principle has been under-considered, if not grossly underestimated.

It is important to reflect on the observation that entropy will strive towards a maximum. With this observation, Clausius triggered one of the most compelling, albeit slippery, principles of dissipation. While the notion that entropy increases — strives — towards a maximum is all but a colloquialism of any high school physics course, the implications for how systems are therefore organized is not as immediately familiar. Open, non-isolated energy systems emerge — think of a tornado, a hurricane, a building, a city — to dissipate available energy gradients. Successful systems will strive to generate maximum entropy. Again, thermodynamic principles state that the systems which prevail in a process of natural selection will be those designs that extract maximal power from an available energy gradient and in so doing maximize the production of entropy, thus re-radiating remaining energy at the lowest possible level.[9] They will have maximized their exergy consumption and in so doing yield maximum entropy. This is the endgame of an open, non-isolated system and it explains how such systems will persist and prevail.

If buildings were closed, isolated systems, the role of dissipation and entropy inherent in the second law would be a very problematic issue for buildings, cities, and life. Their fates would be known: the thermodynamic death of equilibrium. Closed thermodynamics push towards a paralyzing equilibrium state, while open thermodynamics push towards vital states that are far from equilibrium. The role of entropy in the former is rather pessimistic while the role of entropy in the latter is quite optimistic.

To think of entropy as wasted energy would be to greatly misconstrue the topic. It is better to understand entropy as the remainder of an exergy consumption process, a structure of dissipation that mostly consumed the exergy of a particular available energy gradient. In this way, maximum entropy and notions of maximum power are intimately connected. When Clausius observes that systems will strive towards maximum entropy and Boltzmann characterizes life as a struggle for entropy, both are peering into the life-amplifying dynamics that others have characterized as maximum power systems.

A Struggle for Available Energy or Entropy?

Population dynamicist Alfred J. Lotka, after Boltzmann, noticed a fundamental relationship between energetics and evolution. Seeing a connection perhaps between the dynamics of evolution and the behavior of energy systems, Lotka stated:

> "it has been pointed out by Boltzmann that the fundamental object of contention in the life-struggle, in the evolution of the organic world, is available energy. In accord with this observation is the principle that, in

the struggle for existence, the advantage must go to those organisms
whose energy-capturing devices are most efficient in directing available
energy into channels favorable to the preservation of the species."[10]

Efficiency here has a specific meaning. When writing about energy and efficiency, Lotka does not have in mind what would otherwise be easy today to associate with the discourse on energy efficiency. "The problem," ecologists R. E. Ulanowicz and B. M. Hannon observe, "lies in the ambiguities associated with the concept of 'efficiency.' It usually connotes conservative behavior (a first-law notion) which in turn is often paired with minimizing losses, including dissipation."[11] Something more specific in concept and method is embedded in Lotka's assertion that "energy-capturing devices are most efficient in directing available energy into channels favorable to the preservation of the species." In other words, the "efficiency" of a structure in a non-isolated system is quite different from the construal of the efficiency concept in equilibrium thermodynamics.

Ulanowicz and Hannon help parse the more specific and productive meaning of efficiency:

> If two systems receive the same quantity of energy at the same entropy, that system which extracts the most work from its input before releasing it to its environment (as it inevitably must) can be said, in the second law sense of the word, to be the more efficient utilizer.[12]

Efficiency, if used at all, should be tied to this specific usage. Ulanowicz articulates that the maximum consumption of exergy — the portion of energy that can do work per unit energy — is the goal. Construing energy design as the means to minimize energy consumption in non-isolated systems, as in the energy efficiency discourse, is sloppy, misleading, and ultimately ecologically problematic. As such, it summons a range of errant concepts and practices. With the more specific understanding of the variable exergy to entropy ratios of an energy system in mind, Lotka's intent becomes even clearer in a subsequent passage:

> In every instance considered, natural selection will so operate as to increase the total mass of the organic system, to increase the rate of circulation of matter through the system, and to increase the total energy flux through the system, so long as there is presented an unutilized residue of matter and available energy.[13]

At no point does Lotka deem the purpose of a vital energy system to minimize energy. This is not what he has in mind when he uses the word efficiency. Rather, the systems that prevail will maximize energy and matter (and recall that matter is but captured and stored energy). Not only will a system strive to increase the quantity of energy, the velocity of the energy will also increase. In short, maximal energy will be dissipated in the system for any available gradient.

It is this Lotka passage that in turn guided Howard T. Odum towards his construction of the maximum power principle: "During self-organization, system designs develop and prevail that maximize power intake, energy transformation and those uses that reinforce production and efficiency."[14] The use of efficiency in the open thermodynamic sense is likewise maintained here while the focus on maximum energy is refined through maximum intake, transformation, and use that recursively reinforce production and efficiency. Elsewhere Odum and Richard Pinkerton note about efficiency, "natural systems tend to operate at that efficiency which produces a maximum power output."[15] Most importantly, they further observe that, "these systems perform at an optimum efficiency of maximum power output, which is always less than the maximum efficiency."[16] Add to this the Jevons paradox and it is clear that a focus on maximizing energy efficiency will not disclose much about the ideal dissipations of energy in a system and is likely counterproductive.[17] Take care of the power of the system and the necessary efficiencies will generally take care of themselves.

In short, there is a profound difference between energy efficiency and exergy efficacy. Any astute observer of thermodynamic systems — whatever their field — eventually arrives at the latter. So, eventually, will architects as they move beyond the platitudes of the former.

This arc of thought between maximum entropy and maximum power navigates what have become slippery adjacent concepts. What complicates this series of thermodynamic observations is the apparent contradiction between the ends of maximum entropy and maximum power.

Power or Entropy?

As you read, Lotka based his observations about the energetics of evolution — "the fundamental object of contention in the life-struggle, in the evolution of the organic world, is available energy" — on a seemingly contradictory passage from Ludwig Boltzmann, quoted at the beginning of this chapter. Boltzmann claimed that the struggle for existence was a struggle for entropy, not its corollary — available energy. Did Lotka misunderstand? Did Odum, too, misunderstand?

This apparent discrepancy has engendered much debate: is life a struggle for exergy or entropy, a maximum or minimum? It is important to recognize that the two constructions are one in the same principle. Again, for any unit of energy, the aim should be to maximize entropy, which implies that the system has consumed as much exergy as possible from that unit of energy. It would be of no use to maximize available energy unless it is ultimately dissipated as entropy, for little work would be done. "In other words," Schneider and Kay note,

> ecosystems develop in a way which increases the amount of exergy that they capture and utilize. As a consequence, as ecosystems develop, the exergy of the outgoing energy decreases. It is in this sense that ecosystems develop the most power, that is, they make the most effective use of the exergy in the incoming energy while at the same time increasing the amount of energy that they capture... thus we expect a more mature ecosystem to degrade the exergy content of the energy it captures more completely than a less developed ecosystem... more organization emerges to dissipate the energy."

An example from landscape ecology will be instructive here. Consider the energetic analysis of various forest types. Jeffrey Luvall and H. Richard Holbo measured the net flux of solar radiation entering forest types as well as the degradation of that solar gradient.[18] It is readily apparent that a 400-year-old Douglas fir forest degrades — processes and dissipates — 90% of its incoming energy whereas a younger, less complex forest may yield only two thirds of its solar input. This observation points to something essential about the thermodynamics of any self-organized system. The forest degrades the solar gradient in a mature system "design" refined over millions of years, storing some energy here, pumping and transpiring water there, photosynthesizing a very small amount. The measured velocity of each forest type — each dissipation "design" — is essential to the dissipative structure of the forest. The younger forest might spend more on mass growth, for instance, that less fully degrades the solar gradient.

The capacity of the old forest to dissipate its solar income almost completely is an indicator of its thermodynamic depth and complexity. The 400-year-old forest itself no doubt reflects an incredible accumulation of available energy in its biomass alone — Lotka is correct — but it also reflects life's struggle to not just accumulate available energy but to dissipate it through cunning exergy-matching use, the cycling of "waste" reinforcements, and ultimately to degrade the maximized entropy into non-radiative processes. So Boltzmann, too, is correct: the whole endeavor is a struggle for entropy (spent exergy). Lotka did not go far enough but said nothing that would contradict Boltzmann or a contemporary understanding of the non-equilibrium thermodynamics. How far could architects go if they accordingly designed the dissipations inherent to a building and city?

Dissipative Structures

To help more fully understand why successful energy systems are organized the way they are, Prigogine and Stengers explain that "structures are created by the continuous flow of energy and matter from the outside world; their maintenance requires a critical distance from equilibrium, that is, a minimum level of dissipation. For all these reasons we have called them dissipative structures."[19] We have to see any energy system — a tornado, a forest, a building, a city — as a dissipative structure organized to cycle energy in a very particular way. The formation and configuration of any energy system grants increasing access to the flux of energy.

It is essential to keep in mind that architects engage non-isolated thermodynamic systems. These generally exhibit non-linear causes and effects. Therefore, a non-modern agenda for energy will target certain thermodynamic indicators — such as emergy-to-exergy ratios, scaling laws, and maximum entropy production — as much as any specific optimization workflow to design novel thermodynamic formations and figurations for its inevitable dissipations.

Novel structures of entropy and order can, and do, emerge from a context characterized by dissipation that can cycle up to macro-structures, like buildings and cities. This is absolutely essential for understanding the role of a building in its energy hierarchy and thus, its role in its ecology. Again, if buildings and urbanization were closed, isolated systems, they would by law run down through an entropic process of dissipation. But yet, buildings, cities, and life persist. It is precisely their openness, their *non-isolated* thermodynamic constitution that is their energetic core and source of persistence and potential vitality. New orders emerge from both the dissipations in an object/system boundary and the energetic transition s between that object/system boundary and its surroundings.

For buildings and cities, this life-persisting novelty is anything but an isolated process. Theories and practices based on isolation thus offer little insight or direction for thinking about energy in this larger, more totalizing way. "The point is that," as Riccardo Pulselli and Enzo Tiezzi state, "interaction of the system with the outside is the condition for life of the system."[20] To explain dissipative structures, Prigogine offered this equation:

$$dS = d_iS + d_eS$$

Here d_iS is the dissipation inherent in the object/system and d_eS is the dissipation exchanged with the exterior of the system. In the insulation apparatus of buildings, architects, and engineers only consider d_iS. Yet, it must be grasped that d_eS is what produces all the order and novelty latent in the system. Thus, these larger transitions and forms of dissipation must be part of an architectural agenda for energy. Architects must more broadly apprehend and design these dissipations to make any valid claim about energy or the ecology of buildings. Anything less is merely an isolated explanation characterized by curious, convenient system boundaries.

Constructal Law

The slippery concepts of maximum entropy and maximum power that would ideally prevail in dissipative structures, and their connections to both the thermodynamics of life and design, is somewhat clarified in the applied physics of Adrian Bejan. With the Boltzmann/Lotka/Odum observations above in mind, consider this observation by Bejan, what he calls the "Constructal Law:" "for a finite-size flow system to persist in time (to live) it must evolve such that it provides greater and greater access to the currents that flow through it."[21] Bejan eschews the slipperiness of the maximum entropy/exergy semantic tussle by observing that systems will prevail that provide increasing access to dissipation, acknowledging that the essential struggle is for the designed use of both exergy consumption and entropy production.[22] Increasing access to dissipation in a self-organized system points towards increased choice about how to direct and modulate the transition inherent to dissipation.

Manifested in multiple books and articles is Bejan's obsession with the necessity of design that appears with any prevailing flow structure: that is, any dissipative structure. For Bejan, the often dendritic patterning and rhythm of energetic flux, and the consistent appearance of scaling laws for size and configuration in all flow structures are each important indicators of the fitness of design to the non-equilibrium thermodynamics. It should be deeply interesting for architects that applied physics now understands optimal flow structures in the terms and variables of *design*. The rudiments of design — size, shape, location, orientation, treatment, rhythm, sequence — are fundamental in the lexicon of energy dynamics observers. The aim is to dissipate an energy gradient in designed ways. Architecture is a dissipative structure, a flow structure that exists to dissipate available energy gradients in the most powerful way possible. The vitality and role of design in this non-isolated perspective on the non-equilibrium thermodynamics of a building is radically different from the unfounded aims and constrained system boundaries of a zero-energy building, for instance.

From Fourier to Bejan, this historical arc of ideas is essential to understanding the basic dynamics of dissipation. Fourier observed and quantified a process of dissipation — the flux of energy. Clausius and others observed that part of this dissipation, entropy, tends towards a maximum. This led Boltzmann, Lotka, and Odum to see life as a struggle for maximum entropy and ultimately maximum power. Bejan sees design as the process of directing and patterning this dissipation.

Non-classical, Non-modern Design

What is fundamentally at stake in this book is the difference between the perceptions of entropy as waste — something to be minimized through design — versus an understanding of entropy as absolutely inevitable as it is productive. Non-equilibrium thermodynamics offer more optimistic and pointed perceptions about inevitable dissipation: the cycling of energy in a system can and does become a source of order. As Fritjof Capra observed,

> In classical thermodynamics, the dissipation of energy in heat transfer, friction, and the like was always associated with waste. Prigogine's concept of a dissipative structure introduced a radical change in this view by showing that in open systems dissipation becomes a source of order.[23]

Inducing the latter, non-classical perception through design produces an entirely different paradigm for the role of energetic dissipation in architecture. The former classical interpretation of thermodynamics dominates the modern idea of insulation while the latter stands to transform fundamental attitudes and assumptions about the role of energy dissipation in buildings, from form to energy analysis and accounting. Instead of reductive platitudes about R-values, architects should grasp and design the sweeping array of energy transitions inherent to buildings and deduce cogent strategies for modulating and directing that dissipation in powerful ways.

Architecture, today, needs this non-modern approach to the thermodynamic figuration of not only buildings but the emergetic supply chains and next-uses of buildings to make any worthwhile claim about energy and the whole energy hierarchy of a building. This no doubt is foreign to the more isolated practices and pedagogies of architecture that emerged in modernity, but it is essential to more vital futures for the discipline. Non-classical thermodynamics and non-modern notions of energy design are the métier of designed dissipations.

Designed Dissipations

In the introduction to this book, I described buildings as non-isolated, open, dissipative structures. One way to temper the insulating/isolating logics of building science in the twentieth century is to accept and embrace, as an exergy/entropy function, the dissipative reality of buildings and their heat. Rather than seen as a Calvinist source of guilty waste, the inevitability of energy transition and dissipation might be put to work through exergy design rather than be resisted. This nudges design closer to the non-equilibrium thermodynamics of reality.

Designed dissipations in the thermal diffusivity, the thermal storage of heat capacity, and thermal effusivity of building masses are but a few, *small* building-scale examples that just barely point towards some of the larger forms of dissipation that could be engaged by architects. Such an approach resists the tendency to (only) resist heat flow that an insulating habit of mind suggests. Instead, the aim should be to design heat flow in architecturally, thermodynamically, and ecologically powerful ways. Whether at the scale of a building or in the many externalities of modernist architecture, dissipation is a central topic for the future of architecture.

To be architecturally productive in multiple — and simultaneous — temporal and spatial dissipations requires a more recursive approach to the design of a building and its energy hierarchies. In a thermal context, it is not inherently problematic that heat flows from a warm body and a warmed interior into or out of the mass of a building. The issue is how architects can begin to design more sound ways to capture and channel that thermal capacity so as to amplify architecture. To do so more fully engages the great mass of any building and its material geographies, as well as the great quantity of energy that courses through a building, for buildings are non-isolated, open, dissipative structures.

Velocity

In either the short or long-term consideration of dissipation, the central operative concern is the relative rate of dissipation and relative residency time of energy in a system. Whether to speed up, slow down, or store, the fundamental task of design in the context of dissipation is modulation. Modulating dissipation in the context of buildings is an inordinately intricate task, exactly where architecture engages actual complexity. To envisage the many flows of energy involved and their complexity-inducing feedback mechanisms requires reflection to fully discern, much less design. In other words, architects rarely think about the thermodynamics of civilization when they think about energetic dissipation. However, it is at best a grave thermodynamic error to only think about energy at a certain scale over all the others. Precisely because buildings are inherently non-isolated, their great potential complexity warrants studied attention and velocity is one of the soundest ways to gauge this complexity. This is not a simple task but for today's most ecologically ambitious designers, it is nonetheless essential.

Accelerating, slowing down, and storing for later — each a potential mechanism of adaptive feedback — are each essential functions of the velocity modulation of energy design in architecture and urbanization. Regarding insulation, the conventional role of conductivity has obviously been to reduce energy to the velocity of heat conduction through a building envelope. This relatively small role reveals little about what the purpose of the slowed velocity of this very low-quality energy actually is in the design. Insulation materials and practices will dampen the velocity of heat transfer in this narrow way, but they reveal very little about the character of larger energy transitions related to the building. Other forms of dissipation must be considered.

Buildings, it must be understood, have varying velocities of matter and energy. A building as a whole might appear static, but only from a narrow, anthropocentric perspective. The extraction and convergence of material into a building and the weathering and/or obliteration of a building all have velocities attached to them. Sped up, the construction of a city (buildings organized on plots) would look very much like the convection of Bénard cells. Architecture is but the temporary expression of vast amounts of matter and energy that come and go in the nascent fury of civilization. Architects have parochially privileged shorter-term fluxes of energy, but that habit need not persist. As a matter of natural selection, as Lotka indicates, the velocity of matter and energy is the central concern.

Energy Accumulation

The modulation of energy velocity determines rates of dissipation, accumulation, and degradation. Faster rates reflect greater energy intake and dissipation, while slower rates release or accumulate over a longer period. Both will be part of any energy system design as that which grants greater access to the flow and flux of energy. Our concern, however, should be how the design of the dissipations, velocities, and accumulations triggers productive feedback dynamics in the system and what effect they have on the total degradation of energy.

These dissipative dynamics are at the core of two systems of thought that might otherwise appear to be disparate. Both focus on accumulation and dissipation as operative principles in the world and its systems. What is in part compelling is that although they emerge from different contexts — economic critique and ecological energetics — they point towards startlingly similar ends.

On the one hand, Georges Bataille's notion of the General Economy is a focused non-isolated economic perspective based on accumulation and, thus, on surplus rather than scarcity.[24] This shift from culturally-imbued notions of Calvinist frugality in isolated systems to the simple necessity of accumulation and abundance in open energetic systems is essential to grasping the dissipative structure of architecture and the thermodynamics of life. Bataille's point of departure was the superabundance of solar energy in the open system of the planet as the real basis of an economy.[25]

On the other hand, Howard T. Odum, too, uses the superabundance of solar energy as the basis of his ecological accounting system. Bataille and Odum are in many ways the two sides of a solar-driven *oikos*, the Greek root for both economy and ecology. Odum conceptualized energy hierarchies as the convergence of energy into increasing accumulations and feedbacks of emergy.[26] His methodology is the most comprehensive accounting of the dissipations and degradations of energy in any system. It is a recent methodology, not yet a complete picture of an extraordinarily hybrid world, and this fact deeply troubles modern scientists who dismiss the method for more isolated system boundaries that are more readily and knowingly quantified. But it is precisely Odum's breathtaking attempt to delve into our inability to know the difficult whole that is his great contribution. With his energy hierarchies, architects can finally consider buildings within the necessarily messy questions of the energetic velocities, accumulations, and dissipations inherent to architecture. The emergy concept and methodology, if anything, helps re-describe and question relevant system boundaries for architecture. There is no more worthy task today for a non-modern designer than to question their system boundaries and those of their buildings.

Accumulation in Architecture

Because matter is captured and stored energy, the accumulation of matter in a building is a useful example to consider. As one of architecture's primary energy velocities, the accumulation of matter in a building must be seen as a store of certain forms of energy for certain durations to be fed back in non-simple ways. This inevitable, slower energetic velocity of a building and the resultant residency time of energy in the system resonate strongly with the fundamental reality of architecture's accumulation of matter and energy; the "great weight" and "amassment" of architecture as Alberti referred to it:

> Him I call an architect, who, by sure and wonderful art and method, is able, both with thought and invention, to devise, and, with execution, to complete all those works, which, by means of the movement of the great weights, and the conjunction and amassment of bodies, can, with the greatest beauty, be adapted to the use of mankind: And to be able to do this, he must have a thorough insight into the noblest and most curious sciences. Such must be the architect.[27]

If we accept Alberti's definition of architecture, then the inherent thermodynamics of the movement, conjunction, and amassment of buildings as dissipative structures is absolutely essential. This all depends, as Alberti claims, on the most noble and curious of sciences. Today there is no more noble or curious science than the hybrid endeavor of non-equilibrium thermodynamics in the context of architecture.

Otherwise dispersed forms of energy converge, through directed design, from matter to materials to products to components to systems to the final building itself. This is a process of accumulation and dissipation, riddled with questions about velocity. The question, then, is what is done with that massive accumulation and dissipation of energy?

In pre-modern, and now non-modern, modalities, people squeezed as much exergy from any quantity of energy — from use to exterior and interior climate modulation to defense to food storage, etc. — constantly working to eliminate energy gradients by circulating ever more energy in varying velocities in the most ingenious of ways. There is no other thermodynamic explanation for the emergence of buildings, cities, books, libraries, petroleum, industrial production, agriculture, and eventually the Internet, than the deepest-seated necessity to increase energy flux and to abhor a gradient. Somehow, though, by the end of the twentieth century, the only accumulated energy quantified and evaluated in architecture was air temperature, dissipated BTUs, or perhaps the lighting levels. This was perhaps a necessary thermodynamic regression, but there is now so much more to consider.

Great Weight

As the first chapters demonstrated, the great weight of a building was routinely left out of the development of building warmth and insulation strategies. Compelling, intricate examples emerged, but the trajectory of typical energy practice was to bracket out the more unwieldy power of the amassment of a building. In the modern mind, the amassment of a building complicated the affairs of insulation. While the accumulation of mass in a building — as one of architecture's primary ways of dissipating available energy gradients — had been used in multiple cultures throughout the world in the history of building, suddenly building scientists and others in the expert culture of modernity deemed it irrelevant, generally due to its intricacy. What is easy to measure and quantify is too often what gets quantified. This is a strange perversion of the energy systems of buildings. In this way, insulation theory too often became a pillow for ecologically/thermodynamically disconnected, if not unambitious, modern minds.

In the non-modern mind, the inevitability of this amassment in architecture is a source of worthy complexity and thermodynamic depth. The reality is that the accumulation of mass affords multiple ways to dissipate not only thermal energy but also moisture, sound, and other, much larger forms of energy, such as information. Buildings as not only captured energy but, more importantly, as captured information, are impossible to overestimate in the non-linear dynamics of non-isolated systems. If energy residency time in a system is a key indicator of system maturity, then everything from typology to construction technique to adaptive re-use are all inordinately essential thermodynamic functions in architecture.

In contrast to the emergy-to-exergy ratios of a double-skin glass façade, for instance, the knowing, simultaneous exergy design of these multiple forms of energy with a single, perhaps low-emergy material, points towards a very different habit of mind. In terms of accumulation and excess, what a designer does with that excess matters far, far more than the excess itself. If there is any doubt, revisit the Luvall-Holbo observations on 400-year-old Douglas fir forests.

In architecture, the massive accumulation of energy in all its forms and formations in a building is routinely cited as an ecological problem, perhaps *the* ecological problem, with a litany of statistics about the consumption of material and energy.[28] Unfortunately, this techno-theological litany is then habitually coupled with a Calvinist notion of frugality: the excess of architecture must be minimized, or, the aim of the good is to do less bad. Efficiency is the beginning and the end of the consideration of energy in the system. This cripples the latent power of this massive accumulation of matter and energy.

For instance, from the point of view of how architects are trained to consider the thermal behavior of a building, the mass of contemporary buildings are largely understood as thermally inert and it is only the layers of insulation that are operative from an energetic point of view. Thermal diffusivity, concerned with conductivity of time and space, provides a more dynamic habit of mind. At the same time, thermal diffusivity concerns point towards more monolithic approaches to construction as a core energy agenda. An architect can diffuse and dissipate heat through the great thermally active mass of a building as an exergetic function. In such a case, dissipation is not something to be avoided through design; instead, thermal diffusion is the starting point of a far more compelling and exergetically efficacious strategy. It puts the great mass of a building to work. Designing the inherent mass of a building around thermal diffusivity permits a more specific strategy for dealing with the thermal dissipation of heat.

The modern perspective on the excess of architecture is unfortunate because architecture's excess is, in fact, its greatest ecological capacity, not its liability. Despite the reductive development of insulation theories and practices, the accumulated mass of a building is decidedly non-trivial from an exergetic and emergetic point of view. Since matter is but captured energy, the great mass of energy captured and stored — accumulated — in a building should be activated in multiple ways to an end of maximizing its power and the degradation of the solar gradient. In that way, sheer accumulation is not a problem. Again, it is in reality our great opportunity to maximize the use of that massive amount of emergy in countless ways. When it comes to energy concerns, this deserves our most intent and delirious forms of design imagination. Aiming to minimize or eliminate the dissipations and accumulations of architecture rather than aiming to maximize the power of this high-emergy accumulation simply targets the wrong end.

Ex-externalities

The above discussion of amassment considers only the accumulation and dissipations inherent in the great weight of a building itself. Tracking the dissipations and accumulations of energy throughout a building's energy hierarchy, as a cunning designer interested in energy might, becomes a central prerequisite for design. The more one tracks and observes how energy moves through a particular energy hierarchy, the more that designer can identify productive modes of adaptive feedback for the great weight of a building. What architects could achieve in the energy hierarchies of a building will require architects to peer into that hierarchy, tracking the velocities, accumulations, and dissipations of energy. This is no doubt unfamiliar terrain for the discipline, but it is fundamental to an ecological agenda.

While a designer will work to modulate the transfer of heat in a building, this is not the sole form of dissipation to consider. Further, despite its dominance in contemporary practice, it is not necessarily the most important. Regarding the emergy analysis of buildings, urban system dynamics scientist Riccardo Maria Pulselli and his colleagues found that

> Emergy inflow due to building manufacturing corresponds to 49% (considering a building lifetime of 50 years), while maintenance is 35% (maintenance needs material use as well) and building use is 15%. In other words, the choice of building materials, the energy for their production and assembly in the building yard, and their duration have to be taken into account, besides the energy consumption (electricity, fuel) during the building lifespan.[29]

The relative amounts of accumulated matter and energy in a building — its production, construction, operation, and maintenance — and the dissipation required thereof, should re-orient central assumptions about the dissipation of energy in the context of buildings. Moreover, once larger and longer scale accounts of dissipation enter the domain of design concern, the potential for important forms of complex adaptive feedback are suddenly methodologically possible. There is much at stake in this shift from the resistance to thermal dissipation to the complex adaptive feedback that is central to the dissipative structures of reality.

System Boundaries: Emergy, Exergy, & Exchange

The history of building science in general — and the insulation apparatus in particular — reflects an inexplicable focus on the objects and artifacts of buildings, a focus shared by architects in general. This focus on the object and the artifact of architecture is itself an artifact of a bygone era: an era characterized by unreflective assumptions about the system boundaries, energy hierarchies, and milieus of buildings. This is an epistemology driven by sharp distinctions: energy will be construed as separate from matter, matter and energy will be construed as separate from history, and all this will enable the thermodynamically ludicrous assertion that design is somehow possibly autonomous as an endeavor.

This was an age of separations, phantom expert cultures, and the folly of administratively convenient categories of knowledge. In the insulation apparatus of this modern period, architects, engineers, and building scientists tended to associate the boundary and domain of a building with the artifacts of its building envelope, a seemingly obvious but curiously selected system boundary for a building, from a thermodynamic perspective. Even today, the exterior envelope is often deemed to be the boundary of a building's energy system.

The system boundaries of an energy claim today should include both the short *and* long-term and the small and large cycles of energy. The degree to which time, and thus velocity, enters an agenda for energy determines much about its relative efficacy. An agenda for heat in architecture cannot be limited to the short-term — hourly, daily, seasonal — cycles of energy but must simultaneously confront the long-term cycles — decades, centuries, millennia, etc. — of energy.

Likewise, an agenda for dissipation cannot be isolated to a building spatially. The implications of dissipation on multiple spatial scales — building structure and envelope, the effects of buildings on adjacent outdoor comfort, the dissipation of energy in the material geographies of contemporary buildings and its role in patterns of urbanization — are inextricable concerns in a non-isolated agenda for energy as well.

The reality is that far broader boundaries and modes of exchange are at stake in the question of buildings and their energy systems. Addressing these broader boundaries and modes of exchange requires a shift in focus away from the artifact mentality that dominates so much of design, research, analysis, and codes. Careful attention to what system boundaries matter for a building in a particular context is always required.

A shift in focus from artifacts to the behaviors and capacities of large systems, however, must itself not be too narrow. Addressing the state of a building's energy system requires a twofold evaluation of the system: firstly, even if cursory, an emergy evaluation of the history of the energy embedded in the system. Second, an exergy evaluation is necessary knowledge since exergy describes the current state of the system. The twinned focus on the emergy and exergy content of a system describes not only current capacities and behaviors but equally the accumulated capacity of the system to do work and its potential feedback capacities. This twinned mode of analysis links the past and future capacities and behaviors of an energy system.

Elsewhere in the sciences, as in the Luvall-Holbo example cited above, production to biomass ratios are an indicator of system complexity and capacity as a dissipative system. Emergy-to-exergy ratios offer a corollary indicator about the role of buildings and cities in their larger energy hierarchies. Once again, U-values and R-values might describe some narrow aspect of heat transfer, but they hardly indicate the broader capacity of a system. Today, we increasingly need to think about the big, sloppy, and (formerly) externalized whole. We will no doubt still engage discrete forms of building science analysis in the process of design, but those discrete acts must always then be put back into the messy thermodynamics of the whole.

Scale Analysis

Given the great weight and potential complexity of a building, as well as the massive orders of magnitude involved, it is necessary to develop both intuitions and methods to evaluate which scales and orders of energy matter. Discerning (much less astutely, making design assertions for) the mixture of linear and non-linear behaviors in the vast thermodynamics of a non-isolated building is a central problem. Physicist Giulio Boccaletti provides some insight on this matter. As Boccaletti observes, "one of the most robust bridges between the linear and the non-linear, the simple and the complex, is scale analysis, the dimensional analysis of physical systems."[30] Scale analysis of a system helps deduce what matters in a system, a necessity amongst the thermodynamic indicators described above.

Boccaletti also provides some welcome warning: "any time a complicated system is translated into a simpler one, information is lost. Scale analysis is a tool only as insightful as the person using it. By itself, it does not provide answer and is no substitute for deeper analysis. But it offers a powerful lens through which to view reality and to understand 'the order of things.'"[31] It would be misguided to rely too much or too simply on a technique like scale analysis, for other indicators and non-modern intuitions still matter.

Conclusion: Passivhaus or Massivhaus?

With the above principles about the dissipative structure of architecture in mind, it is revealing to now look back at the prevailing paradigms and techniques for energy transition and through design. The Passivhaus paradigm is the logical outcome of the history of insulation instantiated on its own isolated terms. To be sure, the isolated, linear focus on the super-insulation of super-tight construction is not a novel approach. The burst of super-insulated buildings in the northern climates of North America in the 1970s is one example of this recurrent tendency. What is "novel" about the Passivhaus approach today are the techniques used to achieve the stated targets and the means to document and disseminate the behavior. As with the earlier Canadian examples, conduction and infiltration, as the key concerns of the operative equations, must be minimized: the attempt of isolating a non-isolated system is nowhere more clearly manifest than in a Passivhaus. By isolating short-term energy cycles as the predominate consideration, a Passivhaus approach will inevitably perform rather well under its own managerial terms and metrics. But this isolated characterization is, of course, partial and has limited architectural efficacy as an agenda for energy.

In the closing pages of this chapter, as a contrary proposal to concerns about energy and thermodynamics in twenty-first century architecture, I offer a non-isolated paradigm based on the full thermodynamics of a building's energy hierarchies. I refer to this paradigm as the Massivhaus paradigm. The Massivhaus paradigm begins with fundamentally different principles about what constitutes the energy hierarchies of a building in the thermodynamics of life and points to fundamentally different, if not more powerful, outcomes from what architects can do with those energy hierarchies. This non-modern paradigm most certainly includes both short and long-term energy cycles that operate in small and large spatial scales. Further, they do so in a recursive manner. As such, in the Massivhaus approach, accumulation, velocity, diffusivity, dissipation, durability, and design are the central thermodynamic concerns of maximum power design.

It is not trivial or surprising, in my view, that Alberti focused on the weight of the building given the deep obligations he felt architecture had to civilization. In Alberti's pre-modern paradigm, perhaps the ecological power of an extant system was still much more innate and less abstract than our own. Both Alberti's and Vitruvius's attention to these matters were manifest in their recommendations about how to site a city or when to best harvest wood or how to cut stone. Each author distilled empirical knowledge from hundreds and thousands of years of building. This seems to be rather odd and specific, if not foreign, knowledge for a contemporary architect. But seen in another light, if these two non-modern commentators were merely articulating the more refined points that further maximize the self-organized knowledge inherent to pre-modern building systems, then their observations become quite shrewd from ecological and architectural perspectives.

When combined with Alberti's emphasis on the maintenance of buildings, as Luis Fernández-Galiano articulates in great detail, then a compelling non-modern maximal power design agenda is eloquently evident in these early books on architecture.[32] The mass of architecture, from its extraction, convergence, and maintenance, is no doubt one of the most important thermodynamic considerations of a building. The mass of architecture itself, as but captured energy of many manifold forms, deserves extensive non-modern consideration today.

In my continuous search for buildings that exhibit aspects of non-isolated, non-modern architecture, the extremes of architectural amassment return again and again in both pre-modern and contemporary examples. By extremes I refer to, on one end of the spectrum, massive buildings designed to last for centuries, if not longer. The Pantheon is a superb model of this type. On the other extreme, the lowest-emergy, often most ephemeral buildings that nonetheless perform well for years, months, or even days represent another modality. Many vegetal, thatch-based, and animal hide-based buildings and bridges throughout the non-modern world are fine examples. The North America teepee, though, is my most intimate example. In the more extreme cases — both anomalies in contemporary construction — the accumulation, velocity, diffusion, and dissipation designs are deeply compelling to consider.

It should be noted that it was relatively difficult to find non-isolated examples of energy practices in recent modes of architecture, especially as they might pertain to the thermal focus of this book and the Massivhaus agenda discussed above. However, in what follows, I consider a few contemporary non-isolated strategies and examples.

2226

2226 by Baumschlager Eberle is a recently completed office building for the architects' own use in Lustenau, Austria.[33] The name refers to the Celsius comfort range achieved in the 3,200 gross square meter building with, as the architects note, "no heating, ventilation or cooling system." Considering a broader range of dissipations in the built environment than that found in typical building science practices, the design of the building was motivated by a central reflection of contemporary practice: "as buildings require less and less energy, more and more is being spent on the maintenance and service needed to sustain this reduction." To do so, the architects fundamentally reconsidered assumptions about matter, conceiving of it as no longer inert but rather as a convergent set of architectural capacities. So whereas additive technologies have been the predominate paradigm for modernity, this building evinces another tack.

The architects understand that, today, many building designs follow a persistent logic of technological escalation that are highly reminiscent of the promises of modernism. In contrast to these habits of mind and practice in architecture, Baumschlager Eberle sought a building that as a fundamental architectural proposition relied much more substantively on its emergy accumulation to do more work.

The lower-emergy building — accommodating a level of work similar to, if not more than, that of a similar office building — is premised on extracting more work from the massive, load-bearing walls and thermally active, but not mechanically charged, floor slabs. Likewise, the high spaces and tall windows are highly conducive to a civilized work space. In this regard, the spatial composition, when coupled with the occupant-modulated thermal environment, stands as good evidence that more work and more productive employees might be possible from more sanely designed office environments. In this modality, more attention is paid to the actual work of the building rather than merely adding more and more layers to the modernist apparatus of a typical contemporary office building.

2226
Exterior at night

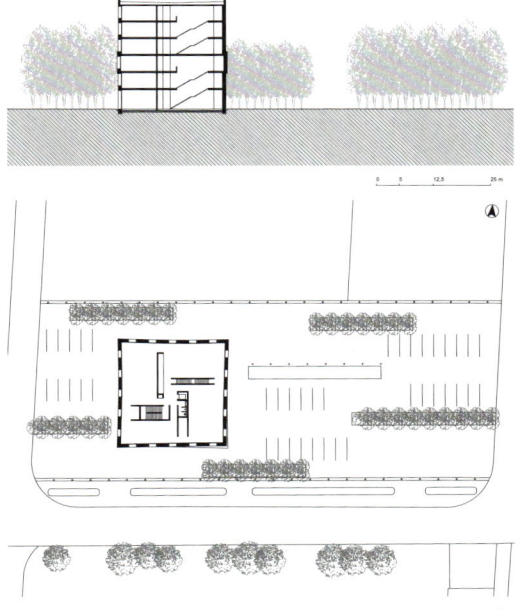

2226
Top: Exterior
Bottom left: Site section, site plan / Right: Office interior

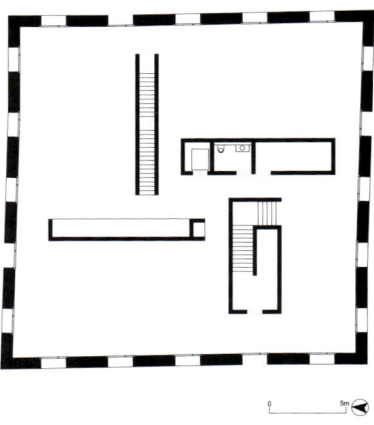

2226
Top left: Load-bearing exterior walls / Right: Floor plan
Bottom: Workspace interior

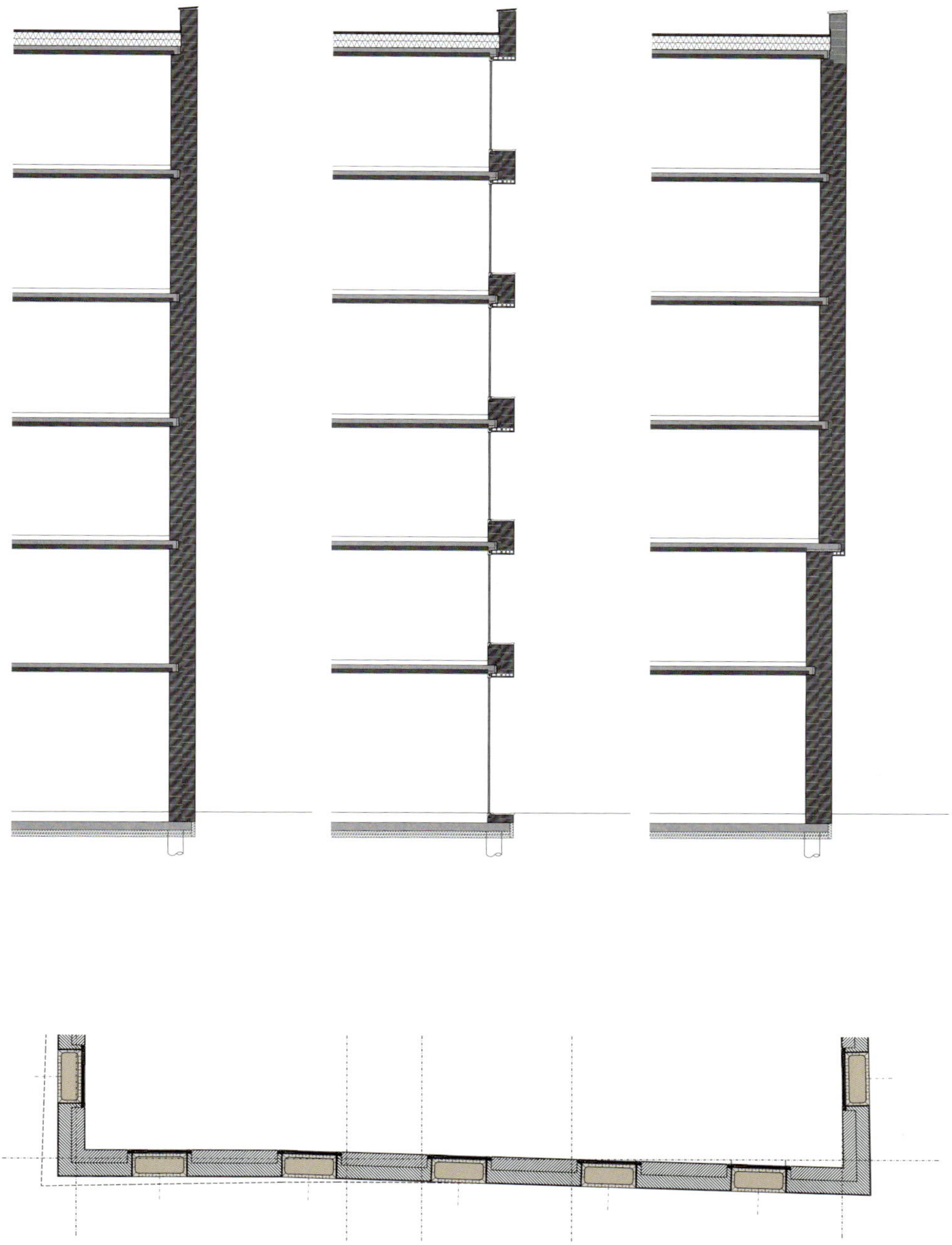

The Architecture of Dissipation

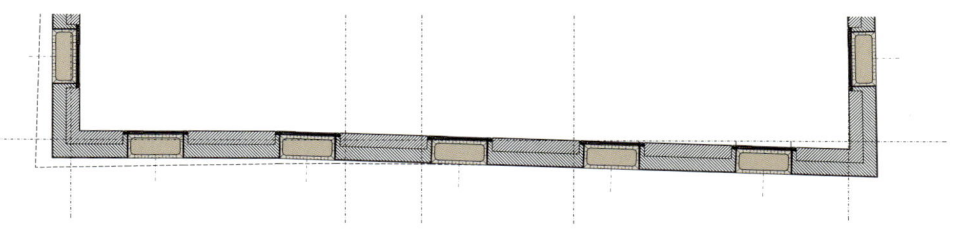

2226
Top: Uninsulated wall sections
Bottom: Plan detail

Rauch House
Front approach

The building's walls are walls, not additive layers or skins. They are load-bearing enclosure and structure as well as the thermal modulation system. They consist of two wythes of 36 centimeter masonry units, the inner intended primarily for the load-bearing structure and the outer for thermal accommodation. The mass of these walls is cooled at night by simply opening the windows. Tall, deeply punched windows are the primary lighting strategy, using the lowest-emergy light possible to the work in the building, which is otherwise typically accomplished with very high-quality electricity. A building management system modulates vents in these fenestrations to moderate fresh air exchanges and thermal exchanges.

The Rauch House

Given the salient topics in this book — non-isolated systems, the relative velocity and power of a building, and dissipative structures — in many ways no building better manifests some of the deep thermodynamic potential of buildings in this dissipative architectural agenda for energy than the Rauch House.[34] No building helps better illustrate key aspects of architecture as a dissipative structure than this building.

A short description of the construction of this building will help situate its various exergy and emergy behaviors.[35] Nearly all the material — 80% — for the construction of this building emerged from the excavation of the site for the building.[36]

Construction commenced from this excavated surface. An unreinforced, low cement ratio concrete was used for the continuous strip foundation. Above this strip foundation, earthen walls rise up at 45 centimeters thick. Projecting stringcourses of fire-baked bricks occur at each lift in the rammed-earth compaction process. These more durable projections slow the velocity of wind-driven rain on the surface of the wall. Trass-lime ring beams occur in the wall where floor structures encounter the vertical walls. The ring beams help to distribute the floor loads more evenly to the wall as well as stabilize the wall horizontally. When trass, a tuff powder resembling pozzolana and of similar volcanic origin, is combined with hydrated lime, it more thoroughly binds the mortar mix. Unlike other binders, trass is neither baked nor requires any other thermal processing. The fenestration openings in the earthen envelope require concealed reinforced trass lintels. The window and door frames are oak, sourced from southern Germany.

Each floor structure extends 30 centimeters into the wall. On most floors and the roof, the horizontal structure is based on the local, traditional "Dippelbaumbalken" method. This uses three-sided, rough-sawn, air-dried timbers installed directly next to each other to form a crude but continuous structural deck. At the ends of the timbers where they meet the exterior wall, the wood is coated with a clay-rich earth mortar that prevents any condensation from occurring where this wood is only 15 centimeters from the face of the exterior wall and thus could otherwise incur condensation. The somewhat rough timber structural deck is topped with a mix of granulated cork from Sardinia (a by-product of wine bottle cork production), sawdust, trass lime, and loam. This subfloor is then finished with a dense 12 centimeter rammed-earth floor that is finished with casein and wax.

Earthen Construction as Heat Pump

While it is possible to determine the U-values and R-values of an earthen wall, this of course does not disclose much about the wall's actual thermal behavior, much less its larger energy hierarchy. Massive earthen construction affords unique sink and source capacities for heat flux above and below ground. The thermal diffusivity of the earthen construction, most important in this case from the outside in (because the majority of the mass is on the exterior of the wall), is one consideration. The thermal effusivity of the earth plaster is another. The penetration depth of the compacted earthen floors is one as well.

Earthen Construction as Humidity Pump

In his review of the energy performance of the Rauch House, Thomas Kamm emphasizes the role of the absorptive capacity of loam.[37] Massive earthen construction affords unique sink capacities for humidity as well as heat. This is one way normal insulative and capacitive approaches to thermal-humidity milieu can converge in architecturally specific ways. As building scientist Margit Pfundstein notes, "in principle, any type of water or moisture absorption is undesirable for any type of insulation material."[38] Water, more dense and conductive than insulation materials, significantly drops the insulating effect of that material. This is one reason why the role of vapor barriers and dew point condensation is so emphasized in the pedagogies of insulation in the second half of the twentieth century. Another reason for this emphasis is the propensity for mold and rot that can appear in the lightweight approach to construction.

Rauch's earthen construction, on the other hand, exhibits non-trivial capacity to absorb and release moisture over diurnal and seasonal cycles. The great mass is much more accommodating regarding humidity flux and engenders far better humidity conditions in both winter and summer. As Kamm notes, "from the frequency diagrams for all the spaces, a distinct trend is discernible regarding the

Rauch House
Top: Exterior walls
Bottom: View from terrace

indoor air humidity. Since loam releases the stored moisture again to the indoor air, the relative indoor air humidity in the winter months, due to the sorption capacity of the earthen walls, is higher than with conventional walls."[39] In the summer months, the earthen walls exhibit a more stable modulation of room relative humidity levels (~45%–58%) while conventional construction in the same climate would fluctuate from ~36% to ~75% relative humidity. The hygroscopic capacity of more massive construction forms is thus both a very, very low technology and a high-performance air-conditioning system. Like its thermal capacities, the massive walls are a kind of humidity pump.

Earthen Construction as Dissipation Pump

In short, the thermal and moisture storage capacity of the earthen wall's mass and material properties is significant. But there is another very important type of storage inherent in this house that is not articulated in Kamm's energy analysis. As a sink of stored emergy, the specification and design of this building is particularly compelling. Kamm discusses the relative embodied energy of the house, but this embodied energy discussion does not reveal enough about the geo-ecological dynamics inherent in Rauch's convictions about earthen construction. While the emergy of the earthen construction is itself rather low, Rauch's studied insistence

Rauch House
Site plan

on the reversion of the building back into the site prompts an interesting emergy-to-exergy ratio.

As Kamm documents, 85% of the materials in this building were sourced directly from the construction site of the house.[40] They were processed at Rauch's work yard that is about 500 meters away from the construction site. Petroleum-powered equipment was used to process and fabricate the earthen construction, but the relative amount of this fuel consumption compared to other transportation and processing footprints is minor. As such, the emergy strategy for this project is ultimately a convergent one in which the necessary exergy from the system is obtained with relatively low-emergy stock. This strategy thus has limited physical feedbacks in the system, save some non-trivial relationships with the managed forests of Vorarlberg for the timber floor structure and as fuel for the kitchen masonry heater and pellet stove, but does serve the energetic functions of a household so its emergy-to-exergy ratio is very productive when compared to a more normative household architecture of the same size.

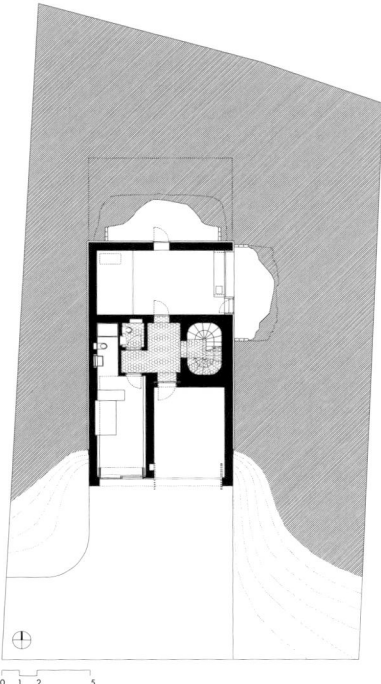 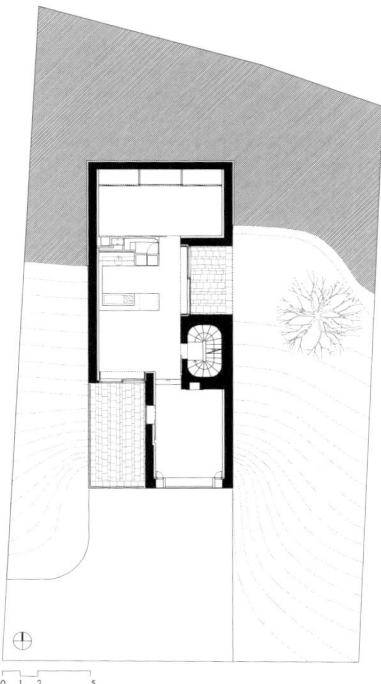

Rauch House
Left: Lower level plan / Right: Middle level plan

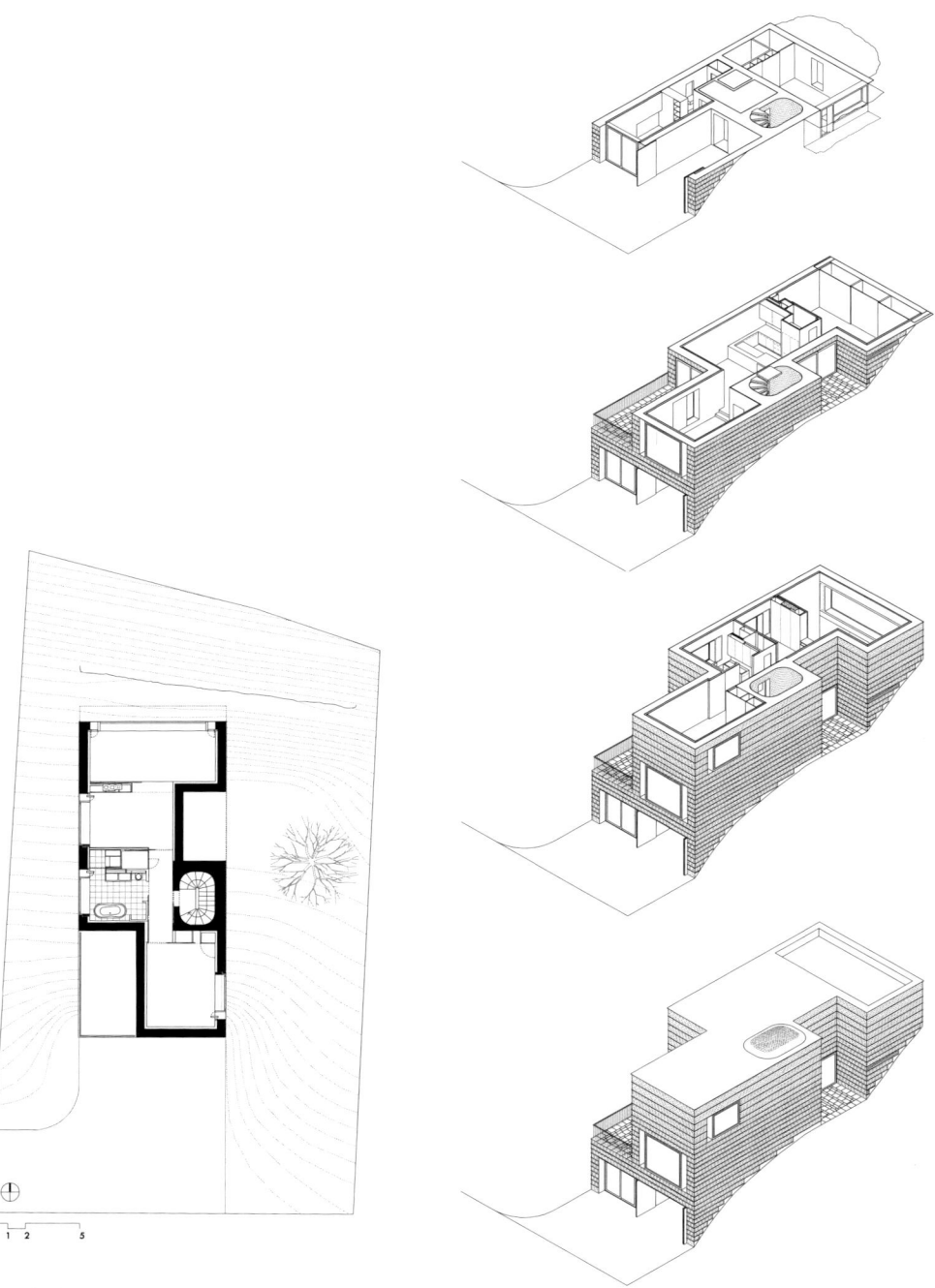

Rauch House
Left: Upper level plan / Right: Section axonometrics

However, as a matter of information-emergy accumulation, this case study house is very compelling in regard to emergy accumulation and flow. As a low-material emergy value and high-information emergy transformity value, this building nonetheless feeds back in important ways. The maximum power design of this building depends on its case-study, experimental role. Rauch and architect Roger Boltshauser rigorously worked to keep the emergy accumulation of the project quite low. The timbers for the horizontal floor structures, for instance, are locally sourced, minimally processed (surfaced on three sides), and air-dried. Reed insulation is used internally as a very low-emergy layer of thermal modulation. Higher-emergy materials such as steel and the foamed glass exterior below grade insulation are kept rather minimal.

The key component of the maximum power design, then, is the degree to which the techniques and lessons of this house feed back into future projects, lectures, exhibitions, and publications. This is a key part of its ecological use and feedback reinforcements, together amplifying the ecological effect of the otherwise low emergy intake.

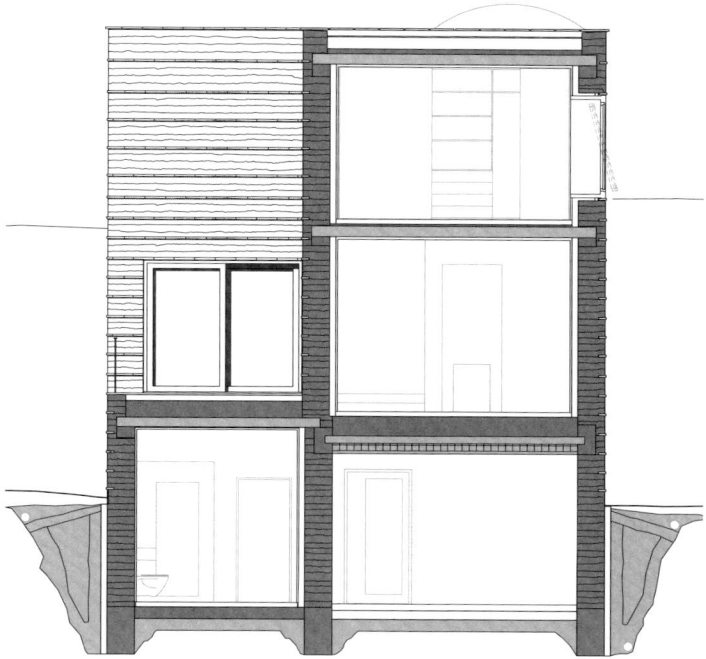

Rauch House
Lateral section

266 Insulating Modernism

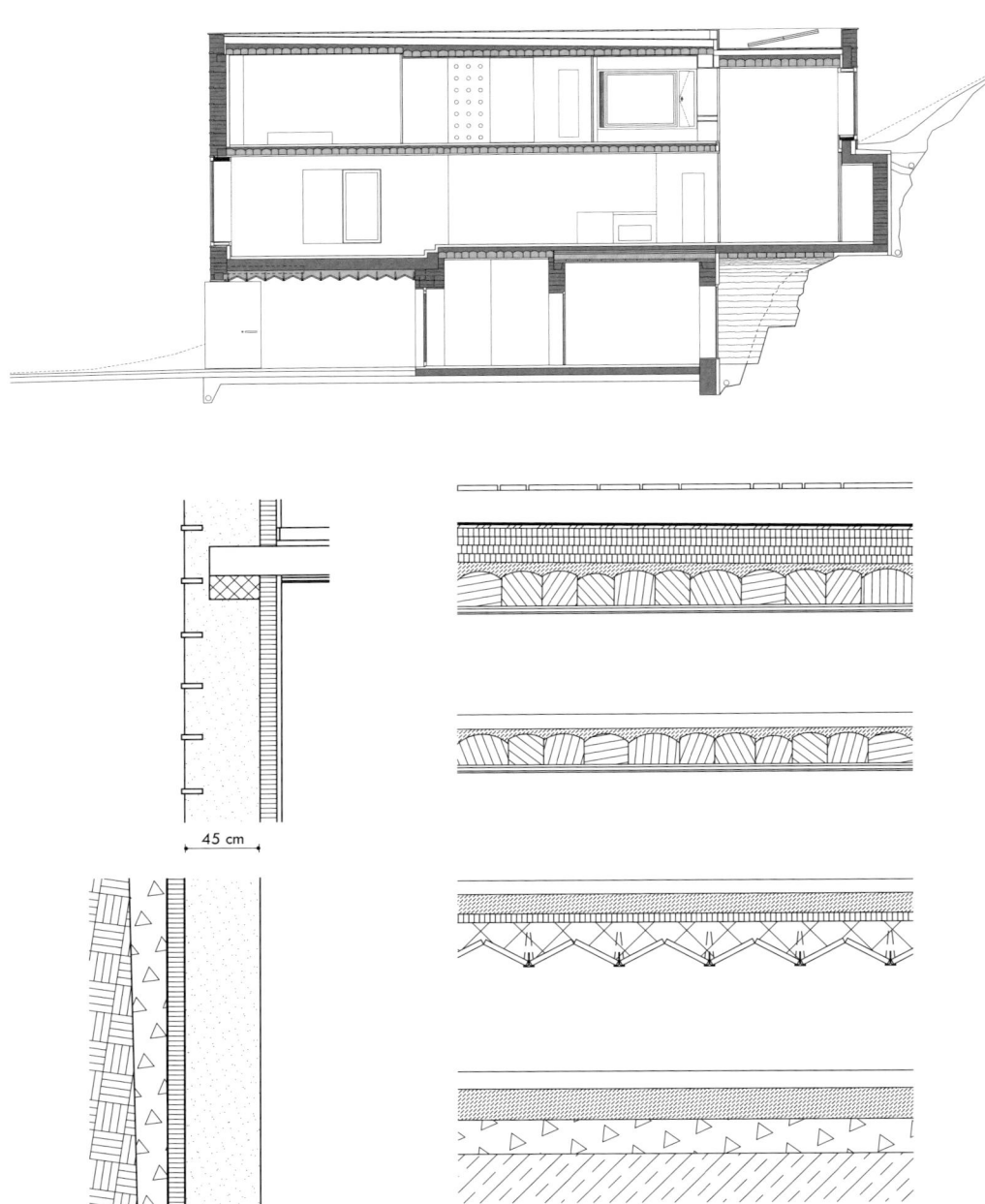

Rauch House
Top: Longitudinal section
Bottom left: Wall section details / Bottom right: Floor section details

Rauch House
Top left: Studio / Right: Lower level pressed earth tiles
Bottom left: View into the site excavation / Right: Fireplace

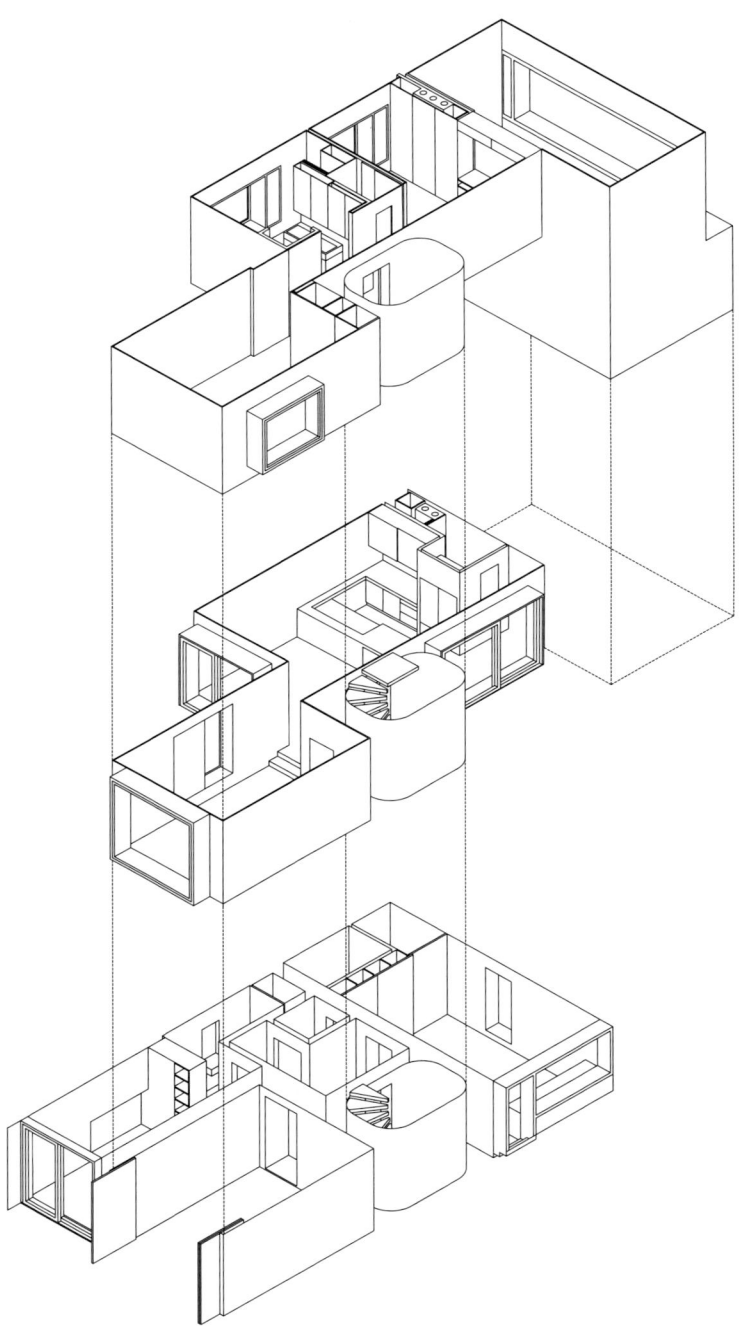

Rauch House
Axonometric

Kamm discusses other forms of feedback in terms of social and economic sustainability. He refers to local skills and knowledge as one manifestation. The project aimed to amplify skills and knowledge in the region. As Kamm notes, "by strengthening the region in this way, a foundation is established for counteracting the exodus from rural communities."[41]

Controlled Erosion

Perhaps the most important contribution this building makes is best expressed in Rauch's own words. Very importantly, Rauch speaks about the emergy and exergy dynamics of his house in his own terms, not the terms of the insulation apparatus. To begin his discussion of earthen construction he reminds us that "earth is fragmented, eroded, decomposed and deposited rock."[42] In short, he reminds us that dirt is the entropic dissipation (by-product) of other geophysical processes. In ecologically and exergetically astute ways, Rauch sees that "loam, silt, sand, crushed stone and pebbles are the direct 'products'" — not the wastes — of these processes. Rauch taps exergy from the entropy of these geophysical processes. In an inversion of typical thought, he understands dissipation in overtly productive terms:

> Today, erosion and sedimentation are usually viewed negatively throughout the world. We seek, for instance, to prevent erosion or to divert it by constructing protective structures...but erosion is so powerful that it cannot be thwarted: in other words, the problem is merely shifted in time. All surfaces are fundamentally subjected to erosion. At the same time, the respective material has an impact on its configuration. It is fascinating to accept erosion and design with it.[43]

It is indeed as fascinating to accept the dissipative structure of erosion and design with it as a fundamental principle, as it is to accept entropy and design with it. They are one and the same in the energetic reality of the world. Within a geological milieu, Rauch aims to maximize his entropy production by consuming more exergy from the earthen energy gradient of the site for construction and the sun for operation. When this larger thermodynamic optimism about the dissipation of energy in matter in short and long terms — expressed here by Rauch and shared throughout this book — is accepted as productive, a unique agenda for energy appears.

Rauch describes the effort of constructing his house as an act of "controlled erosion."[44] Grasping the entropic flow of rock to dirt, from earth building back to

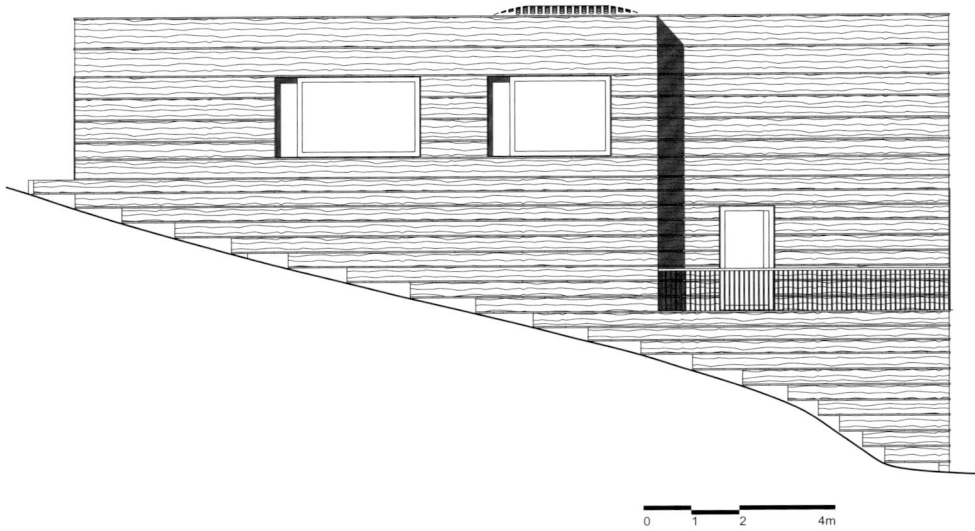

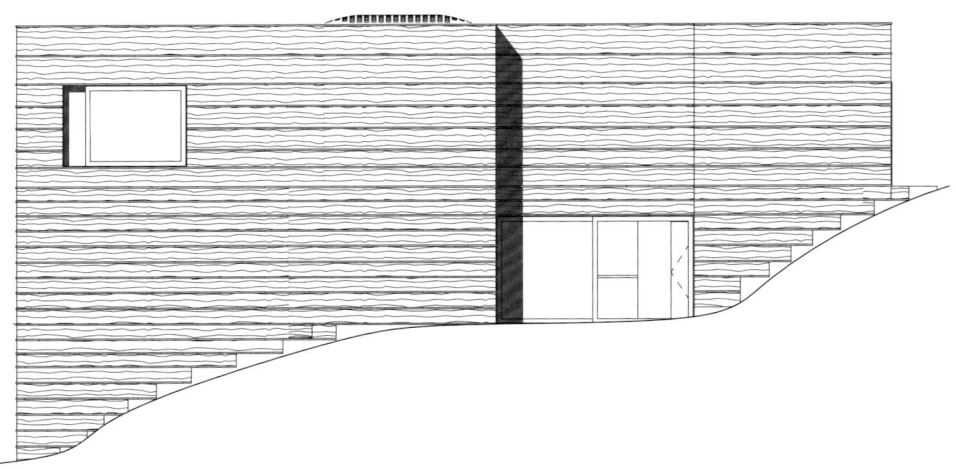

Rauch House
Top: West elevation
Bottom: East elevation

earth, Rauch fundamentally sees earthen construction in temporal, flowing terms. The building represents a structured pause in the velocity of its material/energy cycle. The energy input to re-order the configuration and to modulate the velocity of this sediment in the form of the house temporally arrests this flow, giving it a different shape. This fluvial perspective is central to Rauch's preoccupation with earthen construction. Rauch's description of the house as controlled erosion is a very clear analog for designed dissipation.

Adrian Bejan's Constructal Law is perhaps useful to recall here: "For a finite-size flow system to persist in time (to live) it must evolve such that it provides greater and greater access to the currents that flow through it."[45] Rauch finds uncanny means to access the geophysical dissipations in architecturally powerful ways that yield deeply compelling ecological outcomes. As such, Rauch fundamentally thinks about short and long-term energy cycles as a designer and builder. His earthen means are as specific as his understanding of the energy and potential exergy captured in soil. As a result, his work exhibits very cogent emergy-to-exergy yields that are difficult to dismiss.

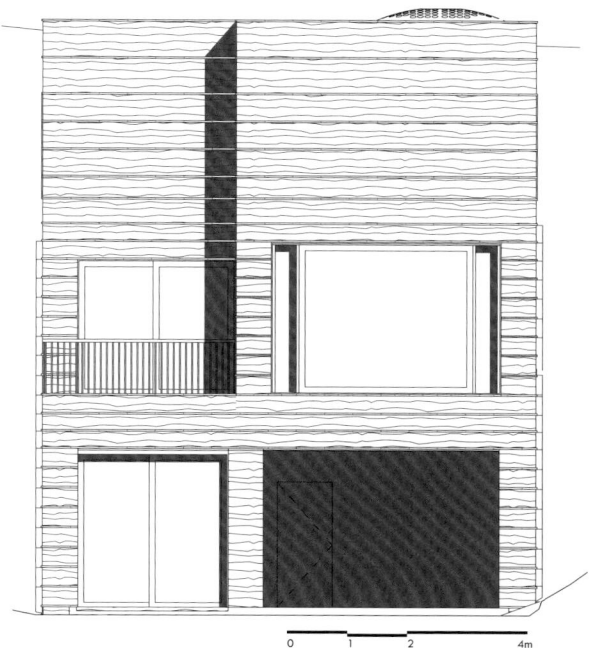

Rauch House
South elevation

A Dissipative Epilogue: Breathing Walls

By Salmaan Craig

Is better insulation really the best solution? In contemporary energy design practice, the drive towards more insulation and greater air-tightness goes largely unchallenged. There is however one alternative approach which deserves to be better articulated and further developed. It involves reconfiguring the way that heat flows through the building envelope. Normally, building envelopes are designed to behave has air-tight insulators. The alternative is to have it operate as an air-permeable heat-exchanger.

Imagine you are in a cold climate inside a warm house. The air outside is fresh but cold and so must be pre-heated before entering the house. Any number of energy sources could be used to do this, but instead you turn your attention to the heat that is conducting out through the perimeter walls. Now imagine that the perimeter walls are porous and air-permeable. You switch on a fan which sucks the fresh air through these porous walls, from outside to inside. On its way through, the temperature of the fresh air stream is elevated by the heat traveling in the opposite direction — the heat that would otherwise be lost to the surrounding environment by conduction. By doing this you have turned the wall into a heat recovery device, one whose performance improves by increasing the intake of fresh air. And this 'heat flow inversion' principle works in hot climates too: hot fresh air from outdoors will be pre-cooled if sucked through a perimeter wall (so long as the room it enters is actively chilled).

Breathing Wall Semantics

In heat transfer engineering this phenomenon is called 'contra-flux' or 'counter-flux' heat exchange. Building technology researchers have sought to apply it in a number of ways. The naming conventions have tended to differ from country to country depending on what functional characteristic or particular application each researcher has chosen to emphasize. Examples include *Dynamic Insulation* in English, *Porenlüftung* (pore ventilation) in German, and *Motstaumstak* (inverted flow roof) in Norwegian. The core idea is to transform the building envelope from an insulating jacket into a distributed heat recovery and ventilation system.

The technology exists mainly in the minds of a small group of curious researchers and has yet to make a notable impact on the construction industry. For it

Top: Cold fresh air, warm interior
Middle: The modernist conceit of isolating interior from exterior
Bottom: Could incoming air be heated through building configuration?

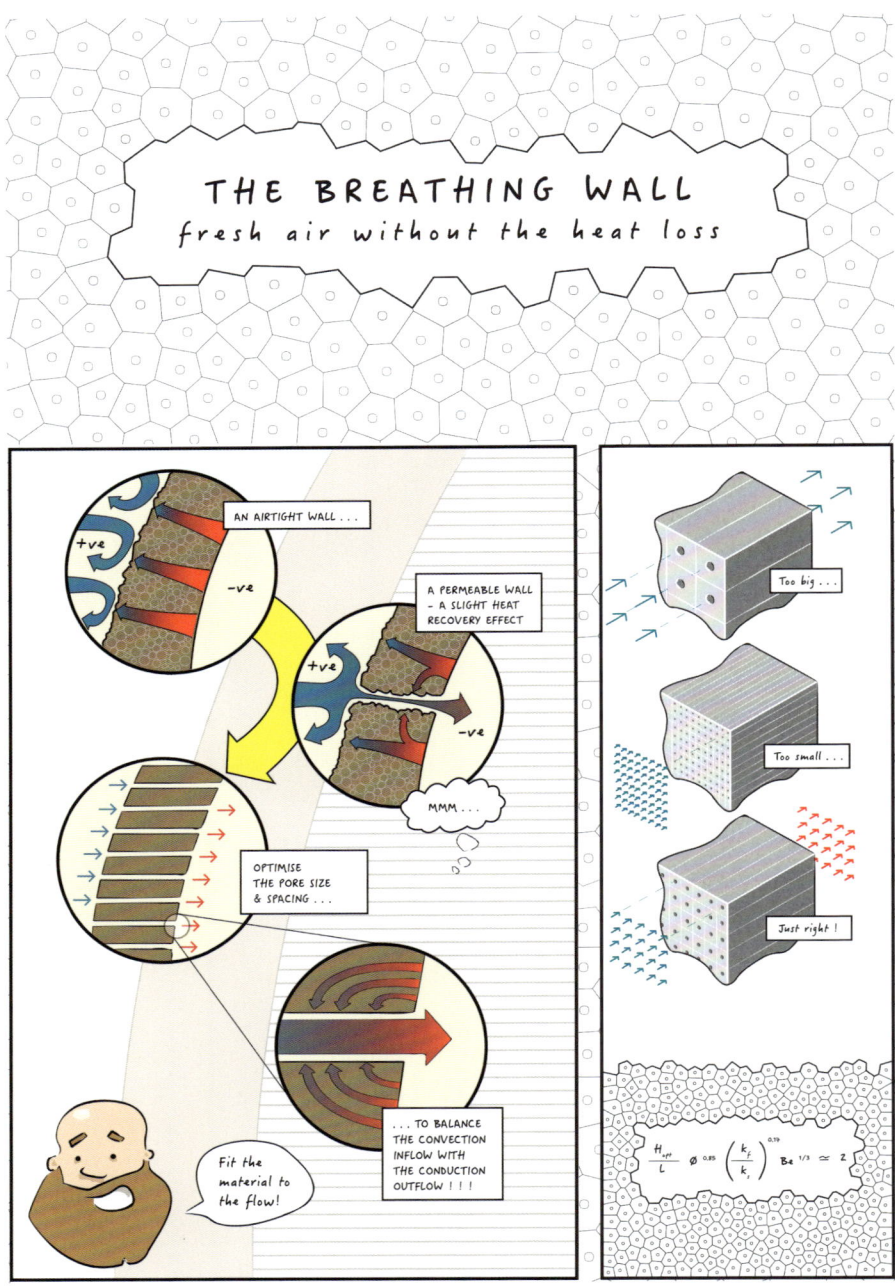

Left: Airtight resistance versus contra-flux design
Right: Pore size design and equation

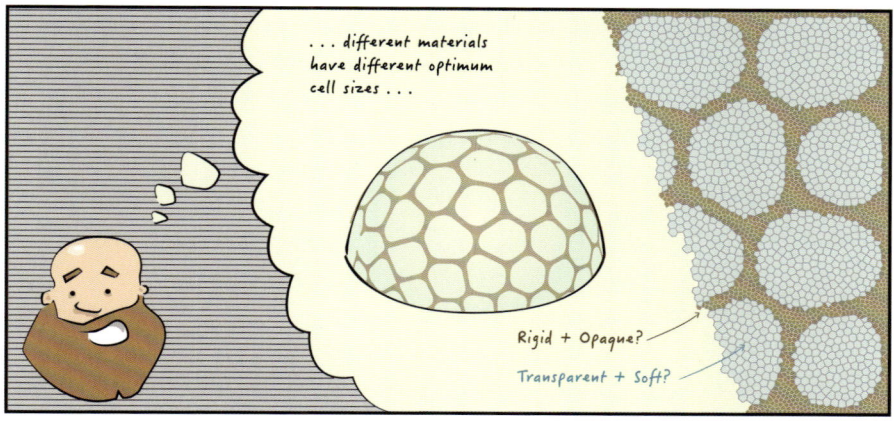

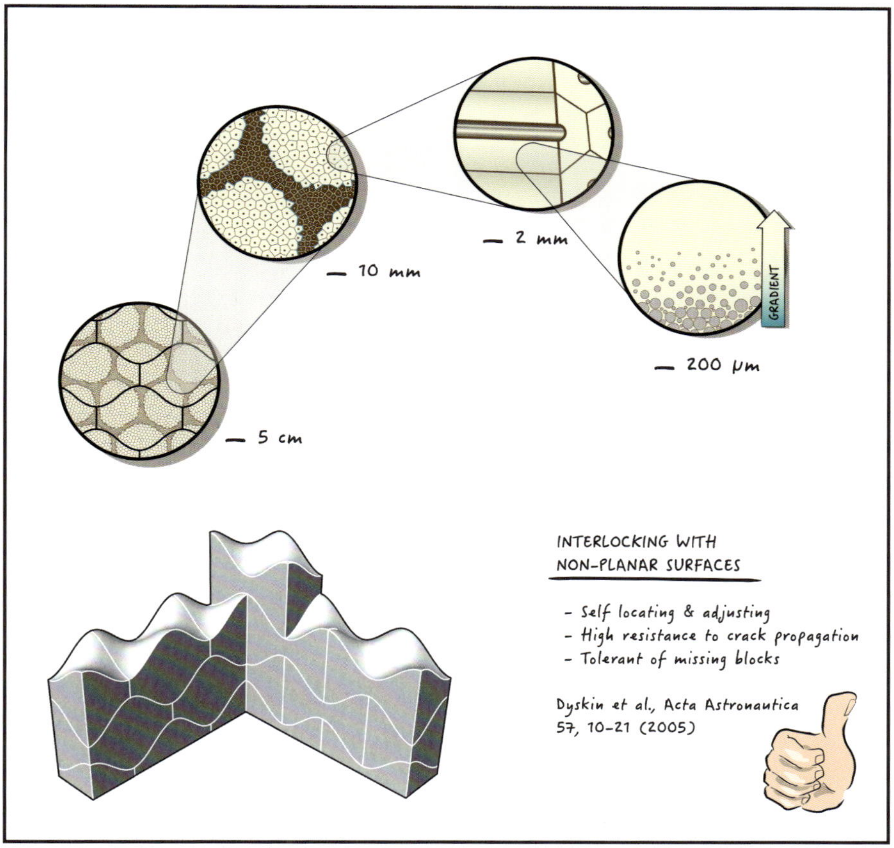

Top: Distribution of porous and non-porous media
Bottom: Manufacturer of porous media

to have any chance of leaving the drawing board there is one fundamental design flaw which must be first remedied. In short, it is this: when the fresh air is sucked through the wall beyond a certain flow rate, the incoming flow pushes away the warm air inside the room. The wall-cum-heat-recovery device is thus largely decoupled from the source of heat it is meant to recover. I argue that the best way to address this flaw is by embedding heating elements directly into the internal surface of the wall (rather than, say using room convectors or a heated air supply). This discussion forms the main body of the essay and its main features are illustrated in the accompanying figures.

The naming conventions used by the research community to describe the technology seem to have added to the confusion. This is an interesting side discussion which has particular relevance to Moe's main argument in this book. *Dynamic insulation* is a label tactically framed for a general construction audience whose idea of thermal control extends only to basic insulation practices. The name introduces it as a sophisticated yet incremental improvement to normal insulation procedures. This choice of name is understandable, but the tactic ultimately backfired. Things came to a head in one particular analysis, which erroneously assumed an undisturbed boundary layer on the internal surface of the wall — a conceptual 'hangover' from insulation calculation and practice. The study set the standard for further research, leading to the propagation of flawed designs which, when measured, did not agree well with theoretical predictions.

The key to unlocking the full potential of the 'breathing wall' technology is to think of it not as a form of insulation, but as a type of heat-exchanger. Which is to say that it should be analyzed, designed, optimized, applied, and operated using heat-exchanger theory. Heat-exchangers are crucial to the metabolism of all sophisticated technologies, both in engineering and in biology. And 'breathing walls' — that is, walls designed explicitly as heat-exchangers — may one day provide a competitive and architecturally rich alternative to today's low-energy design paradigm.

More Fresh Air, More Recovered Heat, Greater Vapor Control

In buildings with 'breathing walls', operational energy use can be low so long as three conditions are met. First, the energy cost of pulling the air through the porous material should be modest. The simplest and most reliable way of doing this is to use fans to depressurize the internal space. This does of course imply an upfront energy cost that without careful attention may be difficult to redeem. The challenging yet tantalizing alternative is to cultivate and harness natural pressure differences, powered either by wind or thermal buoyancy.[46]

Second, the conduction energy used to pre-heat the incoming fresh air should be recovered downstream before it escapes from the main exhaust of the building. This can be done using a standard air-to-air or air-to-water heat-exchanger.

The third condition pertains not to services design but to architectural form. In standard energy-efficient design, the emphasis is on minimizing conduction losses. Buildings with compact forms — that is, with a low ratio of external surface area to internal volume — tend to perform better. However, the rules of the game are changed with 'breathing walls'. In principle, the technology complements buildings with a *high* ratio of surface area to volume. Conduction flux is not something to be minimized, but recuperated and put to work.

Researchers have tended to downplay the significance of this switch in geometrical emphasis. The general consensus is that dynamic insulation "would appear appropriate for only small detached buildings" (which by nature of their size have a high area to volume).[47] But 'dynamic insulation' — i.e. 'breathing walls' — *can in principle* be appropriate for large buildings, so long as these buildings have a large surface area. Thus, there is an opportunity here for a new geometric typology of low-energy buildings: a different kind of compactness — intricate and arresting — with external forms that enhance the rate of exchange between the building and the environment, rather than enhance the state of isolation.

This design opportunity has yet to be properly characterized, let alone explored. Part of the reason for this oversight again lies with the general (but understandable) conservatism of 'breathing wall' researchers. After all, serious questions regarding the potential energy savings remain. Theoretical studies show promise but there is a gap in reporting when it comes to comparing predictions with measured data (a subject I tackle in more detail below).[48]

More recent research has instead tended to emphasize the respiratory benefits for occupants.[49] Higher rates of ventilation, whatever the means, will deliver more oxygen and exhaust more VOCs (Volatile Organic Compounds, from interior fixtures and fittings). But 'breathing walls' can improve indoor air quality further, as the insulation can operate as a filter, ridding the incoming air of particulate pollution from the urban environment. The rate of clogging on fibrous insulation in 'breathing' mode was evaluated, with the conclusion that it can be managed over a period of decades, without a detrimental increase to the pressure drop across the insulation.

The other notable consequence of ventilating through pores has to do with moisture control. In cold climates, 'breathing wall' technology turns the usual problem of interstitial condensation on its head. Moisture produced indoors by people and their activities is ejected at the main exhaust. The tiny jets of air that spout from the pores of the insulation push the water vapor molecules away from the wall. So long as the air flow is fast enough, the push-away will happen despite the presence of a static vapor pressure difference across the wall willing moisture to travel in the other direction. This behavior was confirmed first experimentally then theoretically.[50, 51] The incoming flow, in other words, acts as a de-facto vapor barrier. Moisture from indoors no longer migrates out through the envelope and so cannot condense on any cold surfaces in the exterior envelope assembly.

A Fitful Inquiry

'Breathing wall' research has tended to come in fitful bursts from a variety of sources. Curiosity was first stoked in the 1960s on a Norwegian farm when some engineers noticed something interesting about the air flow in traditional agricultural barns. On its way out, driven up by a warm, tightly packed livestock, the air transferred some of its heat and moisture to stacks of hay stored in the loft. This work inspired several early trials of 'dynamic insulation', yet only in the ceilings of agricultural buildings.[52] In the 1980s and 1990s a better understanding of the behavior led to installations in domestic dwellings in Austria,[53] Norway,[54] and Japan.[55] Measurements at these installations confirmed the key operating principle as it was then (and still is) perceived: lower U-values with increased ventilation. The first installations in the United Kingdom were in Scotland, in a residential building in Aberdeen — which was unsuccessful due to deleterious air leakage through cracks and joints in the building — and in a leisure center above a swimming pool in Callender, which is still in operation today.[56]

Data revealing the performance of 'breathing walls' *in-situ* are scarce, but one full-scale laboratory experiment has been undertaken.[57] Baker, the author of this experimental study, concludes that the theory of 'dynamic insulation', published in the late 1990s in a series of studies by one group,[58] does appear to predict the measured trend of decreasing U-value with increased ventilation well enough. But he stops short of comparing the data explicitly to the theoretical predictions. To date, in fact, not one full validation of 'dynamic insulation' theory has been published.[59]

There is, in fact, good reason to doubt the predictive power of 'dynamic insulation' theory. The key source of uncertainty lies with the temperature of the internal surface. Baker noted that his measurements were markedly lower than expected. Taylor and Imbabi also found the same discrepancy.[60] Until a full validation exercise is undertaken and published, doubt will remain about these systems and whether 'dynamic insulation' theory predicts their performance adequately.

After all, internal surface temperatures matter. If they are too low, they can cause occupants radiant discomfort. This is the point at which most responsible architects and engineers stop their inquiry into 'breathing wall' technology. On the one hand, because of its integrated nature, working with the technology demands an exceptional level of integration between services design and architectural design. On the other hand, the modeling theory appears to be lacking in its ability to predict key aspects of behavior. And on top of that, there is the real risk of thermal dissatisfaction, meaning occupants might be compelled to operate the system in a way that leads to higher energy use than originally intended. Most designers, first

 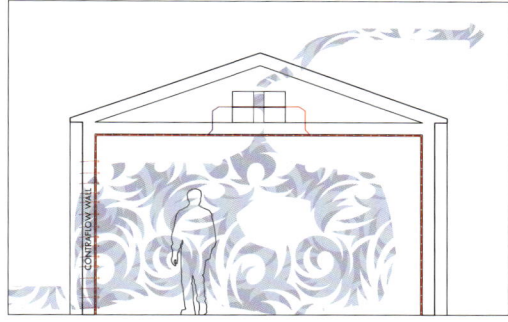

In conventional installations of 'breathing walls', fresh cold air is pulled in through the porous wall with a fan (not indicated), and the wall, in 'contra-flux' mode, is supposed to pre-heat the incoming air with the heat that would otherwise be lost by conduction. However, the fresh air (blue) pushes away the internal air in the space (red), which is warmed by the internal heat sources (in this case a person and a convector). The wall is therefore 'decoupled' from the warm air in the room; it is only coupled to the heat sources by radiation, not by convection. The wall is meant to recover heat, yet it can only recover heat that it first absorbs! This has lead to thermal discomfort (cold interior walls) in existing installations and ambiguities in the literature with regard to predicted and measured performance. Note the air-to-air heat exchanger indicated in the roof space.

One way to solve the issue is to use a different space heating system. In this diagram, the interior surfaces are heated by a network of embedded water pipes. Small-diameter water pipes made of polymer are commercially available. On the interior surface of the 'breathing wall', there are pores penetrating the heated surface to allow fresh air through. In this way, the 'breathing wall' is in direct contact with the heating source. It can be designed explicitly as a heat recovery device – the 'porosity' can be optimized such that the air enters at the same temperature as the interior surface, which in turn can be controlled to ensure thermal comfort. Downstream, an air-water heat exchanger can be used to recover heat from the air stream, and add it back to the water-heating loop.

attracted by curiosity, understandably leave 'breathing walls' on the drawing board. Too much risk, too little reward.

An Uncomfortable Oversight

Yet for researchers the question still remains: why do 'breathing walls' suffer unpredictably low internal surface temperatures? The temperature discrepancy is odd, because basic heat-exchanger theory suggests it ought to be straightforward to master.[61] In principle, it should simply be a case of finding the right balance between the two opposing flows. That is, the right balance between the outgoing conduction on the one hand, and the heat-absorbing capacity of the incoming flow on the other.

There is one possible explanation. It may be that previous installations and experiments have all suffered from one design flaw, borne from one false assumption in the original analysis about the basic phenomenology of the technology. I think that in the examples in the literature, the design is such that the wall is not able to absorb sufficient heat from the room.

Remind yourself of the inverted behavior of vapor transfer in 'breathing walls'. Despite a static vapor pressure difference willing moisture to travel out through the wall from inside to out, the incoming air pushes the water vapor molecules away in the opposite direction. The same could be true with the warm air inside the room. When the fresh air spouts from the highly distributed pores, the warm air molecules inside the room are pushed away from the surface of the wall. In other words, so long as the incoming air is fast enough, *there is no convective heat transfer from the room to the wall.*

The *raison d'être* of a 'breathing wall' is to turn the envelope into a heat recovery device. But to recover the heat, clearly the wall must absorb it first. If convection is the primary mode of heating the room, then the transfer of heat to the heat-exchanger device — the 'breathing wall' — is significantly diminished. The only heat left for the wall to absorb is that emanating from radiant sources, such as from people and equipment. This partial 'decoupling' may explain why measurements of internal surface temperature are consistently (and unpredictably) lower than the room temperature.

If true, then the origin of this oversight can be found in the analysis by Taylor and Imbabi.[62] This is the 'dynamic insulation' theory that the wider research community has thus far accepted and assumed to be accurate, despite indications to the contrary. Taylor et al. assumed that there was an 'air-film resistance' between the wall and the room. In thermal transmittance calculations, this is short-hand for a 'convection boundary layer', the thin layer of air that falls down the surface of the wall when the wall comes in contact with the warmer air. This is the physical space across which convection heat transfer occurs. The jets, however, displace this boundary layer. No boundary layer, no 'air-film resistance', no convective heat transfer (from room to wall).

Fortunately the poor coupling between room and wall (no convection, only radiation) is easily remedied. It is just a case of choosing an alternative heating (and cooling) system. Instead of convectors, use radiators. Even better, spread the heating across the entire internal surface of the 'breathing wall'. Water-fed 'thermally active surfaces' are increasing in popularity,[63] particular in Europe, where it is now common practice to embed small-diameter hydronic tubing (made from plastic) into the ceiling or floor. This technology could be incorporated into the internal surface of a 'breathing wall', so long as there were enough open pores in the surface between the pipes. The surface temperature and the thermal comfort of occupants could then be easily controlled. The design emphasis also changes. The wall may be treated explicitly as a heat-exchanger. Its function is to pre-heat the incoming air with the energy that would otherwise be lost to the external environment. Heat downstream of the air flow may be recovered to the water cycle using an air-to-water heat-exchanger.

Designed Porosity

A crucial aspect of heat-exchanger design is its geometry. Often these devices are intricate, beautiful pieces, with numerous fins packed closely together, their distance apart an expression of the maximum heat transfer density achievable for a given pressure drop. Until now the first step to designing a 'breathing wall' was to look up and specify an insulation material with a known and measured permeability. Now, however, with advances in materials processing and digital fabrication, it may be possible to design the internal architecture of a 'breathing wall', and further optimize its performance. For instance, though the economics might not quite be in place for a real construction project, the idea of a 3D printed 'breathing wall', or 3D printed 'breathing wall' bricks, with an internal geometry optimized for heat recovery, is technically feasible today.

The key question concerning the geometry, therefore, is not how, but what. Tentative answers may be found in other industries. In aerospace engineering, the fight against overheating at high speeds is a critical one — think shuttles entering the earth's atmosphere, or the spinning blades of a turbine engine. Two fairly recent studies are particularly insightful.

The first[64] proposes to 'vascularize' a solid slab subject to intense heating from the side by introducing arrays of parallel channels, and to flood these channels with coolant flowing in the opposite direction to the oncoming heat wave. An analytical argument yielded expressions for the optimal channel spacing (for a given material, thickness, void fraction, fluid, and pressure drop) and the maximum temperature inside this optimal configuration (for a given heating flux). These analytical results were confirmed by numerical simulations. The results are general, which means they could in principle be used to inform the design of new 'breathing walls'.

The second study[65] deals with the same scenario but instead of parallel channels, the 'vasculature' is a fluid tree, flowing from one point (the entry point) to many (where the heat flux emanates). A similar procedure to the first study was followed. In comparing the two studies the aim was to find the tipping point — when tree-like architectures perform better than parallel channels. The authors found that tree designs are more effective when the (dimensionless) pressure drop number and the void fraction were sufficiently large. This tipping point appears too large to justify using trees for dynamic insulation (counterflow heat exchangers) with air as heat recovery medium. They may however be in the range for the some types of hydronic vasculatures within buildings.

The scope of the search can be expanded to include other possible heat exchanger types. Imbabi[66] recently proposed an alternative version of a 'breathing wall', one that operates in cross-flow instead of counter-flow. But again the opportunity to optimize the geometry of the heat-exchanger unit — in this case, a slab of insulation with surface studs to enhance transfer — was missed. Many of the forced or mixed convection correlations available in the literature might apply, for example, parallel plates,[67] staggered plates,[68] or staggered cylinders.[69]

The Vascularization of Building Envelopes

The body of 'breathing wall' research has so far been piecemeal in nature. Most notably the prevailing 'dynamic insulation' theory has yet to be put to the test adequately. Indeed, measurements suggest some flaws in the original analytical reasoning. These false assumptions arguably stem from the false characterization of the technology as a form of insulation. The technology is more accurately characterized as a type of heat-exchanger. This uncovers obvious ways of improving the thermodynamic performance, and opens the door to the application of state-of-the-art results from the world of heat-exchanger engineering. These results are based on the insights of the new thermodynamics, also known as Constructal Law, which has been discussed in various parts of this book. Along with advances in materials processing and digital fabrication technology, newly considered 'breathing walls' may one day provide for a genuine and architecturally rich alternative to today's low-energy design paradigm.

Notes

1. Ludwig Boltzmann, "The Second Law of Thermodynamics," in *Theoretical Physics and Philosophical Problems*, ed. Brian McGuinness, trans. Paul Foulkes, Dordrecht: D. Reidel, **1974**.
2. Eric D. Schneider and James J. Kay, "Order from Disorder: The Thermodynamics of Complexity in Biology," in Michael P. Murphy, Luke A. J. O'Neill, eds., *What is Life: The Next Fifty Years: Reflections on the Future of Biology*, Cambridge, UK: Cambridge University Press, **1995**. pp. 161–172.
3. R. E. Ulanowicz and B. M. Hannon, "Life and the Production of Entropy," in *Proceedings of the Royal Society, London, Series B*, vol. **232**. p. **188**.
4. Ilya Prigogine and Isabelle Stengers, *Order Out of Chaos: Man's New Dialogue with Nature*, New York: Bantam Books, **1984**. p. **104**.
5. Luis Fernández-Galiano, *Fire and Memory: On Architecture and Energy*, trans. Gina Cariño, Cambridge, MA: The MIT Press, **2000**. pp. **49–55**.
6. William Thomson, "On a Universal Tendency in Nature to the Dissipation of Mechanical Energy," in *Proceedings of the Royal Society of Edinburgh* for April 19, **1852**. pp. 139–143.
7. Rudolf Clausius, "Über verschiedene für die Anwendung bequeme Formen der Hauptgleichungen der mechanischen Wärmetheorie," *Annalen der Physik und Chemie*, vol. **125** (7), **1865**. p. **400**.
8. Fernández-Galiano. p. **49**.
9. Eric D. Schneider and James J. Kay, "Life as a manifestation of the second law of thermodynamics," *Mathematical and Computer Modelling*, vol. **19** (6–8), March–April **1994**. pp. 25–48.
10. Alfred J. Lotka, "Contributions to the energetics of evolution," in *Proceedings of the National Academy of Science*, vol. **8** (6), **1922**. p. **147**.
11. Ulanowicz and Hannon. p. **185**.
12. Ibid. p. **185**.
13. Lotka. p. **148**.
14. Howard T. Odum, "Self-Organization and Maximum Empower," in C.A.S. Hall, ed., *Maximum Power: The Ideas and Applications of H. T. Odum*, Boulder, CO: University Press of Colorado, **1995**. p. **311**.
15. Howard T. Odum and Richard C. Pinkerton, "Time's Speed Regulator: The Optimum Efficiency for Maximum Power Output in Physical and Biological Systems," *American Scientist*, vol. **43** (2), April **1955**. p. **331**.
16. Ibid. p. **332**.
17. You cannot make energy more or less efficient. All energy is transferred in a process. You might alter the exergy efficiency of a process, but that is already

a more specific and focused task, one that is overtly related to the designed dissipation of energy.

18 Jeffrey Luvall and H. Richard Holbo, "Thermal remote sensing methods in landscape ecology," in Monica G. Turner and Robert H. Gardner, eds., *Quantitative Methods in Landscape Ecology*, New York: Springer-Verlag, 1991. pp. 127–152.

19 Ilya Prigogine, Gregoire Nicolis, and Agnes Babloyants, "Thermodynamics of Evolution," *Physics Today*, vol. 25, 1972. p. 25.

20 Ricardo Puselli and Enzo Tiezzi, *City Out of Chaos: Urban Self-Organization and Sustainability*, Boston: WIT Press, 2009. p. 16.

21 Adrian Bejan, "Constructal-theory network of conducting paths for cooling a heat generating volume," *International Journal of Heat Mass Transfer*, no. 40, 1996. p. 815.

22 Expressing a bit of exhaustion, Bejan and his colleague Sylvie Lorente lambaste all the various maximum/minimum positions in Adrian Bejan and Sylvie Lorente, "Constructal law of design in nature," *Philosophical Transactions of the Royal Society of Britain*, vol. 365, 2010. p. 1337.

23 Fritjof Capra, *The Web of Life: A New Scientific Understanding of Living Systems*, New York: Anchor Books, 1996. p. 89.

24 Georges Bataille, *The Accursed Share: An Essay on General Economy*, trans. Robert Hurley, New York: Zone Books, 1988–1991.

25 William Braham, "Temptations of Survivalism, or What do you do with your waste?" *Places: Forum of Design for the Public Realm*. http://places.designobserver.com/entry.html?entry=13998 (consulted August, 15, 2011).

26 Howard T. Odum, *Environment, Power, and Society for the Twenty-first Century: The Hierarchy of Energy*, New York: Columbia University Press, 2007.

27 Leon Battista Alberti, *The Ten Books of Architecture: The 1755 Leoni Edition*, New York: Dover Publications, 1986. (See first paragraph of the un-paginated preface.)

28 The opening of nearly every book, article, and presentation on "sustainability" begins with a hand-waiving litany of statistics about the consumptive nature of buildings. For example see: Brenda and Robert Vale, *Towards a Green Architecture*, London: RIBA Publications Ltd., 1991; Richard L. Crowther, *Ecologic Architecture*, Boston: Butterworth Architecture, 1992; Rocky Mountain Institute, *A Primer on Sustainable Building*, Aspen, CO: Rocky Mountain Institute, 1995; Laura C. Zeiher, *The Ecology of Architecture: A Complete Guide to Creating the Environmentally Conscious Building*, New York: Whitney Library of Design, 1996; Klaus Daniels, *The Technology of Ecological Building: Basic Principles and Measures, Examples and Ideas*, trans. Elizabeth Schwaiger, Boston: Birkhäuser Verlag, 1997; Sandra Mendler and William Odell, *The HOK Guidebook to Sustainable Design*, New York: John Wiley & Sons, Inc., 2000; David Nye, *Consuming Power: A Social History of American Energies*, Cambridge, MA: The MIT Press, 1998.

29 Riccardo Maria Pulselli, Eugenio Simoncini, and Nadia Marchettini, "Energy and emergy based cost–benefit evaluation of building envelopes relative to

geographical location and climate," *Building and Environment*, vol. **44** (5), 2009. p. **922**.
30 Giuilio Boccaletti, "Scale Analysis," in John Brockman, ed., *This Will Make You Smarter: New Scientific Concepts*, New York: Harper Perennial, **2012**. p. **185**.
31 Ibid. p. **187**.
32 Fernández-Galiano. pp. **62–125**.
33 2226, Lustenau, Austria, **2013** – Architect: Baumschlager Eberle, Lochau, Austria; Jürgen Stoppel, project architect; Hugo Herrera Pianno, Markus Altmann, project assistance – Structural engineer: Mader & Flatz Ziviltechniker GmbH, Bregenz, Austria – Lighting design: Ingo Maurer, Munich, Germany – Electrical planning: Lars Junghans, PhD, Ann Arbor, Michigan, USA – Fire safety: IBS–Institut für Brandschutztechnik und Sicherheitsforschung, Linz, Austria.
34 Rauch House, Schlins, Austria, **2005–2008** – Planning Team: Roger Boltshauser and Martin Rauch – Offices: Boltshauser Architekten AG (project leader: Thomas Kamm) / Lehm Ton Erde Baukunst GmbH – Civil engineer: Josef Tomaselli – Rammed-earth construction: Martin Rauch and his team, foreman: Johannes Moll – Carpenter: Manfred Bischof – Ceramic work/Art: Marta Rauch-Debevec, Sebastian Rauch.
35 For a complete account of the Rauch house, see: Roger Boltshauser, Martin Rauch, eds., *The Rauch House: A Model of Advanced Clay Architecture*, Basel: Birkhäuser, **2011**.
36 Thomas Kamm, "Energy," in Roger Boltshauser and Martin Rauch, eds., *The Rauch House: a Model of Advanced Clay Architecture*, Basel: Birkhäuser, **2011**. p. **157**.
37 Ibid. p. **157**.
38 Margit Pfundstein, *Insulating Materials: Principles, Materials, Applications.* Basel: Birkhäuser Edition Detail, **2008**. p. **12**.
39 Kamm. p. **157**.
40 Ibid. p. **165**.
41 Ibid. p. **170**.
42 Rauch. p. **113**.
43 Ibid. p. **113**.
44 Ibid. p. **113**.
45 Adrian Bejan, "Constructal-theory network of conducting paths for cooling a heat generating volume," *International Journal of Heat and Mass Transfer*, vol. **40** (4), **1997**. p. **815**.
46 Bruce J. Taylor and Mohammed Imbabi, "Environmental Design Using Dynamic Insulation," *ASHRAE Transactions*, vol. **106** (1), **2000**. Whatever the method — forced or natural — the normal recommendations for maximum pressure difference apply: 5–10 pascals, to avoid any difficulties with opening doors.
47 Bruce J. Taylor and Mohammed Imbabi, "The application of dynamic insulation in buildings," *Renewable Energy*, vol. **15** (4), **1998**.

48 Ibid. Also see Bruce J. Taylor, D. A. Cawthorne, and Mohammed Imbabi, "Analytical investigation of the steady-state behaviour of dynamic and diffusive building envelopes," *Building and Environment*, vol. **31** (6), **1996**.

49 See Mohammed Imbabi, "Modular breathing panels for energy efficient, healthy building construction," *Renewable Energy*, vol. **31** (5), **2006** and Bruce J. Taylor, R. Webster, and Mohammed Imbabi, "The building envelope as an air filter," *Building and Environment*, vol. **34** (3), **1998**.

50 Helmut Bartussek, "Luftdurchlässige Konstruktionen: Eine Übersicht über den Stand der Entwicklung (Air-permeable constructions: An overview of the state of development)," *Schweizer Ingenieur & Architekt,* **30–31, 1986**.

51 Bruce J. Taylor, D. A. Cawthorne, and Mohammed Imbabi, "Analytical investigation of the steady-state behaviour of dynamic and diffusive building envelopes," *Building and Environment*, vol. **31** (6), **1996**.

52 Discussed in T. Græe, *Breathing Building Construction,* Oklahoma: ASEA Stillwater, **1974**.

53 See Bartussek, note **50**.

54 J. T. Brunsell, "The performance of dynamic insulation in two residential buildings," *Air Infiltration Review*, vol. **16** (4), **1996**.

55 Arvid Dalehaug, *Dynamic insulation in walls,* Research report No. **53**, Hokkaido Prefectural Cold Region Housing and Urban Research Institute, Japan, **1993**.

56 P. H. Baker, "The thermal performance of a prototype dynamically insulated wall," *Building Services Engineering Research and Technology,* vol. **24** (1), **2003**.

57 Ibid.

58 Bruce J. Taylor and Mohammed Imbabi, "The effect of air film thermal resistance on the behaviour of dynamic insulation," *Building and Environment*, vol. **32** (5), **1997**. This is an extension of work done by the same group in notes **2** and **3**.

59 Baker (see note **11**) concludes by stating that "a future publication will deal with the development and validation of a model of the wall with full climatic boundary conditions." However, a decade has passed and this has not yet happened. In a research article published in **2012**, Imbabi (see note **21**) refers to a study in preparation entitled "A validated theoretical model of dynamic insulation," but this too has not yet been published.

60 See note **58**.

61 Look in any good heat-transfer textbook at the chapter on heat-exchangers. The performance of the different types is described by a set of analytical models based on two simplified starting or 'boundary' conditions: either the source of heat is of constant temperature or of constant flux. In practice, the reality is usually very close to one of the two scenarios. And even if it is not, the models are still useful: since they represent the two limit cases — the two extremes on either side of a continuum of operation — it is safe to assume that the performance of your system will fall somewhere between. If the interior surface of a 'breathing wall' was at a fixed (heated) temperature, then

it would be easy to control the temperature of the incoming air. It would just be a case of ramping-up the air flow until the temperature of the incoming air converged with that of the internal surface. If instead the heating flux at the internal surface was fixed, it would also be straightforward to have the two temperatures converge at a desirable (and comfortable) point. The trick is in both cases to balance the heat-absorbing capacity of the incoming flow with the heat of the outgoing conduction flux.

62 See note **58**.
63 Kiel Moe, *Thermally Active Surfaces in Architecture,* New York: Princeton Architectural Press, **2010**.
64 Sunwoo Kim, Sylvie Lorente, and Adrian Bejan, "Vascularized materials with heating from one side and coolant forced from the other side," *International Journal of Heat Mass Transfer,* vol. **50, 2007**.
65 Sunwoo Kim, Sylvie Lorente, and Adrian Bejan, "Dendritic vascularization for countering intense heating from the side," *International Journal of Heat Mass Transfer,* vol. **51, 2008**.
66 Mohammed Imbabi, "A passive–active dynamic insulation system for all climates," *International Journal of Sustainable Built Environment,* vol. **1** (2), **2012**.
67 Adrian Bejan and Enrico Sciubba, "The optimal spacing of parallel plates cooled by forced convection," *International Journal of Heat Mass Transfer,* vol. **35** (12), **1992**.
68 Andrew C. Fowler, Gustavo A. Ledezma, Adrian Bejan, "Optimal arrangement of staggered plates in forced convection," *International Journal of Heat Mass Transfer,* vol. **40, 1997**.
69 George Stanescu, Andrew C. Fowler, and Adrian Bejan, "The optimal spacing of cylinders in free-stream cross-flow forced convection," *International Journal of Heat Mass Transfer,* vol. **39, 1996**. See also Tunde Bello-Ochende and Adrian Bejan, "Optimal spacings for mixed convection," *International Journal of Heat Mass Transfer,* vol. **126, 2004**.

"*Not only are these systems open but also they exist only because they are open. They feed on the flux of matter and energy coming to them from the outside world… They form an integral part of the world from which they draw sustenance, and they cannot be separated from the fluxes that they incessantly transform.*"[1] **Ilya Prigogine and Isabelle Stengers**

Conclusion: The Metabolic Rift, Gift & Shift of Architecture's Necessary Excess

In the scientific, material, and physiological history of insulation, architecture has suffered foolishly insufficient and narrow systems boundaries for energy. The system boundaries inherent in the insulation apparatus help demonstrate that buildings have foremost been characterized as objects in architecture and building science, an understandable, perhaps, but nonetheless wholly arbitrary choice given a building's actual networks and hierarchies of energy. The composition of a building as an object is what emphatically matters in the design studio, for example, and heat transfer is often reduced to the R-values of various insulation objects: batts and rigid sheets. Further, the cultural implications of a building's objecthood are common currency of not only architectural journalism and history courses but building science as well. Likewise, the assembly and performance of the systems in the building-object are the focus of technical courses where integration is so often portrayed as a matter of spatial accommodation of systems rather than a more cogent ecological integration of building and energy hierarchy. The role of insulation products/objects in the design construction of building/objects is paramount to this history of insulation. Whence, then, the accumulation of matter and energy inherent in these objects of architecture? Why these excessively parochial system boundaries?

Architects are habitually indifferent to these questions, which they are trained not to consider. The isolating basis of insulation has only exaggerated this posture characterized by peculiarly selected system boundaries: the unwitting refrain of modern energy praxis on the one hand and an isolated formalism on the other. The totality of matter and energy dissipations in any building-as-object is left as an externality to the discipline. As such, the composition of this energy hierarchy is routinely unaccounted for in architecture. With respect to energy, the impossible notion of isolated, autonomous buildings — claims about net-zero energy buildings and self-sufficient architecture, for instance — likewise externalizes consequential formations of energy and matter. It is only possible to consider the concept of "net-zero buildings" if an opportunistic and irrelevant system boundary is operative. It is only possible to aim to "minimize the impact" of a building if a woefully narrow systems boundary is operative. In the non-isolated reality of architecture, architects should instead maximize the (positive) impact of architecture through the maximum power design of maximal energy, intake, use, and feedback design of systems, not objects alone. Anything less is a betrayal of the ecological and thermodynamic potential of architecture; a betrayal of its actual system boundaries. In contrast, it would be nearly impossible to separate formal procedures from that of thermodynamic operations in a relevant, novel practice of thermodynamic formation and figuration fit for this century.

In far too many cases, modern architecture neither teaches nor practices its own geographic, ecologic, and energetic engenderment that points towards its larger system boundaries. Thus, the topics of matter, energy, and formation are isolated and inadequately framed. So the equally important question is: whence this paucity of geo-ecologic curiosity, imagination, and praxis in architecture?

Form Exists to Degrade Energy

Any thermodynamic system, such as an urban metabolism, emerges and exists strictly to dissipate available energy gradients.[2] How systems of matter and energy form, how they behave, as well as what they yield, all depend on their dissipative structure design: their particular formation of energy dissipations and degradations.[3] The thermodynamic constitution of non-isolated, self-organized systems — such as forests, buildings, cities, ecosystems, and life itself — will tend to do three things: circulate and transform the most available energy, at the fastest rate possible, and with the most reinforcing feedbacks.[4] These three dynamics — the volume, velocity, and feedback of energetic flux — govern the power, behavior and fate of any system. Given these dynamics, non-equilibrium systems will prevail that thus produce maximum entropy, having captured, transformed, and consumed the most available energy and thereby re-radiated the most anergy at the lowest energy level possible.[5]

Without a grasp of these maximum entropy production dynamics, little of any ultimate purpose will be discerned, or designed, from urban metabolisms: the struggle for maximum entropy production. This struggle begins with an enormous available energy gradient: the incident solar gradient supplies Earth with some ~160,000 terawatts of exergy annually.[6] (Humans, for scale, consume about 16 terawatts of exergy annually.) It is not possible to make claims about energy shortages in this context; quite the opposite. It is for us to determine how to best degrade this insolation in the most powerful, if not magnificent, way possible. Some of this exergy is used to drive processes while other portions are stored in matter. Given the excess of this solar gradient, the question, as Georges Bataille notes, "is always poised in terms of extravagance. The choice is limited to how the wealth is to be squandered."[7]

The central ecological/thermodynamic problem these modern metabolisms constitute is that they never been powerful enough. They only consider two thirds of our thermodynamic constitution as energy dissipating structures: maximal intake and transformation. They habitually lack a full reckoning of the third critical component of any maximal power system: the requirement for feedback reinforcement. In modern metabolisms, feedback reinforcements include both the "goods" of modern metabolisms (liberating, powerful systems of knowledge, mobility, and industrial production) as well as the "bads" of modern metabolisms (counterproductive, ultimately power-draining systems of capital accumulation, chronic carbon cycles, resource degradation, amongst other externalized costs and effects).[8] The first and second contradictions of capitalism are prime manifestations of this thermodynamic reality.[9] The lack of adequate feedback recognition and design ultimately chokes the modern metabolism; its flow structure design. Consequently, less energy will eventually cycle through such a system. As such, these modern metabolisms have not been as powerful as possible with their particular dissipations of massively available energy gradients.

When understood as a dissipative structure, architecture's oversight of the actual bio-telluric formation of buildings amidst its perpetual formal preoccupations and fetishes is, at face value, dismally ironic at best. Form as object supersedes form as formation and the delirious potential inherent therein. The potential vitality of architecture and qualities of life in urbanization are lost in the process. But, on a deeper plane of analysis, this formal oversight reveals architecture's complicity in a larger rift that separates so many parts of modern society: a rift between the appearance of things and their energetic engenderment. This rift is evident in many master modernization processes of which architecture is but one gear and insulation but one cog. To illuminate the fettering effects of this rift, it is instructive to understand one of its earliest instantiations: in keen observations about early industrialized agriculture that offer insight into a systemic set of problems for contemporary architecture.

The Metabolic Rift

Environmental sociologists Jeremy Bellamy Foster, Jason W. Moore, Brett Clark, and Richard York, amongst others, have robustly expanded upon a rift that Karl Marx perceived in the emerging social dynamics of the nineteenth century

Metabolism, Globalized and Externalized
"The Chincha (Guano) Islands: Middle Island, As Seen from North Island"

The Contradictions of Capitalism
The collapse of Interstate 35W bridge in Minneapolis, August 1, 2007
The externalization and discounting of future costs make the future a colony of the present, one of many contradictions in the enabling fictions and metabolic rifts of designs for unimpeded, short-term growth.

and their ecosystem effects. Marx observed that the increasing divide between town and country in nineteenth century England — the "urbanization of the countryside" — manifested a fundamental disruption in the ecological relations of people and the material/energy dissipations that supported them.[10]

The topic of emblematic concern for Marx was the introduction of fertilizers to counter the soil exhaustion caused by agricultural over-production in that period, a product of the nascent industrialization of farming and land tenure.[11] The sources for British soil fertilization were global: from bones on Napoleonic battlefields to, most famously, guano imports from Peru as harvested by Chinese slave labor.[12] The alienation inherent in this now global "irreparable rift" for Marx was only exacerbated by the geographic extent of the guano trade as well as the food and clothing trade that begat the soil exhaustion in the first place.

Marx's focus on soil re-fertilization drew primarily from the work of German agricultural chemist Justus von Liebig.[13] Liebig used the term *metabolism* to describe the exchanges of consumption and feedback that characterize biological systems. A major problem that Marx observed in the English hinterland dynamics was that the reinforcing feedback nutrients required for the soil's metabolism in the countryside were not fed back from the city in the form of night soils. This break in the feedback reinforcement cycle of the soil metabolism — nutrients out from the rural land in the form of food or textile materials coupled with no refreshment of those nutrients back into the land, an asymmetry of accumulation — marked a problematic rift in this early formation of modernization.

It is no mere historical coincidence that Karl Marx based his critique of the emerging metabolic rifts of the nineteenth century industrial agriculture on the work of Justus von Liebig, who was intimately involved with the elucidation of heat-work equivalence. Marx identified first and foremost a thermodynamic problem as he articulated a metabolic problematic. Architecture suffers a metabolic rift as well. The thermodynamic basis, explanations, and designs of our metabolism today are absolutely central to more sane modes of design and urbanization in the future.

Marx did not directly use the term "metabolic rift," but he did observe that large-scale agricultural practices produce "conditions that provoke an irreparable rift in the interdependent process of social metabolism, a metabolism prescribed by the natural laws of life itself. The result of this is a squandering of the vitality of the soil."[14] Further, the "natural laws" that govern how energy is captured and channeled in matter — soil, in this case — extended, in Marx's mind, to the energy captured and channeled through human labor. The vitality of the soil and the vitality of the labor class — both important qualities of life — were squandered in emerging modernization dynamics. Changing conditions of labor in the town and the countryside were both the source and sink of this rift.

Environmental sociologist James O'Connor has cogently argued that Marx's thoughts on this irreparable rift illuminate a "second contradiction" of capitalism and other master modernization processes.[15] Where Marx observed that through over-production, capitalism does not maintain sufficient conditions for endless accumulation (the first contradiction), O'Connor observes that it also engenders a contradiction of under-production. In its efforts to maintain production and accumulation, O'Connor writes that "when individual capitals attempt to defend or restore profits by cutting or externalizing costs, the unintended effect is to reduce the 'productivity' of the conditions of production and hence to raise average costs."[16] Under-production introduces its own crises in the system.

Put into the terms of contemporary dynamics, Jason W. Moore observes that this second contradiction "might take the form of soil exhaustion or deforestation, but it might also take the form of defunding public education or the deterioration of vital infrastructures."[17] In other words, this rift and its contradictions are not a matter isolated to agriculture but rather a systematic reality of modernization, a condition of depravations and degradations in nearly all aspects of contemporary life, not the least of which are architecture and urbanization. Thus, Moore importantly observes that Marx's capital-driven town-country dynamics did not produce a metabolic rift but rather it *is* a metabolic rift that constitutes capitalism and other master modernization processes.[18] The ignorance that characterizes this rift enables dynamics as problematic as they are systemic. The metabolic rift between master modernization processes and the ecological relations that presuppose them is not specific to soil. Today it remains a vexing issue for qualities of life in general and by no means excludes architecture.

Following Moore's perspective, the current discipline of architecture does not *produce* a rift but — especially as object-construed and object-focused — *is* the rift. The soil metabolism problem that Marx observed is an emblematic example of the perpetually complicated relations of matter, energy, and practices that enact this rift. Focused on the object, and unaware of the system, architects generally have no cognition of, and no agenda for, the systemic, maximal power habitually squandered — to use Marx's and Bataille's shared term — in the vast accumulations of matter, energy, and information that constitute a building. Faced with such great squandered resources, architects who are disposed to such awareness typically employ tropes of efficiency and conservation as a response. A litany of statistics are routinely used to portray architecture's excess as something to minimize rather than as a source of latent ecological power.

However, and no matter how counterintuitive it may seem, the resources accumulated and "squandered" within this rift are not inherently problematic, whereas the tropes of efficiency and conservation are ultimately problematic.[19] The fettering effects of prevalent habits of accumulation and squander that constitute the rift are problematic. But it must be known that the massive, constitutive accumulation of matter and energy in a building is not the problem itself. Far more consequential are the prevalent habits of mind that shape a building's literal material and energetic metabolism. The habits of mind that perpetuate the rift are the central problem.

So while Georges Bataille's observation that "man's disregard for the material basis of his life still causes him to err in a serious way," is certainly poignant, its real importance runs counter to many assumptions about material and energy consumption.[20] This conclusion primarily asserts that pseudo-ecological assumptions about efficiency and conservation risk perpetuating the rift as much as ecologically unmotivated practices. In too many cases, efficiency concerns are but apologist agendas that excuse far more deleterious neoliberal dynamics that architects, again, are trained to ignore.[21] These all-too-common and well-intended practices are perhaps less bad, but they propagate the problematic rift nonetheless.

Necessary Excess

Perhaps surprisingly, the metabolic rift can be best addressed in architecture through its necessary excess rather than practices that fundamentally perceive that excess as a liability (as in efficiency or ambitions for net-zero buildings). The key to addressing architecture's metabolic rift is to in turn reflect more deeply on its metabolic *gift*: the necessary excess of architecture.

Architecture only exists as a surplus to basic requirements of shelter. Even the most basic building requires a great accumulation of matter and energy to exist and persist; architecture requires more. Architecture, as an elevated condition of building, only emerges in a context of even greater accumulated need, ambition, *or* wealth (however the last may be defined). This accumulation of otherwise dissipated energy can be understood as a metabolic gift once matter is understood as but an expression of captured energy. The energy captured in matter — geologic action, solar inputs, labor, information, etc. — is the currency of metabolic gift economies: the donor value of architecture. The massive, constitutive accumulation of matter and energy in a building therefore puts architecture in a non-trivial position in the metabolic economy of material and energy systems. Architecture's metabolic rift is methodologically problematic precisely because it obfuscates its vast accumulation of metabolic gifts and its role in this gift economy.

The great accumulation inherent in architecture's literal metabolism, when considered together with the metabolic rift, poses powerful questions for the necessary excess of architecture. Marx and O'Connor astutely observe that the contradictions of capital accumulation continuously yield squandered resources, labor, and qualities of life. While such processes are deeply problematic, efficiency and conservation, again, only serve to perpetuate the rift. Addressing this metabolic rift requires other perspectives and techniques.

Georges Bataille perceived accumulation, and its squander, in other terms. As he noted about the economy of superabundance and accumulation, "On the surface of the globe, *for living matter in general*, energy is always in excess; the question is always posed in terms of extravagance. The choice is limited to how the wealth is to be squandered."[22] From Bataille's perspective, forms of ultimately beneficial exudation are the desired outcomes of accumulation. He notes that a living organism,

> receives more energy than is necessary for maintaining life; the excess energy (wealth) can be used for the growth of a system (e.g. an organism); if the system can no longer grow, or if the excess cannot be completely absorbed in its growth, it must necessarily be lost without profit; it must be spent, willingly or not, gloriously or catastrophically.[23]

It is useful, again, here to note that in terms of exergy, the planet Earth receives around 160,000 terawatts of gross solar radiation while civilization consumes about 16.5 terawatts of exergy annually.[24] The accumulation of this solar radiation (about 110,000 terawatts on the surface of the Earth), poses radically different views on the character of any so-called energy crisis.

To grasp mutual benefits in this metabolic economy requires a substantially larger perspective than the conventional energy analysis in architecture, which proves far too isolated in this regard. Architecture's necessary excess must be understood in this larger energy system, also characterized by excess. This ultimately constitutes a radical shift in assumptions about the economy of energy dissipations in the terms of architectural discourse. "Changing from the perspectives of *restrictive* economy to those of a *general* economy," Bataille notes, "actually accomplishes a Copernican transformation: a reversal of thinking — and of ethics."[25]

So, fundamentally, the issue is not whether squandering is good or bad, or if dissipation is good or bad. A different focus — and ethic — is at stake. Most contemporary impressions of squander are colored by the ascetic Calvinist ethics that characterized capitalism in Max Weber's view.[26] This negative characterization of squander and dissipation persists in the capital-driven notions of efficiency that determine much environmental discourse. This Puritanical construal of squander and dissipation misses its great energetic potential able to address metabolic rifts. Other more robust and resilient positions about squander understand it as being as mutually productive as it is necessary.[27] Again, it is not a question of minimizing architecture's impact; rather we should maximize its (productive) impact in non-isolated, non-linear thermodynamic systems.

Our solar superabundance is inevitable and requires some form of exudation. The issue, then, is how the superabundance of energy accumulated in the matter and processes of architecture submits to the terms of productive and ambitious design. Self-organizing systems do just this and this is what architecture's metabolic rift obfuscates in design: the latent power of this inherent solar accumulation. As Bataille notes, "our ignorance only has this incontestable effect: It causes us to *undergo* what we could *bring about* in our own way, if we understood. It deprives us

of the choice of an exudation that might suit us."[28] Paraphrasing Bataille, by perpetuating its rift, architecture undergoes what it could otherwise bring about through design, if it understood it.

The Metabolic Gift

The squandered vitality of the soil that Marx observed illuminates a very important question about contemporary architecture's general energy economy of energy systems and their hierarchies. The only way to understand and address architecture's metabolic rift is to recognize the inherent energy hierarchies of architecture. Buildings are lodged in a hierarchical cascade of energy driven almost entirely by a superabundant solar input — a metabolic gift.

Ecosystem scientists map energy systems to articulate the ecological functions of an object, process, or system. These energy hierarchies track the accumulation of energy that is initially as abundant as it is diffuse: solar energy. Because of this solar abundance, Odum used solar inputs as the basis of his energy system mapping methodology. This allows all objects and processes to be compared, a unique and powerful component of this methodology. The resulting systems track how and how much solar energy (solar emjoules) are captured and channeled by the system. The position of anything, such as a building, in an energy hierarchy indexes the number of transformities of solar energy that are embedded in it. What matters in this kind of emergy evaluation is not exacting accounts of emergy quantities but rather finally grasping the relative scales and magnitudes of buildings in energy hierarchies.

Based on this account of energy systems as a general economy, Odum offers what he calls the maximum power principle: "in the competition among self-organizing processes, network designs that maximize empower will prevail."[29] Optimal systems maximize, not minimize, their power (energy) intake, their transformities, and their feedback reinforcement. Energy systems are, it turns out, not at all as squander-phobic as most energy practices and pedagogies in architecture construe them to be. Rather, energy systems evolve to maximize their power: their capacity to do work on their milieu. To do so, energy systems do not simply take in as much power as possible but select networks that reinforce conditions that will feed back in productive ways. The results are systems of great mutuality and accumulation.

In the context of energy hierarchies it is important to understand the role of high-transformity objects and processes (such as buildings, people, and information). Objects and processes in an energy system of high transformity are characterized by Odum as high-quality because of their capacity for feedback reinforcement.

Odum describes feedback reinforcement in a specific way: "the action of a unit or process to enhance production and survival of a contributing unit or process, thereby enhancing itself; a loop of mutually enhancing interactions."[30] This means that high-transformity objects and processes have great capacity to select reinforcement of the larger collective manifest in the energy hierarchy. Architects, as specifiers of great amounts of matter, energy, and information, hold a high-quality position in their energy hierarchies, the importance of which is equaled by their lack of awareness of it.

Odum's energy hierarchies and energy system diagrams force the question of system boundaries. This is perhaps his most significant contribution and import for design. He provides a methodology that is by definition non-isolated and thus serves to counter the limitations of isolated perspectives on energy. Once exposed to the concept of emergy, the role of the design's system boundary becomes paramount to any claim about energy, or, more importantly, the power of a system. In this important way, Odum's methodology helps close the metabolic rift in contemporary architecture.

Emergy is not deterministic. It will not necessarily demand or determine one design over another. However, understanding the energy hierarchy of any building will forever transform how an architect thinks about buildings and energy. From a non-isolated perspective of the energy inherent in architecture, a few key questions emerge from emergy analysis:

- **What is your system boundary?**
- **What do you externalize through design?**
- **What is the power of your object and its system?**

These questions help shape a more totalizing agenda for energy in architecture. In contrast, the isolating habits of mind that emerged with insulation — evident equally in modernist building assemblies, modern buildings as a whole, and the supply chains and material geographies that presuppose modernist buildings — have ultimately constrained what architects think energy is, how it operates, and what they can do with it.

In the realm of building envelopes, isolating conductivity as the only parameter of heat flux systematically limited the thermal power of buildings. In the realm of buildings as a whole, isolating buildings as autarkical, autonomous entities has systematically constrained the ecological power of buildings, and isolating building design to the composition of buildings-as-objects has systemically inhibited the praxis of architects.

Necessary Excess of Energy Systems

Architecture and energy systems both thrive on surplus, abundance, and excess. Architects need to contend with this necessary excess not through efficiency or optimization, but through maximal ecological power and its architectural potential. Before any claim on performance, an architect must first know the inputs, throughputs, and outputs of the overall system: the energy hierarchy of any object or process. Without this larger view of energy hierarchies, there is no way to evaluate the efficacy of a particular design with respect to the overall system: a building's general economy. As part of a non-modern architectural agenda for energy, this concern for the whole far exceeds the energy simulation of any particular component of the design. With the planetary flow of energy that constitutes the general economy of the world in mind, Bataille wrote, "woe to those who, to the very end, insist on regulating the movement that exceeds them with the narrow mind of the mechanic who changes a tire."[31]

For designers to operate cogently within these energy hierarchies requires alternate agendas for architectural formation and different agendas for energy in architecture. This metabolic gift economy of maximal intake and maximal feedback reinforcement should exert great demands of the necessary excess of architecture. Architecture could be designed in a way that makes a building *and* the overall energy system more powerful.

Autarky and other Non-architectural/ Ecological Agendas for Energy

Like so many other capital-driven systems, too many agendas for objects and energy in architecture narrowly bind the area of concern, neglecting the power of maximal power intake, use, and feedback. Many types of energy analysis — energy simulations, embodied energy analysis, Passivhaus, and Life Cycle Analysis (LCA) — while cogent for certain bounded purposes, neglect to include important bio-geophysical inputs as well as different qualities of energy inherent in the energy hierarchy of any object or process. The systems boundary considered in these methods is simply too narrowly defined.

As such, they methodologically maintain the rift even as they intend to address its symptoms. Bio-geophysical inputs are not free from an energy system point of view, they are so only from a capitalist point of view. As Ravi Srinivasan, William Braham, Daniel Campbell, and Charlie Curcija collectively note, "The notion that raw materials for building construction are plentiful and can be extracted 'at will' from Earth's geobiosphere, and that these materials do not undergo any degradation or related deterioration in energy performance while in use is alarming and entirely inaccurate."[32] They continue: "The most significant inadequacy that relates to this research is that LCA lacks a rigorous thermodynamic framework which is elemental for analyzing ecosystems and in certain situations it may even violate thermodynamic laws."[33] And so it goes with a range of energy simulation approaches in the building industry today that perpetuate the metabolic rift.

The persistent rift in architecture's metabolism, even in overtly ecological work, enables some problematic perspectives on the role of energy in architecture. Rather than a cogent response to the rift, the impulse for autarky, for example, is but another poignant symptom of the rift.[34] The thermodynamically, economically, socially, and ecologically incongruous notion that a building could be self-sufficient, self-sustaining, or achieve net zero-energy is as physically impossible as it is epistemically disturbing in regard to architecture's metabolic rifts and gifts. Net-zero energy claims suffer from the enabling narrowness of their system boundary. Such narrow system boundaries, and, by extension, system aspirations, block the capacity of feedback reinforcement through design: a profoundly un-ecological motivation. In reality, buildings are highly dependent on large scale material and energy flows. This precisely is the great ecological potential of building, not its ecological liability.

In this respect, architecture does not need to become more ecological. Rather, architects need to become aware of just how systemic, and thus ecological, architecture is in reality. Architecture is constitutively ecological. Rather than autarky, a cogent response to architecture's metabolic rift is radical contingency: finally situating architecture in the energy hierarchies of the world. In doing so, it finally becomes methodologically possible for buildings to become more ecologically powerful through design. Further, *and most importantly*, since both energy systems and architecture thrive on abundance, excess, and affordance, it also becomes methodologically possible for buildings to become ecologically powerful in ways that amplify the motivations and ambitions of architecture itself.

Many contemporary objects and systems maximize bio-geophysical throughput by using an abundance of resources. This is not inherently problematic from an energetic or resource point of view. However, it is extremely important to recognize that today these high-emergy-intake objects and systems are generally not maximum power designs. Their emergy, exergy, and energy ratios are awry. While high power systems, they do not maximize their feedback reinforcement as they maximize their throughputs as the primary means to maximize power in the system. As such, this rift creates a conundrum for the architectural formation of matter and energy. With this metabolistic oversight — these unconsidered system boundaries — architects and buildings cannot fully engage the architectural and ecological potential of their own necessary excess and exuberance.

In the end, the necessary excess of architecture and energy systems can and should be self-reinforcing. The necessary excess of architecture aligns neatly with the propensity of energy systems to maximize (not minimize) their power. Architecture's ecological capacities are directly correlated to these large scale material and energy systems that presuppose any building.

Conclusion: The Metabolic Shift

Taken together, architecture's metabolic rift and gift demand a sharp metabolic shift in the direction of maximum power and maximum entropy designs. Instead of objects alone, contemporary life demands maximal power objects *and* systems. Contemporary life demands fluxable systems boundaries from designers. Rather than the narrow systems boundaries and the autarkical, Calvinist impulses of efficiency, contemporary life needs the affordances and abundance that emerge from mutuality and reinforcement in maximal power energy systems. Well beyond the pervasive preoccupations with efficiency and optimization, the thermodynamics and energy hierarchy design of real systems require very different ethics, theories, techniques, and designs.

The metabolic gift of architecture's necessary excess suggests that architects should be radically more invested in the dynamics and multifarious dissipative pathways for *solar accumulation* rather than as efficient cogs of modern *capital accumulation* alone. Whereas Marx was concerned about the effects of accumulation would have on soil and labor, it is absolutely critical to recognize simultaneously another type of super-accumulation to move beyond the fetters of this rift. The accumulation of solar energy — the cascade of solar energy in the energy hierarchy of any object, system, or process — is essential to addressing this rift and designing more exuberant architectural and ecological outcomes. But if, as you read this, you begin thinking about photovoltaic systems rather than the solar-driven energy hierarchies of architecture, then the essential lessons still evade you.

Vast resources are squandered in architecture. It is axiomatic that architecture's necessary excess requires significant accumulation for its mere engenderment as well as its construction and operation. As ecological systems theory establishes, this massive accumulation is not any more inherently problematic than it is necessary or potentially powerful, but only when submitted to the terms of expansive and ambitious design. As George Bataille noted, "The choice is limited to how the wealth is to be squandered."[35] It is a question of how to most powerfully frame the necessary excess of architecture. It is a question of which type of accumulation — capital or solar — ultimately amplifies the mutual purpose and ambitions of architecture, ecology, and contemporary life.

A shift in perspective from the metabolic rift economy to the metabolic gift economy that presupposes architecture presents a radically different understanding of the excessive accumulation and dissipation required for architecture by definition. The metabolic rift economy induces practices that habitually capitulate to capital accumulation: preoccupations that reduce the topic of formation to mere objects and shapes, pervasive obsolescence in the planning and construction of buildings, technological determinism and naïve progressive ideologies, and errant ethics of efficiency that together ultimately have little ecological power in reality. Architects, besieged by a century of acquiescence and capitulation, exert little to no agency in supply chains. But if architects decided to direct a small fraction of the effort spent on its parlor games to the problem of a building's actual engenderment rather than its appearance alone, new forms of agency and new manifestations of design would be powerfully evident.

Insulated Rift

Insulation has been a primary enabling, constitutive example of the isolation inherent in this rift: a habit of mind that keeps system boundaries narrow and isolated. The metabolic gift economy, on the other hand, situates architecture in the energy hierarchies of the world and as such identifies multiple pathways for architects to amplify not only the robustness of the larger collective but architecture itself in this century. The donor value of architecture's necessary excess presumes enlarged and fluxable system boundaries for architecture.

To address this metabolic rift, a recursive process of design characterized by radical contingency is necessary. Such a process would cycle back and forth through specific aspects of a building and its respective bio-geographic implications, yields, and thus potential power. In turn, metabolic observations about these constitutive contingencies will inflect the specificity of a building. Hardly an issue of ecological concern alone, addressing the metabolic rift of contemporary buildings finally allows architecture to fulfill the terms of form and formation that remain so central to the discourse of design.

A shift from object composition to maximum power object *and* system composition is consequential. Less revolutionary than evolutionary, a shift to a recursive object-and-system approach to composition would be a signal advancement of design in this century. The sustained, if not misplaced, virtuosity of object composition can today only be amplified by the terms of its own actual engenderment rather than perpetuating the fabricated claims of autonomy that suited the terms of discourse. In short, design might yet see a profound influx of thermodynamic vitality, but in the most worthy of examples it will only whet and amplify our enthusiasm for the salient task of design: formation.

The contradictions inherent in the rift between architecture's preoccupations with the topic of formation coupled with dismissed and externalized but literal formations can no longer be intellectually — or physically — sustained. A recursive object-system composition of architecture's necessary excess is far more ambitious intellectually and ecologically: an advancement and liberation of otherwise fettered, contradictory practices in architecture.

To move beyond these contradictions, metabolism must become a deliberate master process of the non-modernization and composition of architecture today. The metabolic rift, gift, and shift in architecture constitute an essential framework towards this end. Such a framework enables novel formations and figurations for architecture in contemporary life. To fulfill its own ambitions, architecture needs to become ever more formal, ever more engaged in the processes of actual formation of buildings. Modern architecture, though, has never been formal enough, especially about its necessary excess.

Buildings are not insulated or isolated from large scale energy systems. The necessary excess inherent in these energy systems is in fact architecture's greatest ecological opportunity. To isolate the heat flux of a wall from these larger concerns profoundly diminishes the capacity of architecture. A much more recursive consideration and design of these small and large, short-term and long-term energy fluxes is essential to an architectural engagement with energy. One without the other is a thermodynamic, ecological, architectural, and *formal* failure.

The history of insulation clearly characterizes the missions and omissions of architecture's agenda for energy flux. By more deeply interrogating the terms of insulation's history and dynamics, we can more deeply engage a much larger set of energy dynamics in this century. Reconsidering heat flux and insulation is by now absolutely essential to conceptualizing, if not designing, buildings that maximize energy intake, exergy matching and use, and feedback reinforcement as part of a maximum power approach to design. Given the circumstances of this century, these ecological/thermodynamic characteristics of optimized energy systems demand great attention today. The implications of maximal power systems are not easily grasped but when they are, they promise profound adjustments to assumptions about energy, design, architecture, and urbanization.

Hammer

To grip a hammer makes everything look like a nail. Architecture's current grip on R-values and insulation has made energy exchange seem like a matter of conduction and its resistance. This is of course only partially valid and yet does not at all afford any insight into the totality of dissipations involved in the thermodynamics of architecture. To grip insulation theories and practices this way is to draw peculiarly parochial systems boundaries that overemphasize the spatial and temporal scale of the building as an object and neglect it as an open thermodynamic system. It is only with such a parochial system boundary that the thermodynamic quackery of "zero-energy buildings" could be even uttered without irony. There is no thermodynamic, ecological, or architectural justification to design with such unquestioned system boundaries. To do so perpetuates the metabolic rift of architecture.

To grip simultaneously the diffusion and dissipation of energy at the scale of the building envelope and at the scale of its energy hierarchy gives the design proposition of "optimization" a radically different, more totalizing character. In my view, this more totalizing mobilization of dissipation is the most potent ecological design catalyst of this century because it affords the broadest praxis, the most penetrating insights and claims, and, most importantly, novel formations of architecture. To simultaneously grip energy at these multiple scales is, like a hammer, at once a tool and a specific prescription for action.

Insulation taught architects to resist the flow of energy with astonishing and overreaching effectiveness. But an architect equipped with a thermodynamic understanding of energy systems knows that energy just flows and this dissipation, not flux obstacles, is what must be deduced and designed. Far more than an R-value instinct, the architect's task is the purposeful modulation of the velocity of energy in an energy hierarchy. In some cases this might include resistance but in other cases the key exergetic functions of accelerating, mixing, slowing, delaying, or storing might matter more and have greater eco-thermodynamic efficacy. Whether you are considering the heat flux in a wall or the qualities of life in civilization, it is the mindful modulation of dissipating available gradients that should be designed.

Over-insulated

In this regard, insulation over-trained architects. Too quick to resist and make efficient, we can no longer really see — much less design — the multiple forms of energy that pulse through our bodies, buildings, cities, and civilizations. Too quick to quantify, architects have systematically forgotten to first qualify different forms and qualities of energy. Buildings, when carefully partitioned as isolated and closed systems, cleave off great ecological potential and exuberance. In doing so, the reductive, Calvinist values and mathematics of consumption cynically supplant the donor value of architecture and the affordances of its latent gift economy. When reduced to the utility of heat or light, energy and its astonishing transformations and formations lose their vitality and potential exuberance not only in architecture but in life at large. When reduced to utility and use, the richness and literal power of feedback and recurrence disappears, flattening out the now-latent vitality of design in the process and its metabolic rifts.

Buildings are always and only non-isolated systems. To better engage the non-isolated energy system of a building and its hierarchies in this century, the concepts of energy velocity, diffusion, and dissipation are perhaps the most productive concepts to consider. As a habit of mind, these concepts enable a productive set of practices dealing with a cascade of exergy-matching designs at small scales to the dissipative structures that produce great novelty out of entropic diffusion. In all cases, diffusion introduces a temporal dimension into architectural energy systems, shifting boundary conditions of reality and leading to a far more systemic and connected engagement with the totalizing energy systems of buildings.

The narrow system boundaries of conventional heat transfer practices — from R-values to the Passivhaus paradigm — simply exclude too much to make substantive claims about energy in architecture and the thermodynamics that feed life. There is much to learn from these modern paradigms, but the first and most important thing to understand about these is how their system boundaries are defined.

When it comes to energy, architects need a different hammer and, as Latour has noted, different investigative tools. We need novel building science concepts and novel designs based on a very broad view of energy dissipation and its multiple forms. We need recursive design practices that cycle back and forth through an energy hierarchy as one type of formation. We need new protocols for the thermodynamic formation and figuration of buildings fit for this century and beyond. The paucity of ambition inherent in the parlor tricks and shell games of our purportedly most advanced formations of architecture is no longer a sufficient explication of the topics of form and formation. So very little is composed, ultimately, in the knowing nods and misplaced virtuosity of these perennial examples.

To paraphrase historian Perry Miller, given all the precious, beautiful, initially insupportable but necessary, and wholly irrational blessings of architecture's excess — with all the myriad quandaries of obligation and opportunity therein — the responsibility of mind to preserve its own integrity amid the terrifying operations of building science, green building certification programs and formal agendas for the open thermodynamic systems of architecture is both an exasperation and an ecstasy.[36]

Designed Dissipation, Not Resistance

Any isolated perspectives of architecture cannot persist for long. As but one starting point, I offer dissipation as a time-imbued, multi-scalar parameter that can help inculcate a non-isolated habit of mind about energy in architecture. Simply put, the dissipation of multiple, multivariate forms of energy, when considered recursively, indicates a set of practices that can ultimately make design more architecturally and ecologically powerful.

Architecture in this century needs an evolved agenda for energy. It needs a model of thought and practice that is far less isolated and acquiescent than its modernist antecedent. In short, it needs a non-modern agenda not just for energy but for architecture itself. Finally incorporating the deep implications of thermodynamics is one way around our modernist corners.

The rift between insulation and larger energy systems in modernism is technically easy to mend. The modern rift between isolated and non-isolated epistemologies will require greater but necessary intellectual effort. In both cases, the non-isolated, dissipative structure of buildings and their energy hierarchies is an absolutely necessary starting point for a non-modern architectural agenda for energy in this century and, as such, is the concluding point of this book.

Notes

1. Ilya Prigogine and Isabelle Stengers, *Order Out of Chaos: Man's New Dialogue with Nature*, New York: Bantam Books, **1984**. p. **127**.
2. Axel Kleidon and Ralph D. Lorenz, *Non-equilibrium Thermodynamics and the Production of Entropy: Life, Earth, and Beyond*, Berlin: Springer, **2005**.
3. Eric Schneider and James Kay, "Order from Disorder: The Thermodynamics of Complexity in Biology," *Futures*, vol. **26 (6)**, **1994**. pp. **626–647**.
4. Howard T. Odum as quoted in David Rogers Tille, "Howard T. Odum's Contribution of the Laws of Energy," *Ecological Modelling*, **178**, **2004**. pp. **121–125**.
5. R. E. Ulanowicz and B. M. Hannon, "Life and the Production of Entropy," *Proceedings of the Royal Society, London, Series B,* vol. **232**. pp. **181–192**.
6. Weston A. Hermann, "Quantifying Global Exergy Resources," *Energy*, vol. **31(12)**, **2006**. pp. **1685–1702**.
7. Georges Bataille, *The Accursed Share: An Essay on General Economy*, trans. Robert Hurley, New York: Zone Books, **1988–1991**.

8   Anthony Giddens, *The Consequences of Modernity*, Stanford, CA: Stanford University Press, **1991**; Ulrich Beck, *World at Risk*, London: Polity Press, **2010**.
9   James O'Connor, *Natural Causes: Essays in Ecological Marxism*, New York: The Guilford Press, **1998**. pp. **158–177**.
10 John Bellamy Foster, *Marx's Ecology: Materialism and Nature*, New York: Monthly Review Press, **2000**.
11 John Bellamy Foster, "Marx's Theory of Metabolic Rift: Classical Foundations for Environmental Sociology," *American Journal of Sociology*, vol. **105** (2), Sep. **1999**. pp. **366–405**.
12 Brett Clark and John Bellamy Foster, "Ecological Imperialism and the Global Metabolic Rift: Unequal Exchange and the Guano/Nitrates Trade," *International Journal of Comparative Sociology*, vol. **50**, **2009**. pp. **311–334**.
13 Justus von Liebig, *The Natural Laws of Husbandry*, New York: D. Appleton, **1863**; and *Letters on Modern Agriculture*, London: Walton & Maberly, **1859**.
14 Karl Marx, *Capital, Volume III*, New York: Vintage, **1981**. pp. **949–50**.
15 O'Connor. pp. **158–177**.
16 Ibid. p. **245**.
17 Jason W. Moore, "Transcending the metabolic rift: a theory of crises in the capitalist world ecology," *Journal of Peasant Studies*, vol. **38** (1), **2011**. pp. **12–13**.
18 Ibid. p. **7**.
19 There is a non-trivial arc of consideration of exudation through the notion of accumulation and squandering in the thought of Karl Marx, Friedrich Engels, Thorstein Veblen, Marcel Mauss, Georges Bataille, and Howard T. Odum. Each of these thinkers understands the inevitability of squander in alternate ways. Although of disparate fields, when juxtaposed, these alternate understandings of accumulation and squandering become necessary tools of thought on this topic.
20 Bataille. p. **21**.
21 Erik Swyngedouw, "Apocalypse Forever? Post-political Populism and the Spectre of Climate Change," *Theory Culture Society*, vol. **27**, **2010**. pp. **213–232**.
22 Bataille. p. **23**.
23 Ibid. p. **21**.
24 Mark T. Brown and Sergio Ulgiati, "Updated evaluation of exergy and emergy driving the geobiosphere: A review and refinement of the emergy baseline," *Ecological Modelling*, vol. **221** (20), **2010**. p. **2503**.
25 Bataille. p. **25**.
26 Max Weber, *The Protestant Ethic and the Spirit of Capitalism: and Other Writings* (**1904/05**), trans. Peter Baehr and Gordon C. Wells, New York: Penguin Books, **2002**.
27 Marcel Mauss's "gift economy" and Bataille's "accursed share" both articulate alternate theories of accumulation and squander. Marcel Mauss, *The Gift*, New York: W. W. Norton & Co., **2000**.
28 Bataille. pp. **23–24**. Emphasis his.
29 Howard T. Odum, *Environmental Accounting: Emergy and Environmental Decision Making*, New York: John Wiley & Sons, Inc., **1996**. p. **16**.

30 Ibid. p. **289**.
31 Bataille. p. **26**.
32 Ravi S. Srinivasan, William W. Braham, Daniel E. Campbell, and Charlie D. Curcija, "Re(De)fining Net Zero Energy: Renewable Emergy Balance in Environmental Building Design," *Building and Environment*, vol. **47**, **2012**. pp. **301**.
33 Ibid. p. **301**.
34 William Braham, "Temptations of Survivalism, or What do you do with your waste?" *Places: Forum of Design for the Public Realm*: http://places.designobserver.com/entry.html?entry=13998 (consulted August, **15**, **2011**).
35 Bataille. p. **23**.
36 His original quotation is: "Like the precious, beautiful, insupportable and wholly irrational blessing of individuality, with all the myriad quandaries of responsibility therein involved, the responsibility of mind to preserve its own integrity amid the terrifying operations of the machine is both an exasperation and an ecstasy." See the last sentence of Perry Miller, *The Responsibility of Mind in a Civilization of Machines: Essays by Perry Miller*, Amherst, MA: University of Massachusetts Press, **1979**. p. **213**.

Acknowledgements

Aspects of this book were initially developed for an article (Kiel Moe, "Insulating North America," *Journal of Construction History*, vol. 27, 2012, pp. 87–106). The article benefited from the suggestions of the reviewers and editors of that journal. Bill Addis was an especially generous editor. Jos Tomlow responded to the article with collegial enthusiasm, pointing to some distinctive European sources. This book is better for each of their insights and recommendations. Amy Kulper kindly read an early draft of the article and her comments both advanced aspects of the text as well as encouraged an expansion of the claims inherent in the central thesis. I am grateful for her gracious observations and sustained conviviality.

Iñaki Abalos not only wrote a generous foreword to this book but also importantly offered a number of consequential insights that greatly improved the book in a late stage in its development. I am grateful not just for his contribution to this book but, moreover, for his conviction that thermodynamics represents one of the most radical and important epistemological shifts in design culture today. Both during the research and production of this book specifically, and in general as well, my interactions with Iñaki Abalos, Sanford Kwinter, Bill Braham, Bill Sherman, Steven Moore, and Sal Craig routinely confirmed the vitality of thinking about the thermodynamics of buildings anew. Architects needs alternate pathways of thought and practice in this most non-trivial domain of architecture and its formal, social, and technical capacities. Sal read much of the manuscript in a late stage, offering pointed review and advice. I am grateful for his reading, his "Breathing Wall" contribution to this book, and his always fresh contributions to this field that help architects envision other methods and means for designing cogent flow structures and dissipations of energy in architecture. I also benefited from not only information about the RFG from Ivan Rupnik, but more generally over the past years from Rupnik's fastidious reflections on, and convictions about, architecture.

Several students served as research assistants in the Energy, Environments, & Design (EE&D) research lab at the Harvard University Graduate School of Design (GSD) in the course of this project. Lance Smith waded through archival material and helped shape an insulation material taxonomy. George Gard and Rex Ten both assisted with digital models and drawings of the case studies included in this volume. Maz Kahali assisted with archival images from the Cabot Science Library. Jo

Staudt was invested in issues of insulation and building while in the EE&D lab. Thomas Sherman and Saurabh Shrestha developed aspects of this work on insulation as well. Parallel to the development of the manuscript in 2012–13, GSD thesis student Michael Smith and I both benefited from an extended reflection on the concept of isolation in architecture. This proved productive both in the terms and uses of it in his independent thesis project as well as for me, in separate ways, in its implications for insulation and architecture. I am grateful to each of these students who eagerly ask new questions, and obliterate tired assumptions, about energy in architecture. They confirm the axiom that cognition is distributed.

Gert Walden of Baumschlager Eberle generously provided documentation of the 2226 project. Roger Bolthauser's office provided reference drawings used to produce materials for the Rauch House case study. I was grateful to converse with Martin Rauch and observe the intensity of conviction evident in his most telluric of practices during his Mud Works project with Anna Herringer at the GSD in the spring of 2012. It is with great respect for these practices that I include their work, which both motivates and buoys my own practices in and out of the academy of architecture.

This book was initiated at the MacDowell Colony for the Arts and finished at the American Academy in Rome. In their own completely wonderful ways, each of these inimitable institutions provided not only time and concentration but the most collegial and inspiring of atmospheres for work. I am grateful to both institutions for the manifold opportunities and connections they afford. Mary Medlin, fellow Fellow of the MacDowell Colony, copyedited this book and helped make even my writing readable.

Andreas Müller, editor for the publisher, and the book's designer, Miriam Bussmann, each elevated this book in terms of content and as a visual object. I am indebted to both their rigor and their conviviality.

The research for this book was supported by a Junior Faculty Research Grant awarded by GSD Dean Mohsen Mostafavi. His support of this project is greatly appreciated. The GSD, more generally, provides an intellectual milieu — a collection of many people, projects, and passions for design — that every day inspires my work in ways that are daily appreciated but not easily isolated.

Illustration Credits

Drawings courtesy Baumschlager Eberle — 255–257

Drawings courtesy Boltshauser Architekten AG — 262–266, 270–271, 288

Orphan work, from: Andreas Ragnar Bugge, *Amerikas små hjem, deres planlegning, konstruksjon og utførelse*, Oslo: Grøndahl & Søn, 1927. — 91 left (from page 95 of source), 91 right (from page 96), 92 top left (from page 97), 92 top right (from page 98), 92 bottom left (from page 99), 92 bottom right (from page 100), 94 (from page 92), 96 (page 115)

Orphan work, from: Andreas Ragnar Bugge, ed., *Test Houses: Erected by Norges Tekniske Hoiskole, Trondheim; Results of Tests with Wall-Constructions and Materials for Building Warm and Cheap Dwelling-Houses*, trans. J. Craig, Trondheim: F. Bruns Bokhandels Forlag, 1924. — 80 (from page 7 of source), 81 top (from page 5), 82 (from page 27), 84 (from pages 9–11), 85 (from pages 12–14), 86 (from pages 15–17), 87 (from pages 18–20), 88–89 top (from pages 21–22), 90 (from unpaginated foldout spread), 94 (from page 58–59)

Orphan work: Anders Ragnar Bugge, *Varme og Billige Bolige*, Oslo: Grøndahl, 1932 — 81 bottom (from page 8 of source)

Photograph © Beat Bühler fotografie — 258, 261, 267

Courtesy Salmaan Craig — 273–275, 279

Orphan work, from: *De 8 en Opbouw*, 1934. — 100–101

Courtesy Keith Ewing — 170

David Falconer, US National Archives, National Archives Identifier: 555513 (in the public domain) — 10

Photographs © Eduard Hueber + Ines Leong / archphoto — 254, 255 top and bottom, 256 top and bottom

Courtesy George Gard — 174–175, 177

Photomontage of Campus NTNU Gløshaugen, Trondheim, Norway. Architect: Karl Grevstad. 1960. Creative Commons Attribution-Share Alike 3.0 Unported license — 82 bottom left

The Illustrated *London News*, February, 21st, 1863, p. 200. Creative Commons Attribution-Share Alike 3.0 Unported license — 292

Courtesy of Dr. Carsten Jäger — 26

Kiel Moe — 4, 69, 89 (redrawn based on Andreas Ragnar Bugge, ed., *Test Houses: Erected by Norges Tekniske Hoiskole, Trondheim; Results of Tests with Wall-Constructions and Materials for Building Warm and Cheap Dwelling-Houses*, trans. J. Craig, Trondheim: F. Bruns Bokhandels Forlag, 1924, page 300), 95 (based on "TABLE OF TEST DATA" based on "Architectural Engineering: Heat transmission through dwelling house walls," *American Architect and the Architectural Review*, vol. 126, Sept. 2, 1924, p. 300.), 102,

116, 131–133 (with Lance Smith), 133 bottom, 178–180 (with George Gard), 183, 211 left, 211 right (based on Daniel Krencker, *Die Trierer Kaiserthermen*, Augsburg: Dr. Benno Filser Verlag, 1929), 212 top and bottom, 215–216, 217 top and bottom, 230

Courtesy of the National Institute of Standards and Technology (in the public domain) — 67 top and bottom

NASA Solar Dynamics Observatory (SDO), M6 solar flare on Nov. 12, 2012. Public Domain Image. — 297

The Norwegian Institute of Technology, 1930. Creative Commons Attribution-Share Alike 3.0 Unported license — 82 top left

Courtesy of Rachel Popielarz. — 171

Thomas Preston, *Theory of Heat, London,* New York: Macmillan, 1894 (in the public domain) — 62 top (from page 611 of source), 62 middle left (from page 109), 62 middle row center (from page 112), 62 middle row right (from page 225), 62 bottom (from page 111), 63 (from page 237)

Charles George Ramsey, Harold Reeve Sleeper, *Architectural graphic standards for architects, engineers, decorators, builders, and draftsmen*, New York: Wiley; London: Chapman & Hall, 1989 (facsimile reproduction of 1932 edition). (Published with permission from the publisher.) — 154 (from page 44 of source), 157 (from page 497)

Charles George Ramsey, Harold Reeve Sleeper, *Architectural graphic standards for architects, engineers, decorators, builders, and draftsmen*, 2nd edition, New York: Wiley; London: Chapman & Hall, 1936. (Published with permission from the publisher.) — 158 (from page 184 of source), 159 (from page 185)

Charles George Ramsey, Harold Reeve Sleeper, *Architectural graphic standards for architects, engineers, decorators, builders, and draftsmen*, 5th edition, New York: Wiley, 1956. (Published with permission from the publisher.) — 160 (from page 500 of source), 161 (from page 493)

Samuel Cabot Inc., "Build Warm Houses with Cabot's Quilt: The Original Heat Insulator – In Successful Use for Over 30 Years Resists the Passage of Heat and Cold," 1928. Creative Commons license: Public Domain Mark 1.0 — 148 (from page 3), 150 (form page 5), 151 (from page 4)

Photograph © George Steinmetz — 27

Photograph courtesy of Alyssa Umsawasdi — 147, 149

United States Navy photo by Mass Communication Specialist Seaman Joshua Adam Nuzzo, August 14, 2007. Public Domain image. — 292

About the Author

Kiel Moe is an architect and Associate Professor of Architecture & Energy at the Harvard University Graduate School of Design Department of Architecture. He is co-director of the MDes design research program, the MDes Energy & Environments group, the Energy, Environments, & Design research Lab and the First Semester Architecture Design Studio. He is a Fellow of the American Academy in Rome and the MacDowell Colony. He is also author of *Convergence: An Architectural Agenda for Energy* (2013) and *Thermally Active Surfaces in Architecture* (2010), among other books.

Index of Persons, Firms, and Institutions

Abalos, Iñaki 8–10, 310
Abramowitz, Max 199–200
Abramson, Daniel M. 163, 186
Addington, A. Michelle 167, 187
Addy, C. E. 186
Adlam, T. Napier 223
Agamben, Giorgio 31–32, 53
Agrippa 210
Aiello, Leslie 135, 185
Alabama Polytechnic Institute 197
Alberti, Leon Battista 45, 246, 252, 284
Altmann, Markus 285
American Architect and Building News 72
American Society for Testing and Materials (ASTM) 68, 111
American Society of Heating and Ventilating Engineers 68
American Society of Heating, Refrigeration, and Air Conditioning Engineers 167, 206, 219
American Society of Refrigeration Engineers 68
Arbuthnot, John 192, 222
Architectural Graphic Standards 154, 156, 157–161, 162
Armstrong Cork Company 52
Aronin, Jeffrey Ellis 203, 225
Axelbank, Louis 199–200, 206, 224
Aylward, David A. 186

Babloyants, Agnes 284
Baehr, Peter 308
Baker, P. H. 278, 286
Banham, Reyner 9, 169, 187
Barnes, C. S. 124
Bartussek, Helmut 286
Basaran, Tahsin 226–227
Bataille, Georges 232, 245, 284, 291, 294–296, 298, 302–303, 307–309
Bateson, Gregory 18, 52
Bauhaus Dessau 196
Baumschlager Eberle 253, 285, 311
Beck, Ulrich 53, 129, 185
Bedford, Thomas 104, 121, 193, 203–204, 222–223, 225
Bejan, Adrian 28, 53, 140–141, 155, 185, 232, 241–242, 271, 284–285, 287
Bello-Ochende, Tunde 287
Bénard, Henri 26–27, 28
Bernan, Walter 65, 118
Biot, Jean-Baptiste 61
Bijvoet, Bernard 103
Billington, Neville 107, 112, 123
Bischof, Manfred 285

Black, Joseph 61, 118
Boccaletti, Giulio 251, 285
Boëthius, Axel 226
Boltshauser Architekten AG 285
Boltshauser, Roger 265, 285, 311
Boltzmann, Ludwig 63, 198, 228, 231–232, 236, 238–242, 283
Bozsaky, Dávid 185–187
Bracken, J. H. 119, 186
Braham, William 284, 300, 309–310
Bréguet, Abraham 61
British Research Establishment 103
Brockman, John 285
Brown, Mark T. 308
Brunsell, J. T. 286
Brush, S. G. 283
Bugge, Andreas Fredrik 78–81, 88–90, 93, 95–97, 120
Building Research Station (BRS) 103–104, 193
Burch, D. M. 124
Burkhardt, Walter 197–198
Burnett, Eric 76, 119

Cabot, Samuel 146
Caldwell, Michael 172–173, 187
Calvin, John 10–11, 25, 71–72, 76, 115, 206, 243, 245, 247, 296, 302, 306
Campbell, Daniel E. 300, 309
Cantani, Mario J. 124
Capra, Fritjof 242, 284
Caracalla 210, 213–214, 216, 218
Carnot, Nicolas Léonard Sadi 235
Caudill, William W. 201–202, 224
Cawthorne, D. A. 286
Celotex 96, 142
CIAM 99
Clark, Brett 291, 308
Clausius, Rudolf 235, 236, 242, 283
Close, Paul Dunham 117, 123
Cohen, Ruth Schwartz 74, 119, 186
Commonwealth Experimental Building Station 105
Conrads, Ulrich 223
Constantine I. 210
Coppersmith, Jennifer 60, 118
Corazza, Angelo 226
Corser, Rob 186
Couch, Tristian 227
Courtney, Roger 121
Cox, Paul Alan 186
Cowan, Henry J. 205, 223, 225
Craig, J. 120
Craig, Salmaan 48, 272–283, 310
Crowther, Richard L. 284
Curcija, Charlie D. 300, 309

Dahl, Torben 185
Dalehaug, Arvid 286

Daniels, Klaus 284
Danter, E. 107, 123
Davies, Morris 123
Davis, K. L. 124
De 8 en Opbouw 99–101, 103, 121
De Jonge, Wessel 102, 121
DeLaine, Janet 213, 218, 226
De Landa, Manuel 139–141, 185
Deleuze, Gilles 30, 53, 152, 186
De Ridder, Jan Jacobus 99, 102–103
Dickinson, Hobart C. 66, 68, 118
Diocletian 210
Dionysus 42, 72
Domin, Christopher 224
Drysdale, J. W. 122
Dufton, A. F. 102–104, 121, 193, 223
Duiker, Jan 99, 102–103

Egli, Ernst 202–203, 224
Emery, C. E. 118
Engels, Friedrich 308
Environmental Protection Agency (EPA) 166
ETH Zurich 146
Evans, Jane DeRose 227

Fahrenheit, Gabriel Daniel 61
Federal Trade Commission (FTC) 110, 111
Fernández-Galiano, Luis 176, 181, 187, 235, 252, 283, 285
Fischer, Friedrich Wilhelm Hermann 196
Fitch, James Marston 197–198, 204–205, 224–225
Fitzmaurice, Robert 121
Ford, Ed 176, 187
Forrester, Jay Wright 109, 123
Forschungsheim für Wärmewirtschaft 98
Foster, Jeremy Bellamy 291, 308
Foucault, Michel 31, 53, 74, 119
Fourier, Jean Baptiste Joseph 33, 59–60, 64–65, 118, 233–235, 242
Fowler, Andrew C. 287
Fraunhofer Institute 98
Furuta, Yuzo 226

Galilei, Galileo 60
Galison, Peter 33–34, 53, 97, 223
Gard, George 310
Gardner, Robert H. 284
Gauger, Nicholas 192
Geiger, Rudolf 203
Gellert, Roland 120
Giddens, Anthony 308
Goodwin, Stanley E. 124
Græe, T. 286
Grant, George 45, 53
Grattan-Guinness, I. 117
Greve, Bredo 78

Griffiths, E. 119
Gropius, Walter 98
Guattari, Félix 30, 53, 152, 186
Guilfords Limited 146
Gutberlet, C. 119
Gwilt, Joseph 65, 118

Hall, C. A. S. 283
Hannon, B. M. 237, 283, 307
Harvey, William 191, 222
Haves, Philip 123
Hawkes, Dean 201, 224
Hays, K. Michael 223
Heberden, William 192, 223
Hechler, F. G. 118
Hegel, Georg Wilhelm Friedrich 33
Helmholtz, Hermann von 209
Henderson, Raeana 223
Hering, Carl 109, 123
Hermann, Weston A. 307
Herodotus 166
Herringer, Anna 311
Herrington, Lovic Pierce 191, 204, 222, 225
Heylighen, Ann 225
Hippocrates 191, 222
Historic American Building Survey 197
Holbo, H. Richard 239, 247, 250, 284
Houghten, F. C. 119
Hughes, Lisa 226
Humphrey, John 226
Huntington, Ellsworth 203
Hurley, Robert 284

Institut für Brandschutztechnik und Sicherheitsforschung (IBS) 285
Ice & Refrigeration 75
Ilken, Zafer 227
Illich, Ivan 74–76, 119
Imanishi, Hiroshi 226
Imbabi, Mohammed 285–287
Institution of Heating and Ventilation Engineers 107

Jacobs, Herbert 172, 181–182, 187
Jacobsen, T. 124
Johns, W. L. 124
Johnson, Jim 53
Johnston, Clifton 226
Joule, James Prescott 61, 63, 235
Journal of Architectural Education 198, 200
Junghans, Lars 285

Kagata, Kakeru 226
Kahali, Maz 310
Kahn, Louis Isidore 199
Kamm, Thomas 260, 262–263, 269, 285
Kanayama, Kozo 226

Kay, James J. 23, 25, 28, 52–53, 229, 232, 239, 283, 307
Kelvin, William Thomson 1st Baron 61, 63, 235, 283
Kent, W. 109, 123
Kim, Sunwoo 287
King, Joseph 197, 224
Kleidon, Axel 307
Knoblauch, Oskar 98
Königliche Technische Hochschule Berlin 196
Krencker, Daniel 211, 213, 226
Krintz, D. F. 124
Kulper, Amy 310
Kwinter, Sanford 310

Landsberg, Helmut 203, 225
Laplace, Pierre-Simon 61, 118
Latour, Bruno 34–35, 53, 140, 185, 188, 209, 222, 226
Lavin, Sylvia 223
Lavoisier, Antoine Laurent de 61, 118
Lea, F. M. 121
Le Corbusier 193
Ledezma, Gustavo A. 287
Lefever, René 5–6
Lehm Ton Erde Baukunst GmbH 285
Liebig, Justus von 209, 293, 308
Limperg, Koen 121
Lloyd, Seth 52
Loghem, Johannes Bernardus van 99, 103, 121
Lombardi, Leonardo 226
Loos, Adolf 137–138, 139, 185
Lorente, Sylvie 28, 53, 284, 287
Lorenz, Ralph D. 307
Lotka, Alfred J. 126, 185, 231–232, 236–242, 244, 283
Luvall, Jeffrey 239, 247, 250, 284

MacKenzie, Donald 119, 186
Mackey, Charles Osborn 106–107, 122–123
Mader & Flatz Ziviltechniker GmbH 285
Magie, W. F. 118
Malcolm, S. A. 124
Marangoni, Carlo 26, 26
Marchettini, Nadia 285
Marín, Edgar 208, 226
Markham, Sidney F. 203, 225
Martinsen, David 120
Marx, Karl 291, 293–294, 296, 298, 302, 308
Massachusetts Institute of Technology (MIT) 109, 146, 203
Maurer, Ingo 285
Mauss, Marcel 308
Maxwell, James Clark 61, 63
May, Ernst 98
May, John 52, 54, 117
Mayer, Siegmund 209
McCarthy, John F. 187

McCutchan, Gordon 201–202, 224
Meadows, Donella H. 123
Medlin, Mary 311
Mendler, Sandra 284
Meyer, Hannes 194–196
Middleton, George Frederick 105, 122
Miliaresis, Ismini 226
Miller, Perry 306, 309
Moe, Kiel 8, 120, 186, 222, 276, 287, 310
Moholy-Nagy, Sibyl 198, 224
Moll, Johannes 285
Moore, Jason W. 291, 294, 308
Moore, Steven 310
Morgan, Morris Hicky 222, 226
Mostafavi, Mohsen 311
Müller, Ingo 50, 53
Mumford, Lewis 15, 52
Murphy, Jane 162, 186

Narasimhan, T. N. 118
NASA 202
National Bureau of Standards 74, 113
National Institute of Standards and Technology 73, 113
National Research Council, USA 68
Nessi, A. 106, 123
Neutra, Richard 195–196, 223
New Mexico Energy Research and Development Institute (NM-ERDI) 114
Nicolis, Grégoire 52, 284
Nielsen, Inge 226
Nisolle, L. 106, 123
Nolan, Robert P. 186
Norwegian University of Science and Technology 78
Nye, David 284

Obata, Yoshiro 208, 226
O'Connor, James 294, 296, 308
Odell, William 284
Odum, Howard T. 5–6, 9, 11–12, 27, 52–53, 141, 155, 232, 238, 241–242, 245, 283–284, 298–299, 307–308
Oetelaar, Taylor 226–227
Ohm, Georg Simon 61, 63, 109
Olgyay, Aladar 202–204, 225
Olgyay, Victor 121, 202–204, 225
OPEC 11
Oschendorf, John 23
Osman, Michael 66, 118
Ostwald, Michael J. 223

Pagels, Heinz 52
Péclet, Jean Claude Eugène 65, 118
Pennsylvania State University Building Research Institute 109
Perry, Edward 153
Pfundstein, Margit 185, 260, 285

Pianno, Hugo Herrera 285
Pickering, Andrew 53
Pinkerton, Richard C. 238, 283
Pliny the Elder 142, 185
Popper, Karl 22, 52
Popular Mechanics Magazine 52, 187
Prigogine, Ilya 5–6, 11, 21, 25–27, 52, 232–235, 240, 283–284, 288, 307
Princeton University Architectural Laboratory 203
Pulselli, Riccardo Maria 240, 249, 284, 285
Pynchon, Thomas 5–6, 11, 46, 206

Ramsey, Charles George 154, 186
Rauch-Debevec, Marta 285
Rauch, Martin 260, 262–263, 269, 271, 285, 311
Rauch, Sebastian 285
Rayleigh, John William Strutt 3rd Baron 26, *26*, 27, 28
Rees, Simon 123
Reeve, C. P. 124
Regnault, Henri Victor 61, *63*
Reichsforschungsgesellschaft für Wirtschaftlichkeit im Bau- und Wohnungswesen 97–98, 120, 194
Reid, David Boswell 192, 222
Remmert, W. E. 124
Renn, Charles E. 186
Rietschel, Hermann 196
Ring, James W. 226
Robertson, David K. 124
Robinson, John 118
Rosenberg, A. 119
Ross, Malcolm 186
Rudolph, Paul 197, 224
Rupnik, Ivan 120

Sæland, Sem 79, 90, 91, 93, 96, 97, 120
Samet, Jonathan M. 187
Samonski, Frank H. 224
Samuel Cabot Inc 186
Santorio, Santorio 60–61
Scamozzi, Vincenzo 144, 186, 191, 222
Scheepmaker, Henk 186
Schifferstein, Hendrik N. J. 225
Schneider, Eric D. 25, 28, 52–53, 229, 232, 239, 283, 307
Schwaiger, Elizabeth 284
Sciubba, Enrico 287
Serres, Michel 15, 52
Sherman, Bill 310
Sherman, Thomas 311
Shrestha, Saurabh 311
Shuman, Everett 109, 123
Simoncini, Eugenio 285
Six, James 61
Sleeper, Harold Reeve 154, 186
Sloterdijk, Peter 32–33, 53, 184–185, 187
Smith, Elmer Gilliam 105, 122

Smith, Lance 310
Smith, Merritt Roe 53
Smith, Michael 311
Smith, Ryan 120
Smithson, Robert 5, 6, 11
Spain, R. S. 124
Spengler, Jack D. 187
Spitler, Jeffrey 123
Srinivasan, Ravi 300, 309
Stanesucu, George 287
Staudt, Jo 310
Stefan, Jožef 63
Stengers, Isabella 21, 25–27, 52, 233–235, 240, 283, 288, 307
Stephenson, D. G. 107, 123
Stewart, J. P. 106, 123
Stoppel, Jürgen 285
Straaten, J. F. van 112, 113, 124
Straube, John 76, 109, 119, 123
Stubbins, Hugh Jr. 199
Swyngedouw, Erik 308
Syrkus, Szymon and Helena 99, 121

Tafuri, Manfredo 5–6, 11
Takeuchi, Kazutoshi 226
Taylor, Bruce J. 285, 286
Taylor, TC. S. 119
Technische Universität München 98
Technische Hochschule Hannover 196
Ten, Rex 310
Tiezzi, Enzo 240, 284
Texas Engineering Research Station 105–106, 201–202
Thatcher, Edwin Daisley 222, 227
The Manufacturer and Builder 73, 119
Thompson, Emily 223
Tille, David Rogers 307
Tomaselli, Josef 285
Tomaszewski, Lech 99
Tomlinson, Charles 192
Tomlow, Jos 120–121, 185, 196, 223
Trajan 210
Tredgold, Thomas 64, 118, 191–192, 222
Turner, Monica G. 284
Twitchell 224

Ulanowicz, Robert E. 232, 237, 283, 307
Ulgiati, Sergio 308
United States Congress 68
United States Department of Energy (DOE) 113–114
United States Department of Housing and Urban Development (HUD) 113
United States Mineral Wool Company 73
University of Texas 105
Urbanik, Jadwiga 121

Vale, Brenda and Robert 284
Van Duesen, M. S. *67*, 118
Veblen, Thorstein 308
Vitruvius 45, 136–140, 185, 191, 202, 215, 222, 226, 252

Wacjman, Judy 119,186
Walden, Gert 311
Walton, G. N. 124
Ward-Perkins, John Bryan 226
Wastiels, Lisa 225
Weber, Max 71, 119, 206, 225, 296, 308
Webster, R. 286
Wedebrunn, Ola 185
Wells, Gordon C. 308
Wheeler, Peter 135, 185
Whitman, R. B. 119
Wilkes, G. B. 119
Winslow, Charles Edward Amory 191, 204, 222, 225
Wood, A. J. 118
Wood, David 226
Wouters, Ine 225
Wrangham, Richard 135, 185
Wright, Frank Lloyd 48, 169–170, 172–173, 176, 181–184, 194, 205
Wright, Lawrence T. Jr. 106–107, 122–123
Wright, Orville 123
Wundt, Wilhelm Maximilian 195
Wyllie-Echeverria, Sandy 186

Yamada, Tetsuya 226
Yegül, Fikret 214, 226, 227
Yildiz, Sevin 117
York, Richard 291
Yoshida, Atsumasa 226

Zarr, Robert 68, 118, 119
Zehender, Horst 120
Zeiher, Laura C. 284

Index of Buildings and Objects

2226, Lustenau 253–254, *254–257*, 259
Baths of Caracalla, Rome *211–212*, 213–214, 216, *217*–218
Beni Isguen, Algeria 25, *27*
Betondorp, Amsterdam 99
Broadacre City 170
Building Research Station, Garston 104
Cabot's Quilt 146, *147–151*, 152, 155, 186
Commonwealth Experimental Building Station, Sydney 105
eupatheoscope 193
Gas station, Potlatch, Washington *10*
Gooiland Hotel 103
Holzkirchen building test station 98
Interstate 35W bridge, Minneapolis *292*
Jacobs House, Madison, Wisconsin 48, 169–170, *170–171*, 172–173, *174–175*, 176, *177–180*, 181–184
Joule's insulate thermometer *62*
kata thermometer 193
Lavoisier-Laplace's ice calorimeter *62*
Leisure Centre, Callender 278
Mezhyrich huts, Ukraine 135
NIST guarded hot-plate apparatus by M.S. Van Duesen *67*
Norges Tekniske Høiskole, main building, Trondheim 78–79
Occidental Chemical Corporation, Buffalo, New York 115, *116*
Old Pierce House, Dorchester, Massachusetts 145, 146
Open Air School, Amsterdam 103
Pantheon, Rome 253
Praunheim housing settlement, Frankfurt 98
Rauch House 48, *258*, 259–260, *261–264*, 262, 265, *265–268*, 269–271, *271*
Rayleigh-Bénard convection cells 26–28, 244
Regnault's apparatus *63*
Rocca Pisana, Lonigo 144
Roman baths 209–211, *215–216*, 218–222
SC Johnson and Son Administration Building, Racine, Wisconsin 184
Seafelt 146
Six's max/min thermometer *62*
Stabian Baths, Pompeii 220
Temple of Aesculapius, Insula Tiberina, Rome 14
Test Huts, Trondheim 78–81, *80–82*, 83, *84–92*, 88–91, 93, *94–95*, 95–96
Usonian houses 170
Zonnestraal Sanatorium, Hilversum 10